Mathematical Thought
from Ancient to Modern Times

Mathematical Thought
from Ancient to Modern Times

Volume 1

MORRIS KLINE

New York   Oxford   OXFORD UNIVERSITY PRESS

Oxford University Press

Oxford New York Toronto
Delhi Bombay Calcutta Madras Karachi
Petaling Jaya Singapore Hong Kong Tokyo
Nairobi Dar es Salaam Cape Town
Melbourne Auckland

and associated companies in
Berlin Ibadan

Library of Congress Cataloging-in-Publication Data
Kline, Morris, 1908–
Mathematical thought from ancient to modern times / Morris Kline.
p. cm. Includes bibliographical references.
ISBN 0-19-506135-7 (PBK) (v. 1)
1. Mathematics—History.   I. Title.
QA21.K516 1990 510'.9—dc20 89-25520

10 9 8 7 6 5 4 3 2 1

Printed in the United States of America

To my wife, Helen Mann Kline

Preface to the Three-Volume Paperback Edition of Mathematical Thought

The reception accorded the original edition of this book is most gratifying. I am flattered, if not a penny richer for it, by a pirated Chinese translation. Even more satisfying is a forthcoming authorized Spanish translation.

This work is part of my long-time efforts to humanize the subject of mathematics. At the very beginning of my career I banded with a few colleagues to produce a freshman text that departed from the traditional dry-as-dust mathematics textbook. Later, I wrote a calculus text with the same end in view. While I was directing a research group in electromagnetic theory and doing research myself, I still made time to write *Mathematics In Western Culture,* which is partly history and partly an exploration of the influence of mathematics upon philosophy, religion, literature, art, music, economic theory, and political thought. More recently I have written with the general reader in mind a book on the philosophical foundations of mathematics and a book on the underlying mathematical structure of a good deal of science, most especially cosmogony and physics.

I hope that students, teachers, as well as the general reader will profit from this more affordable and accessible three-volume paperback edition of *Mathematical Thought.* I wish to acknowledge the helpful suggestions made by Harold Edwards, Donald Gillis, and Robert Schlapp among others. My very special thanks go to Fred Pohle for his time, interest, and generosity. Having over the years taught a course based on this book, he saw a need for a multi-volume paperback version and provided the impetus for this edition. Beyond this he gave unstintingly of his time and knowledge in helping me correct errors. I am truly in his debt, as I am to my wife Helen, who undertook much of the work involved in preparing this edition.

Preface

> If we wish to foresee the future of mathematics our proper
> course is to study the history and present condition of the
> science.   HENRI POINCARÉ

This book treats the major mathematical creations and developments
from ancient times through the first few decades of the twentieth century.
It aims to present the central ideas, with particular emphasis on those
currents of activity that have loomed largest in the main periods of the life
of mathematics and have been influential in promoting and shaping subsequent mathematical activity. The very concept of mathematics, the
changes in that concept in different periods, and the mathematicians' own
understanding of what they were achieving have also been vital concerns.

This work must be regarded as a survey of the history. When one
considers that Euler's works fill some seventy volumes, Cauchy's twenty-six
volumes, and Gauss's twelve volumes, one can readily appreciate that a
one-volume work cannot present a full account. Some chapters of this work
present only samples of what has been created in the areas involved, though
I trust that these samples are the most representative ones. Moreover, in
citing theorems or results, I have often omitted minor conditions required for
strict correctness in order to keep the main ideas in focus. Restricted as this
work may be, I believe that some perspective on the entire history has been
presented.

The book's organization emphasizes the leading mathematical themes
rather than the men. Every branch of mathematics bears the stamp of its
founders, and great men have played decisive roles in determining the course
of mathematics. But it is their ideas that have been featured; biography is
entirely subordinate. In this respect, I have followed the advice of Pascal:
"When we cite authors we cite their demonstrations, not their names."

To achieve coherence, particularly in the period after 1700, I have
treated each development at that stage where it became mature, prominent,
and influential in the mathematical realm. Thus non-Euclidean geometry is
presented in the nineteenth century even though the history of the efforts to

replace or prove the Euclidean parallel axiom date from Euclid's time onward. Of course, many topics recur at various periods.

To keep the material within bounds I have ignored several civilizations such as the Chinese,[1] Japanese, and Mayan because their work had no material impact on the main line of mathematical thought. Also some developments in mathematics, such as the theory of probability and the calculus of finite differences, which are important today, did not play major roles during the period covered and have accordingly received very little attention. The vast expansion of the last few decades has obliged me to include only those creations of the twentieth century that became significant in that period. To continue into the twentieth century the extensions of such subjects as ordinary differential equations or the calculus of variations would call for highly specialized material of interest only to research men in those fields and would have added inordinately to the size of the work. Beyond these considerations, the importance of many of the more recent developments cannot be evaluated objectively at this time. The history of mathematics teaches us that many subjects which aroused tremendous enthusiasm and engaged the attention of the best mathematicians ultimately faded into oblivion. One has but to recall Cayley's dictum that projective geometry is all geometry, and Sylvester's assertion that the theory of algebraic invariants summed up all that is valuable in mathematics. Indeed one of the interesting questions that the history answers is what survives in mathematics. History makes its own and sounder evaluations.

Readers of even a basic account of the dozens of major developments cannot be expected to know the substance of all these developments. Hence except for some very elementary areas the contents of the subjects whose history is being treated are also described, thus fusing exposition with history. These explanations of the various creations may not clarify them completely but should give some idea of their nature. Consequently this book may serve to some extent as a historical introduction to mathematics. This approach is certainly one of the best ways to acquire understanding and appreciation.

I hope that this work will be helpful to professional and prospective mathematicians. The professional man is obliged today to devote so much of his time and energy to his specialty that he has little opportunity to familiarize himself with the history of his subject. Yet this background is important. The roots of the present lie deep in the past and almost nothing in that past is irrelevant to the man who seeks to understand how the present came to be what it is. Moreover, mathematics, despite the proliferation into hundreds of branches, is a unity and has its major problems and goals. Unless the various specialties contribute to the heart of mathematics they are likely to be

1. A fine account of the history of Chinese mathematics is available in Joseph Needham's *Science and Civilization in China*, Cambridge University Press, 1959, Vol. 3, pp. 1–168.

sterile. Perhaps the surest way to combat the dangers which beset our fragmented subject is to acquire some knowledge of the past achievements, traditions, and objectives of mathematics so that one can direct his research into fruitful channels. As Hilbert put it, "Mathematics is an organism for whose vital strength the indissoluble union of the parts is a necessary condition."

For students of mathematics this work may have other values. The usual courses present segments of mathematics that seem to have little relationship to each other. The history may give perspective on the entire subject and relate the subject matter of the courses not only to each other but also to the main body of mathematical thought.

The usual courses in mathematics are also deceptive in a basic respect. They give an organized logical presentation which leaves the impression that mathematicians go from theorem to theorem almost naturally, that mathematicians can master any difficulty, and that the subjects are completely thrashed out and settled. The succession of theorems overwhelms the student, especially if he is just learning the subject.

The history, by contrast, teaches us that the development of a subject is made bit by bit with results coming from various directions. We learn, too, that often decades and even hundreds of years of effort were required before significant steps could be made. In place of the impression that the subjects are completely thrashed out one finds that what is attained is often but a start, that many gaps have to be filled, or that the really important extensions remain to be created.

The polished presentations in the courses fail to show the struggles of the creative process, the frustrations, and the long arduous road mathematicians must travel to attain a sizable structure. Once aware of this, the student will not only gain insight but derive courage to pursue tenaciously his own problems and not be dismayed by the incompleteness or deficiencies in his own work. Indeed the account of how mathematicians stumbled, groped their way through obscurities, and arrived piecemeal at their results should give heart to any tyro in research.

To cover the large area which this work comprises I have tried to select the most reliable sources. In the pre-calculus period these sources, such as T. L. Heath's *A History of Greek Mathematics*, are admittedly secondary, though I have not relied on just one such source. For the subsequent development it has usually been possible to go directly to the original papers, which fortunately can be found in the journals or in the collected works of the prominent mathematicians. I have also been aided by numerous accounts and surveys of research, some in fact to be found in the collected works. I have tried to give references for all of the major results; but to do so for all assertions would have meant a mass of references and the consumption of space that is better devoted to the account itself.

The sources have been indicated in the bibliographies of the various chapters. The interested reader can obtain much more information from these sources than I have extracted. These bibliographies also contain many references which should not and did not serve as sources. However, they have been included either because they offer additional information, because the level of presentation may be helpful to some readers, or because they may be more accessible than the original sources.

I wish to express thanks to my colleagues Martin Burrow, Bruce Chandler, Martin Davis, Donald Ludwig, Wilhelm Magnus, Carlos Moreno, Harold N. Shapiro, and Marvin Tretkoff, who answered numerous questions, read many chapters, and gave valuable criticisms. I am especially indebted to my wife Helen for her critical editing of the manuscript, extensive checking of names, dates, and sources, and most careful reading of the galleys and page proofs. Mrs. Eleanore M. Gross, who did the bulk of the typing, was enormously helpful. To the staff of Oxford University Press, I wish to express my gratitude for their scrupulous production of this work.

New York   M. K.
May 1972

Contents

Publisher's Note

to this Three-Volume Paperback Edition

Mathematical Thought from Ancient to Modern Times was first published by Oxford University Press as a one-volume cloth edition. In publishing this three-volume paperback edition we have retained the same pagination as the cloth in order to maintain consistency within the Index, Subject Index, and Notes. These volumes are paginated consecutively and, for the reader's convenience, both Indexes appear at the end of each volume.

I

Mathematics in Mesopotamia

> Logic can be patient for it is eternal.
> OLIVER HEAVISIDE

1. *Where Did Mathematics Begin?*

Mathematics as an organized, independent, and reasoned discipline did not exist before the classical Greeks of the period from 600 to 300 B.C. entered upon the scene. There were, however, prior civilizations in which the beginnings or rudiments of mathematics were created. Many of these primitive civilizations did not get beyond distinguishing among one, two, and many; others possessed, and were able to operate with, large whole numbers. Still others achieved the recognition of numbers as abstract concepts, the adoption of special words for the individual numbers, the introduction of symbols for numbers, and even the use of a base such as ten, twenty, or five to denote a larger unit of quantity. One also finds the four operations of arithmetic, but confined to small numbers, and the concept of a fraction, limited, however, to 1/2, 1/3, and the like, and expressed in words. In addition, the simplest geometric notions, line, circle, and angle, were recognized. It is perhaps of interest that the concept of angle must have arisen from observation of the angle formed by man's thigh and lower leg or his forearm and upper arm because in most languages the word for the side of an angle is either the word for leg or the word for arm. In English, for example, we speak of the arms of a right triangle. The uses of mathematics in these primitive civilizations were limited to simple trading, the crude calculation of areas of fields, geometric decoration on pottery, patterns woven into cloth, and the recording of time.

Until we reach the mathematics of the Babylonians and the Egyptians of about 3000 B.C. we do not find more advanced steps in mathematics. Since primitive peoples settled down in one area, built homes, and relied upon agriculture and animal husbandry as far back as 10,000 B.C., we see how slowly the most elementary mathematics made its first steps; moreover, the existence of vast numbers of civilizations with no mathematics to speak of shows how sparsely this science was cultivated.

3

2. *Political History in Mesopotamia*

The Babylonians were the first of these two early civilizations to contribute to the main course of mathematics. Since our knowledge of the ancient civilizations of the Near East and of Babylonia in particular is largely the product of archaeological research of the last hundred years, this knowledge is incomplete and subject to correction as new discoveries are made. The term "Babylonian" covers a series of peoples, who concurrently or successively occupied the area around and between the Tigris and Euphrates rivers, a region known as Mesopotamia and now part of modern Iraq. These peoples lived in independent cities such as Babylon, Ur, Nippur, Susa, Aššur, Uruk, Lagash, Kish, and others. About 4000 B.C. the Sumerians, racially distinct from the Semitic and Indo-Germanic peoples, settled down in part of Mesopotamia. Their capital city was Ur, and the land area they controlled was called Sumer. Though their culture reached its height about 2250 B.C., even before this, about 2500 B.C., the Sumerians came under the political control of the Akkadians, a Semitic people whose major city was Akkad and who were led at that time by the ruler Sargon. Sumerian civilization was submerged in the Akkadian. A period of high culture occurred during the reign of King Hammurabi (about 1700 B.C.), who is known as the formulator of a famous code of law.

Around 1000 B.C. migrations and the introduction of iron resulted in further changes. Later, in the eighth century B.C., the area was ruled by the Assyrians, who settled primarily in the upper Tigris region. To our knowledge, they did not add anything new to the culture. A century later the Assyrian empire was shared by the Chaldeans and the Medes, the latter being close racially to the Persians further east. This period in Mesopotamian history (7th cent. B.C.) is often referred to as Chaldean. The Near East was conquered by the Persians under Cyrus about 540 B.C. Persian mathematicians such as Nabu-rimanni (*c.* 490 B.C.) and Kidinu (*c.* 480 B.C.) became known to the Greeks.

In 330 B.C., Alexander the Great, the Greek military leader, conquered Mesopotamia. The period from 300 B.C. to the birth of Christ is called Seleucid after the Greek general who first took control of the region following the death of Alexander in 323 B.C. However, the flowering of Greek mathematics had already taken place, and from the time of Alexander until the seventh century A.D., when the Arabs arrived on the scene, the Greek influence predominated in the Near East. Most of what the Babylonians contributed to mathematics predates the Seleucid period.

Despite the numerous changes in the rulers of Mesopotamia, there was, in mathematics, a continuity of knowledge, tradition, and practice from ancient times at least to the time of Alexander.

3. *The Number Symbols*

Our main information about the civilization and mathematics of Babylonia, ancient and more recent, comes from texts in the form of clay tablets. These tablets were inscribed when the clay was still soft and then baked. Hence those that survived destruction are well preserved. They date mainly from two periods, some from about 2000 B.C., and a larger number from the period 600 B.C.–A.D. 300. The earlier tablets are the more important in the history of mathematics.

The language and script of the older tablets is Akkadian, which was superimposed on the older Sumerian language and script. The words of the Akkadian language consisted of one or more syllables; each syllable was represented by a collection of what are essentially line segments. The Akkadians used a stylus with a triangular cross-section held at an angle to the clay, which produced wedge-shaped impressions that could be oriented in different ways. From *cuneus*, the Latin word for "wedge," the script became known as cuneiform.

The most highly developed arithmetic of the Babylonian civilization is the Akkadian. Whole numbers were written as follows:

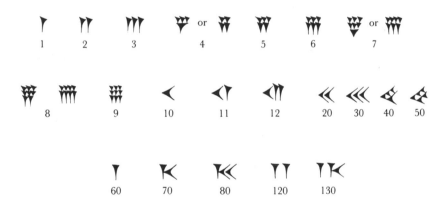

The striking features of the Babylonian number system are the base 60 and positional notation.

At first the Babylonians had no symbol to indicate the absence of a number in any one position, and consequently their numbers were ambiguous. Thus 𝕂𝕂 could mean 80 or 3620, depending upon whether the first symbol meant 60 or 3600. Spacing was often used to indicate no quantity in a given position, but of course this might be misinterpreted. In the

Seleucid period a special separation symbol was introduced to indicate the absence of a number. Thus ⟦symbol⟧ $= 1 \cdot 60^2 + 0 \cdot 60 + 4 = 3604$. However, even in this period no symbol was used to indicate the absence of a quantity at the right-hand end as in our 20. In both periods the content had to be relied upon to indicate the precise value of the entire number.

The Babylonians used positional notation to represent fractions also. Thus ⟦symbol⟧ intended as a fraction meant 20/60 and ⟦symbol⟧, intended as a fraction, could be 21/60 or 20/60 + 1/60². The ambiguity in their number system is therefore even greater than previously indicated.

A few fractions had special symbols. Thus one finds:

These special fractions, 1/2, 1/3, and 2/3, were to the Babylonians "wholes," in the sense of measures of quantities, and not a division of unity into parts, though they arose as measures of quantity having these respective relations to another quantity. Thus we might write 10 cents as 1/10 in relation to the dollar but think of the 1/10 as a unit in itself.

Actually the Babylonians did not use base 60 exclusively. Sometimes the years were written 2 *me* 25, where *me* stood for hundred—in our symbols, 225. Also *lîmu* was used for 1000, generally in nonmathematical texts, though it occurs even in mathematical texts of the Seleucid period. Sometimes 10 and 60 were mixed, as in 2 *me* 1, 10, which means $2 \times 100 + 1 \times 60 + 10$ or 270. Mixed systems involving units of 60, 24, 12, 10, 6, and 2 were used for dates, areas, measures of weight, and coinage, just as we use 12 for hours, 60 for minutes and seconds, 12 for inches, and 10 for ordinary counting. The Babylonian systems, like ours, were a composite of many historical and regional customs. However, in the mathematical and astronomical texts they did use base 60 quite consistently.

We do not know definitely how base 60 came to be used. The suggestion may have come from the systems of weight measures. Suppose we have a system of weight measures with values in it having the ratios

$$1/2, \ 1/3, \ 2/3, \ 1, \ 10.$$

Now suppose there is another system with a different unit but the same ratios, and political or social forces compel the fusing of the two systems. (We have meters and yards, for example.) If the larger unit were 60 times the smaller, then 1/2, 1/3, and 2/3 of the larger unit would be integral multiples of the smaller one. Thus the larger unit might have been adopted because it was so convenient.

As to the origin of positional notation, there are two likely explanations. In an older scheme of writing numbers 1 multiplied by 60 was written with a larger ⟦symbol⟧ than the same symbol for 1. When the writing was simplified the larger ⟦symbol⟧ was reduced in size but kept in the usual place for 60. Hence position became the denotation of a multiple of 60. Another possible explanation comes from the coinage system. One talent and 10 mana may have been written $\blacktriangleleft$, where the $\mathbf{\Upsilon}$ meant 1 talent, which is 60 mana. We do the same when we write \$1.20 and mean 100 cents by the 1. The scheme for writing amounts of money may then have been taken over to arithmetic generally.

4. *Arithmetic Operations*

In the Babylonian system the symbols for 1 and 10 were basic. Numbers from 1 to 59 were formed by combining fewer or more of these symbols. Hence the processes of addition and subtraction were merely a matter of adding or taking away symbols. To indicate addition the Babylonians joined numbers together as in $\blacktriangleleft \mathbf{\mathit{W}}$, which indicates 16. Subtraction was often indicated by the symbol $\mathbf{\Gamma}$. Thus $\mathbf{\mathit{\&}}\ \mathbf{\Gamma\mathit{W}}$ is $40 - 3$. In astronomical texts of a later period the word *tab* appears, signifying addition.

Multiplication of integers was also performed. To multiply by 37, say, would mean multiplying by 30, then by 7, and adding the results. The symbol for multiplication was $\mathbf{\mathit{W}\ \mathbf{\mathit{\rightleftharpoons}}}$, pronounced *a-rá*. It meant "to go."

The Babylonians divided one whole number by another. Since to divide by an integer a is to multiply by the reciprocal $1/a$, to this extent fractions were involved. The Babylonians converted the reciprocals to sexagesimal "decimals" and, except for the few mentioned above, did not use special symbols for fractions. They had tables showing how numbers of the form $1/a$, where $a = 2^\alpha 3^\beta 5^\gamma$, could be written as terminating sexagesimal numbers. Some tables gave approximate values for $1/7$, $1/11$, $1/13$, etc., because these fractions led to infinite repeating sexagesimals. Where fractions involving denominators other than 2, 3, or 5 occurred in the older problems, the same troublesome factors occurred in the numerator and were cancelled.

The Babylonians relied entirely upon the tables of reciprocals. Their tables show, for example,

igi 2 *gál-bi* 30	*igi* 8 *gál-bi* 7, 30
igi 3 *gál-bi* 20	*igi* 9 *gál-bi* 6, 40
igi 4 *gál-bi* 15	. . . .
igi 6 *gál-bi* 10	*igi* 27 *gál-bi* 2, 13, 20

obviously meaning $1/2 = 30/60$, $1/3 = 20/60$, etc. The precise meanings of *igi* and *gál-bi* are not known. Sexagesimal fractions, that is, numbers less than 1, expressed in inverse powers of 60, 60^2, etc., but with denominators merely understood, continued to be used by the Greeks Hipparchus and Ptolemy and in Renaissance Europe up to the sixteenth century, when they were replaced by decimals in base 10.

The Babylonians also had tables expressing squares, square roots, cubes, and cube roots. When the root was a whole number it was given exactly. For other roots, the corresponding sexagesimal numbers were only approximate. Of course, irrationals cannot be expressed in a finite number of decimals or sexagesimals. However, there is no evidence that the Babylonians were aware of this. They might well have believed that the irrationals could be converted exactly to sexagesimal numbers if more places were used. An excellent Babylonian approximation to $\sqrt{2}$ gives $\sqrt{2} = 1.414213\ldots$ instead of $1.414214\ldots$.

Roots occur in their calculations of the diagonal d of a rectangle of height h and width w. One problem asks for the diagonal of a rectangular gate with given height and width. The answer is given without explanation and amounts to using the approximate formula for the diagonal d, namely,

$$d \approx h + \frac{w^2}{2h}.$$

The formula is a good approximation to d for $h > w$. Thus for $h > w$, as is the case in one problem, we can see that the answer is reasonable by noting that

$$d = \sqrt{h^2 + w^2} = h\sqrt{1 + \frac{w^2}{h^2}} = h\left(1 + \frac{w^2}{h^2}\right)^{1/2}.$$

If we now expand the binomial and retain only the first two terms we get the formula. Other approximate answers to square-root problems were given, obtained presumably by using numbers in Babylonian tables.

5. Babylonian Algebra

Distinct from the table texts, which yield much information on the Babylonian number system and operations with numbers, are texts that deal with algebraic and geometric problems. A fundamental problem of the older Babylonian algebra asks for a number which, added to its reciprocal, yields a given number. In modern notation, the Babylonians sought x and $\bar{x}$ such that

$$x\bar{x} = 1, \qquad x + \bar{x} = b.$$

These two equations yield a quadratic equation in x, namely, $x^2 - bx + 1 = 0$. They formed $\left(\frac{b}{2}\right)^2$; then $\sqrt{\left(\frac{b}{2}\right)^2 - 1}$; and then

$$\frac{b}{2} + \sqrt{\left(\frac{b}{2}\right)^2 - 1} \quad \text{and} \quad \frac{b}{2} - \sqrt{\left(\frac{b}{2}\right)^2 - 1},$$

which yield the answers. In effect, the Babylonians had the quadratic formula. Other problems, such as finding two numbers whose sum and product are given, were reduced to the above problem. Since the Babylonians had no negative numbers, negative roots of quadratic equations were neglected. Though only concrete examples were given, many were intended to illustrate a general procedure for quadratics. More complicated algebraic problems were reduced by transformations to simpler ones.

The Babylonians were able to solve special problems involving five unknowns in five equations. One problem, which arose in connection with the adjustment of astronomical observations, involved ten equations and ten unknowns, mostly linear. The solution used a special method of combining the equations until the unknowns were evaluated.

The algebraic problems were stated and solved verbally. The words *uš* (length), *sag* (breadth), and *aša* (area) were often used for the unknowns, not because the unknowns necessarily represented these geometric quantities, but probably because many algebraic problems came from geometric situations and the geometric terminology became standard. An illustration of the way in which these terms were employed for the unknowns and of the way in which problems were stated may be gathered from the following: "I have multiplied Length and Breadth and the Area is 10. I have multiplied the Length by itself and have obtained an Area. The excess of Length over Breadth I have multiplied by itself and this result by 9. And this Area is Area obtained by multiplying the Length by itself. What are the Length and Breadth?" It is obvious here that the words *Length*, *Breadth*, and *Area* are merely convenient terms for two unknowns and their product, respectively.[1]

Today we would write this problem as

$$xy = 10$$
$$9(x - y)^2 = x^2.$$

The solution, incidentally, leads to a fourth-degree equation in x, with the x and x^3 terms missing so that it can be and was solved as a quadratic in x^2.

Problems leading to a cube root also occurred. The modern formulation of one such problem would be

$$12x = z, \qquad y = x, \qquad xyz = V,$$

where V is some known volume. To find x here we must extract a cube root. The Babylonians calculated this root from the cube-root tables we mentioned earlier. They also did compound-interest problems that called for finding the value of an unknown exponent.

The Babylonians sometimes employed symbols for unknowns, but the

---

1. Many examples of algebraic problems can be found in van der Waerden, pp. 65–73. See the bibliography at the end of the chapter.

symbolism came about inadvertently. In some problems two special Sumerian words (somewhat modified by Akkadian endings) were used to represent two unknowns that were reciprocals of each other. Moreover, the ancient Sumerian pictorial symbols were used for these words, and since these pictorial symbols were no longer in the current language, the effect was to use special symbols for unknowns. These symbols were used repeatedly and one could identify them without even knowing how they were pronounced in Akkadian.

The algebraic problems were solved by describing only the steps required to execute the solution. For example, square 10 obtaining 100; subtract 100 from 1000; this gives 900; and so forth. Since no reasons for steps were given, we can only infer how they knew what to do.

They summed arithmetic progressions and geometric progressions in concrete problems; for the latter we find, in our notation,

$$1 + 2 + 4 + \cdots + 2^9 = 2^9 + (2^9 - 1) = 2^{10} - 1.$$

Also, the sum of the squares of the integers from 1 to 10 was given as though they had applied the formula

$$1^2 + 2^2 + \cdots + n^2 = \left(1 \times \frac{1}{3} + n \times \frac{2}{3}\right)(1 + 2 + 3 + \cdots + n).$$

No derivation accompanied the special cases treated in their texts.

Babylonian algebra included a bit of the theory of numbers. Many sets of Pythagorean triples were found, probably by the correct rule; that is, if $x = p^2 - q^2$, $y = 2pq$, $z = p^2 + q^2$, then $x^2 + y^2 = z^2$. They also solved $x^2 + y^2 = 2z^2$ in integers.

6. *Babylonian Geometry*

The role of geometry in Babylonia was insignificant. Geometry was not a separate mathematical discipline. Problems involving division of a field or the sizes of bricks needed for some construction were readily converted into algebraic problems. Some computations of area and volume were given in accordance with rules or formulas. However, the figures illustrating geometric problems were roughly drawn and the formulas may have been incorrect. In the Babylonians' calculations of areas, for example, we cannot tell whether their triangles were right triangles or whether their quadrilaterals were squares and, hence, whether their formulas were the right ones for the figures concerned. However, the Pythagorean relationship, the similarity of triangles, and the proportionality of corresponding sides in similar triangles were known. The area of a circle was obtained, apparently by following the rule that $A = \dfrac{c^2}{12}$ where c is the circumference. This rule amounts to using 3 for π. However, another of their results giving the relation between the circumference of a regular hexagon and its circumscribed circle implies a value of 3 1/8 for π. A few volumes were computed, some correctly and some incorrectly, in the course of solving particular physical problems.

Apart from a few special facts, such as the calculation of the radius of a circle circumscribing a known isosceles triangle, the substance of Babylonian geometry was a collection of rules for the areas of simple plane figures, including regular polygons, and the volumes of simple solids. The geometry was not studied in and for itself but always in connection with practical problems.

7. The Uses of Mathematics in Babylonia

Despite the limited extent of the Babylonians' mathematics, it entered into many phases of their lives. Babylonia was in the path of trade routes and commerce was extensive. The Babylonians used their knowledge of arithmetic and simple algebra to express lengths and weights, to exchange money and merchandise, to compute simple and compound interest, to calculate taxes, and to apportion shares of a harvest to the farmer, church, and state. The division of fields and inheritances led to algebraic problems. The majority of the cuneiform texts involving mathematics (exclusive of the tables and problem texts) were concerned with economic problems. There is no doubt as to the influence of economics on the development of arithmetic in the older period.

Canals, dams, and other irrigation projects required calculations. The use of bricks raised numerous numerical and geometrical problems. Volumes of granaries and buildings and the areas of fields had to be determined. The close relationship between Babylonian mathematics and practical problems is typified by the following: A canal whose cross-section was a trapezoid and whose dimensions were known was to be dug. The amount of digging one man could do in a day was known, as was the sum of the number of men employed and the days they worked. The problem was to calculate the number of men and the number of days of work.

Because the connection between mathematics and astronomy became vital from Greek times on, we shall note what the Babylonians knew and did in astronomy. Nothing is known about Sumerian astronomy, and the astronomy of the Akkadian period was crude and qualitative; the development of mathematics preceded the development of any significant astronomy. In the Assyrian period (about 700 B.C.), astronomy did begin to include mathematical description of phenomena and a systematic compilation of observational data. The use of mathematics increased in the last three centuries B.C. and was directed especially to the study of lunar and planetary motion. Most astronomical texts are from this Seleucid period. They fall into two groups, procedural texts and ephemerides, tables of positions of the heavenly bodies at various times. The procedural texts show how to compute the ephemerides.

The arithmetic behind the lunar and solar observations shows that the Babylonians calculated first and second differences of successive data, observed the constancy of the first or second differences, and extrapolated or interpolated data. Their procedure was equivalent to using the fact that the data can be fit by polynomial functions and enabled them to predict the daily positions of the planets. They knew the periods of the planets with some accuracy, and also used eclipses as a basis for calculation. There was, however, no geometrical scheme of planetary or lunar motion in Babylonian astronomy.

The Babylonians of the Seleucid period did have extensive tables on the motions of the sun and moon which gave variable velocities and positions. Also special conjunctions and eclipses of sun and moon were either in the data or readily obtained from them. Astronomers could predict the new moon and eclipses to within a few minutes. Their data indicate that they knew the length of the solar or tropical year (the year of the seasons) to be $12 + 22/60 + 8/60^2$ months (from new moon to new moon) and the length of the sidereal year (the time for the sun to regain the same position relative to the stars) to within 4 1/2 minutes.

The constellations that lent their names to the twelve signs of the zodiac were known earlier but the zodiac itself first appears in a text in 419 B.C. Each sector of the zodiac was a 30° arc. Positions of planets in the sky were fixed by reference to the stars and also by position in the zodiac.

Astronomy served many purposes. For one thing, it was needed to keep a calendar, which is determined by the positions of the sun, moon, and stars. The year, the month, and the day are astronomical quantities, which had to be obtained accurately to know planting times and religious holidays. In Babylonia, partly because of the connection of the calendar with religious holidays and ceremonies and partly because the heavenly bodies were believed to be gods, the priests kept the calendar.

The calendar was lunar. The month began when the crescent first appeared after the moon was fully dark (our new moon). The day began in the evening of the first appearance of the crescent and lasted from sunset to sunset. The lunar calendar is difficult to maintain because, although it is convenient to have the month contain an integral number of days, lunar months, reckoned as the time between successive conjunctions of sun and moon (that is, from new moon to new moon), vary from 29 to 30 days. Hence a problem arises in deciding which months are to have 29 and which 30. A more important problem is making the lunar calendar agree with the seasons. The answer is quite complicated because it depends upon the paths and velocities of the moon and sun. The lunar calendar contained extra months intercalated so that 7 such intercalations in each 19 years just *about* kept the lunar calendar in time with the solar year. Thus 235 lunar months were equal to 19 solar years. The summer solstice was systematically computed, and the winter solstice and the equinoxes were placed at equal intervals. This calendar was used by the Jews and Greeks and by the Romans up to 45 B.C., the year the Julian calendar was adopted.

The division of the circle into 360 units originated in the Babylonian astronomy of the last centuries before the Christian era. It had nothing to do with the earlier use of base 60; however, base 60 was used to divide the degree and the minute into 60 parts. The astronomer Ptolemy (2nd cent. A.D.) followed the Babylonians in this practice.

Closely connected with astronomy was astrology. In Babylonia, as in many ancient civilizations, the heavenly bodies were thought to be gods and so were presumed to have influence and even control over the affairs of man. When one takes into account the importance of the sun for light, heat, and the growth of plants, the dread inspired by its eclipses, and such seasonal phenomena as the mating of animals, one can well understand the belief that the heavenly bodies do affect even the daily events in man's life.

Pseudoscientific schemes of prediction in ancient civilizations did not always involve astronomy. Numbers themselves had mystic properties and could be used to make predictions. One finds some Babylonian usages in the Book of Daniel and in the writings of the Old and New Testament prophets. The Hebrew "science" of *gematria* (a form of cabbalistic mysticism) was based on the fact that each letter of the alphabet had a number value because the Hebrews used letters to represent numbers. If the sum of the numerical values of the letters in two words was the same, an important connection between the two ideas or people or events represented by the words was inferred. In the prophecy of Isaiah (21:8), the lion proclaims the fall of Babylon because the letters in the Hebrew word for lion and those in the word for Babylon add up to the same sum.

8. *Evaluation of Babylonian Mathematics*

The Babylonians' use of special terms and symbols for unknowns, their employment of a few operational symbols, and their solution of a few types of equations involving one or more unknowns, especially quadratic equations, constituted a start in algebra. Their development of a systematic way of writing whole numbers and fractions enabled them to carry arithmetic to a fairly advanced stage and to employ it in many practical situations, especially in astronomy. They possessed some numerical and what we would call algebraic skill in the solution of special equations of high degree, but on the whole their arithmetic and algebra were very elementary. Though they worked with concrete numbers and problems, they evidenced a partial grasp of abstract mathematics in their recognition that some procedures were typical of certain classes of equations.

The question arises as to what extent the Babylonians employed mathematical proof. They did solve by correct systematic procedures rather complicated equations involving unknowns. However, they gave verbal instructions only on the steps to be made and offered no justification of the steps. Almost surely, the arithmetic and algebraic processes and the geometrical rules were the end result of physical evidence, trial and error, and insight. That the methods worked was sufficient justification to the Babylonians for their continued use. The concept of proof, the notion of a logical structure based on principles warranting acceptance on one ground or another, and the consideration of such questions as under what conditions solutions to problems can exist, are not found in Babylonian mathematics.

Bibliography

Bell, E. T.: *The Development of Mathematics*, 2nd ed., McGraw-Hill, Chaps. 1–2.

Boyer, Carl B.: *A History of Mathematics*, John Wiley and Sons, 1968, Chap. 3.

Cantor, Moritz: *Vorlesungen über Geschichte der Mathematik*, 2nd ed., B. G. Teubner, 1894, Vol. 1, Chap. 1.

Chiera, E.: *They Wrote on Clay*, Chicago University Press, 1938.

Childe, V. Gordon: *Man Makes Himself*, New American Library, 1951, Chaps. 6–8.

Dantzig, Tobias: *Number: The Language of Science*, 4th ed., Macmillan, 1954, Chaps. 1–2.

Karpinski, Louis C.: *The History of Arithmetic*, Rand McNally, 1925.

Menninger, K.: *Number Words and Number Symbols: A Cultural History of Numbers*, Massachusetts Institute of Technology Press, 1969.

Neugebauer, Otto: *The Exact Sciences in Antiquity*, Princeton University Press, 1952, Chaps. 1–3 and 5.

———: *Vorgriechische Mathematik*, Julius Springer, 1934, Chaps. 1–3 and 5.

Sarton, George: *A History of Science*, Harvard University Press, 1952, Vol. 1, Chap. 3.

———: *The Study of the History of Mathematics and the History of Science*, Dover (reprint), 1954.

Smith, David Eugene: *History of Mathematics*, Dover (reprint), 1958, Vol. 1, Chap. 1.

Struik, Dirk J.: *A Concise History of Mathematics*, 3rd ed., Dover, 1967, Chaps. 1–2.

van der Waerden, B. L.: *Science Awakening*, P. Noordhoff, 1954, Chaps. 2–3.

2

Egyptian Mathematics

> All science, logic and mathematics included, is a function
> of the epoch—all science, in its ideals as well as in its
> achievements.   E. H. MOORE

1. *Background*

While Mesopotamia experienced many changes in its ruling peoples, with
resulting new cultural influences, the Egyptian civilization developed un-
affected by foreign influences. The origins of the civilization are unknown but
it surely existed even before 4000 B.C. Egypt, as the Greek historian Herodotus says, is a gift of the Nile. Once a year this river, drawing its water from
the south, floods the territory all along its banks and leaves behind rich soil.
Most of the people did and still do make their living by tilling this soil. The
rest of the country is desert.

There were two kingdoms, one in the north and one in the south of what
is present-day Egypt. Some time between 3500 and 3000 B.C., the ruler,
Mena, or Menes, unified upper and lower Egypt. From this time on the
major periods of Egyptian history are referred to in terms of the ruling
dynasties, Menes having been the founder of the first dynasty. The height of
Egyptian culture occurred during the third dynasty (about 2500 B.C.),
during which period the rulers built the pyramids. The civilization went its
own way until Alexander the Great conquered it in 332 B.C. Thereafter,
until about A.D. 600, its history and mathematics belong to the Greek
civilization. Thus, apart from one minor invasion by the Hyksos (1700–
1600 B.C.) and slight contact with the Babylonian civilization (inferred from
the discovery in the Nile valley of the cuneiform Tell al-Amarna tablets of
about 1500 B.C.), Egyptian civilization was the product of its native people.

The ancient Egyptians developed their own systems of writing. One
system, hieroglyphics, was pictorial, that is, each symbol was a picture of
some object. Hieroglyphics were used on monuments until about the time of
Christ. From about 2500 B.C. the Egyptians used for daily purposes what is
called hieratic writing. This system employed conventional symbols, which
at first were merely simplifications of the hieroglyphics. Hieratic writing is

syllabic; each syllable is represented by an ideogram and an entire word is a collection of ideograms. The meaning of the word is not tied to the separate ideograms.

The writing was done with ink on papyrus, sheets made by pressing the pith of a plant and then slicing it. Since papyrus dries up and crumbles, very few documents of ancient Egypt have survived, apart from the hieroglyphic inscriptions on stone.

The main surviving mathematical documents are two sizable papyri: the Moscow papyrus, which is in Moscow, and the Rhind papyrus, discovered in 1858 by a Britisher, Henry Rhind, and now in the British Museum. The Rhind papyrus is also known as the Ahmes papyrus after its author, who opens with the words "Directions for Obtaining the Knowledge of All Dark Things." Both papyri date from about 1700 B.C. There are also fragments of other papyri written at this time and later. The mathematical papyri were written by scribes who were workers in the Egyptian state and church administrations.

The papyri contain problems and their solutions—85 in the Rhind papyrus and 25 in the Moscow papyrus. Presumably such problems occurred in the work of the scribes and they were expected to know how to solve them. It is most likely that the problems in the two major papyri were intended as examples of typical problems and solutions. Though these papyri date from about 1700 B.C., the mathematics in them was known to the Egyptians as far back as 3500 B.C., and little was added from that time to the Greek conquest.

2. *The Arithmetic*

The hieroglyphic number symbols used by the Egyptians were | for 1, ∩ for 10, ℰ and ℰ for 100, ⚡ for 1000, ⚡ for 10,000, and other symbols for larger units. These symbols were combined to form intermediate numbers. The direction of the writing was from right to left, so that IIII∩∩ represented 24. This system of writing numbers uses the base 10 but is not positional.

Egyptian hieratic whole numbers are illustrated by the following symbols:

| | || ||| — ′′| ⁗‴ ℓ = ⦠ ∧
1   2   3   4   5   6   7   8   9   10

The arithmetic was essentially additive. For ordinary additions and subtractions they could simply combine or take away symbols to reach the proper result. Multiplication and division were also reduced to additive processes. To calculate 12 times 12, say, the Egyptians did the following:

$$
\begin{array}{cc}
1 & 12 \\
2 & 24 \\
4 & 48 \\
8 & 96.
\end{array}
$$

Each line was derived from the preceding one by doubling. Now since $4 \cdot 12 = 48$ and $8 \cdot 12 = 96$, adding 48 and 96 gave the value of $12 \cdot 12$. This process was, of course, quite different from multiplying by 10 and by 2 and adding. Multiplication by 10 was also performed and consisted of replacing unit symbols by the symbol for 10 and replacing the 10 symbol by the symbol for 100.

Division of one whole number by another as carried out by the Egyptians is equally interesting. For example, 19 was divided by 8 as follows:

$$
\begin{array}{cc}
1 & 8 \\
2 & 16 \\
1/2 & 4 \\
1/4 & 2 \\
1/8 & 1
\end{array}
$$

Therefore, the answer was $2 + 1/4 + 1/8$. The idea was simply to take the number of eights and parts of eight that totaled 19.

The denotation of fractions in the Egyptian number system was much more complicated than our own. The symbol $\bigcirc$, pronounced *ro*, which originally indicated 1/320 of a bushel, came to denote a fraction. In hieratic writing the oval was replaced by a dot. The $\bigcirc$ or dot was generally placed above the whole number to indicate the fraction. Thus, in hieroglyphic writing,

$$
\underset{|||}{\bigcirc} = \frac{1}{5}, \qquad \underset{\cap}{\bigcirc} = \frac{1}{10}, \qquad \underset{\cap \ ||}{\bigcirc \ |||} = \frac{1}{15}.
$$

A few fractions were denoted by special symbols. Thus the hieroglyph $\longmapsto$ denoted 1/2; $\mathcal{T}$, 2/3; and $\times$, 1/4.

Aside from a few special ones, all fractions were decomposed into what are called unit fractions. Thus Ahmes writes 2/5 as $1/3 + 1/15$. The plus sign did not appear but was understood. The Rhind papyrus contains a table for expressing fractions with numerator 2 and odd denominators from 5 to 101 as sums of fractions with numerator 1. By means of this table a fraction such as our 7/29, which to Ahmes is the integer 7 divided by the integer 29, could also be expressed as a sum of unit fractions. Inasmuch as $7 = 2 + 2 + 2 + 1$, he proceeds by converting each 2/29 to a sum of fractions with numerator 1. By combining these results and by further conversion he ends up with a sum of unit fractions, each with a different denominator. The final expression for 7/29 is

$$\frac{1}{6} + \frac{1}{24} + \frac{1}{58} + \frac{1}{87} + \frac{1}{232}.$$

It so happens that 7/29 can also be expressed as $1/5 + 1/29 + 1/145$, but because Ahmes' $2/n$ table leads to the former expression, this is the one used. The expression of our a/b as a sum of unit fractions was done systematically according to age-old procedures. Using unit fractions, the Egyptians could carry out the four arithmetic operations with fractions. The extensive and complicated computations with fractions were one reason the Egyptians never developed arithmetic or algebra to an advanced state.

The nature of irrational numbers was not recognized in Egyptian arithmetic any more than it was in the Babylonian. The simple square roots that occurred in algebraic problems could be and were expressed in terms of whole numbers and fractions.

3. Algebra and Geometry

The papyri contain solutions of problems involving an unknown that are in the main comparable to our linear equations in one unknown. However, the processes were purely *arithmetical* and did not, in Egyptian minds, amount to a distinct subject, the solution of equations. The problems were stated verbally with bare directions for obtaining the solutions and without explanation of why the methods were used or why they worked. For example, problem 31 of the Ahmes papyrus, translated literally, reads: "A quantity, its 2/3, its 1/2, its 1/7, its whole, amount to 33." This means, for us:

$$\frac{2}{3}x + \frac{x}{2} + \frac{x}{7} + x = 33.$$

Simple arithmetic of the Egyptian variety gives the solution in this case.

Problem 63 of the papyrus runs as follows: "Directions for dividing 700 breads among four people, 2/3 for one, 1/2 for the second, 1/3 for the third, 1/4 for the fourth." For us this means

$$\frac{2}{3}x + \frac{1}{2}x + \frac{1}{3}x + \frac{1}{4}x = 700.$$

The solution, as given by Ahmes, is: "Add $\frac{2}{3}, \frac{1}{2}, \frac{1}{3}, \frac{1}{4}$. This gives $1\ \frac{1}{2}\ \frac{1}{4}$. Divide 1 by $1\ \frac{1}{2}\ \frac{1}{4}$. This gives $\frac{1}{2}\ \frac{1}{14}$. Now find $\frac{1}{2}\ \frac{1}{14}$ of 700. This is 400."

In some solutions Ahmes uses the "rule of false position." Thus to determine five numbers in arithmetic progression subject to a further condition and such that the sum is 100, he first chooses d, the common difference, to be 5 1/2 times the smallest number. He then picks 1 as the smallest and gets the progression: 1, 6 1/2, 12, 17 1/2, 23. But these numbers add up to 60 whereas they should add up to 100. He then multiplies each term by 5/3.

Only the simplest types of quadratic equations, such as $ax^2 = b$, are considered. Even when two unknowns occur, the type is

$$x^2 + y^2 = 100, \qquad y = \frac{3}{4}x,$$

so that after eliminating y, the equation in x reduces to the first type. Some concrete problems involving arithmetic and geometric progressions also can be found in the papyri. To infer general rules from all these problems and solutions is not very difficult.

The limited Egyptian algebra employed practically no symbolism. In the Ahmes papyrus, addition and subtraction are represented respectively by the legs of a man coming and going, $\diagup\!\!\!\Box$ and $\diagdown\!\!\!\diagdown$, and the symbol $\lceil$ is used to denote square root.

What of Egyptian geometry? The Egyptians did not separate arithmetic and geometry. We find problems from both fields in the papyri. Like the Babylonians, the Egyptians regarded geometry as a practical tool. One merely applied arithmetic and algebra to the problems involving areas, volumes, and other geometrical situations. Egyptian geometry is said by Herodotus to have originated in the need created by the annual overflow of the Nile to redetermine the boundaries of the lands owned by the farmers. However, Babylonia did as much in geometry without such a need. The Egyptians had prescriptions for the areas of rectangles, triangles, and trapezoids. In the case of the area of a triangle, though they multiplied one number by half another, we cannot be sure that the method is correct because we are not sure from the words used whether the lengths multiplied stood for base and altitude or just for two sides. Also the figures were so poorly drawn that one cannot be sure of just what area or volume were being found. Their calculation of the area of a circle, surprisingly good, followed the formula $A = (8d/9)^2$ where d is the diameter. This amounts to using 3.1605 for π.

An example may illustrate the "accuracy" of Egyptian formulas for area. On the walls of a temple in Edfu is a list of fields that were gifts to the temple. These fields generally have four sides, which we shall denote by a, b, c, d, where a and b and c and d are pairs of opposite sides. The inscriptions give the area of these various fields as $\frac{(a + b)}{2} \cdot \frac{(c + d)}{2}$. But some fields are triangles. In this case, d is said to be nothing and the calculation is changed to $\frac{(a + b)}{2} \cdot \frac{c}{2}$. Even for quadrangles the rule is just a crude approximation.

The Egyptians also had rules for the volume of a cube, box, cylinder, and other figures. Some of the rules were correct and others only approximations. The papyri give as the volume of a truncated conical clepsydra (water clock), in our notation,

$$V = \frac{h}{12} \left(\frac{3}{2} (D + d) \right)^2,$$

where h is the height and $(D + d)/2$ is the mean circumference. This formula amounts to using 3 for π.

The most striking rule of Egyptian geometry is the one for the volume of a truncated pyramid of square base, which in modern notation is

$$V = \frac{h}{3} (a^2 + ab + b^2),$$

where h is the height and a and b are sides of top and bottom. The formula is surprising because it is correct and because it is symmetrically expressed (but of course not in our notation). It is given only for concrete numbers. However, we do not know whether the pyramid was square-based or not because the figure in the papyrus is not carefully drawn.

Neither do we know whether the Egyptians recognized the Pythagorean theorem. We know there were rope-stretchers, that is, surveyors, but the story that they used a rope knotted at points to divide the total length into parts of ratios 3 to 4 to 5, which could then be used to form a right triangle, is not confirmed in any document.

The rules were not expressed in symbols. The Egyptians stated the problems verbally; and their procedure in solving them was essentially what we do when we calculate according to a formula. Thus an almost literal translation of the geometrical problem of finding the volume of the frustum of a pyramid reads: "If you are told: a truncated pyramid of 6 for the vertical height by 4 on the base, by 2 on the top. You are to square this 4, result 16. You are to double, result 8. You are to square 2, result 4. You are to add the 16, and 8, and the 4, result 28. You are to take one-third of 6, result 2. You are to take 28 twice, result 56. See, it is 56. You will find it right."

Did the Egyptians know proofs or justifications of their procedures and formulas? One belief is that the Ahmes papyrus was written in the style of a textbook for students of that day and hence, even though no general rules or principles for solving types of equations were formulated by Ahmes, it is very likely that he knew them but wanted the student to formulate them himself or have a teacher formulate them for him. Under this view the Ahmes papyrus is a rather advanced arithmetic text. Others say it is the notebook of a pupil. In either case, the papyri almost surely recorded the types of problems that had to be solved by business and administrative clerks, and the methods of solution were just practical rules known by experience to work. No one believes that the Egyptians had a deductive structure based on sound axioms that established the correctness of their rules.

4. *Egyptian Uses of Mathematics*

The Egyptians used mathematics in the administration of the affairs of the state and church, to determine wages paid to laborers, to find the volumes of granaries and the areas of fields, to collect taxes assessed according to the land area, to convert from one system of measures to another, and to calculate the number of bricks needed for the construction of buildings and ramps. The papyri also contain problems dealing with the amounts of corn needed to make given quantities of beer and the amount of corn of one quality needed to give the same result as corn of another quality whose strength relative to the first is known.

As in Babylonia a major use of mathematics was in astronomy, which dates from the first dynasty. Astronomical knowledge was essential. To the Egyptian the Nile was his life's blood. He made his living by tilling the soil which the Nile covered with rich silt in its annual overflow. However, he had to be well prepared for the dangerous aspects of the flood; his home, equipment, and cattle had to be temporarily removed from the area and arrangements made for sowing immediately afterwards. Hence the coming of the flood had to be predicted, which was done by learning what heavenly events preceded it.

Astronomy also made the calendar possible. Beyond the need for a calendar in commerce was the need to predict religious holidays. It was believed essential, to ensure the goodwill of the gods, that holidays be celebrated at the proper time. As in Babylonia, keeping the calendar was largely the task of the priests.

The Egyptians arrived at their estimate of the length of the solar year by observing the star Sirius. On one day in the summer this star became visible on the horizon just before sunrise. On succeeding days it was visible for a longer time before the sun's growing light blotted it out. The first day on which it was visible just before sunrise was known as the heliacal rising of Sirius, and the interval between two such days was about 365 1/4 days; so the Egyptians adopted, supposedly in 4241 B.C., a civil calendar of 365 days for the year. The concentration on Sirius is undoubtedly accounted for by the fact that the waters of the Nile began to rise on that day, which was chosen as the first day of the year.

The 365-day year was divided into 12 months of 30 days, plus 5 extra days at the end. Because the Egyptians did not intercalate the additional day every four years, the civil calendar lost all relation to the seasons. It takes 1460 years for the calendar to set itself right again; this interval is known as the Sothic cycle, from the Egyptian name for Sirius. Whether the Egyptians knew of the Sothic cycle is open to question. Their calendar was adopted by Julius Caesar in 45 b.c., but changed to a 365 1/4-day year on the advice of the Alexandrian Greek Sosigenes. Though the Egyptian determination of the year and the calendar were valuable contributions, they did not result from a well-developed astronomy, which in fact was crude and far inferior to Babylonian astronomy.

The Egyptians combined their knowledge of astronomy and geometry to construct their temples in such a manner that on certain days of the year the sun would strike them in a particular way. Thus some were built so that on the longest day of the year the sun would shine directly into the temple and illuminate the god at the altar. This orientation of temples is also found to some extent among the Babylonians and Greeks. The pyramids too were oriented to special directions of the heavens, and the Sphinx faces east. While the details of the construction of these works are unimportant for us, it is worth noting that the pyramids represent another application of Egyptian geometry. They are the tombs of kings; and because the Egyptians believed in immortality they believed that the proper construction of a tomb was material for the dead person's afterlife. In fact, an entire apartment for the future residence of king and queen was installed in each pyramid. They took great care to make the bases of the pyramids of the correct shape; the relative dimensions of base and height were also highly significant. However, one should not overemphasize the complexity or depth of the ideas involved. Egyptian mathematics was simple and crude and no deep principles were involved, contrary to what is often asserted.

5. *Summary*

Let us review the status of mathematics before the Greeks enter the picture. We find in the Babylonian and Egyptian civilizations an arithmetic of integers and fractions, including positional notation, the beginnings of algebra, and some empirical formulas in geometry. There was almost no symbolism, hardly any conscious thought about abstractions, no formulation of general methodology, and no concept of proof or even of plausible arguments that might convince one of the correctness of a procedure or formula. There was, in fact, no conception of any kind of theoretical science.

Apart from a few incidental results in Babylonia, mathematics in the two civilizations was not a distinct discipline, nor was it pursued for its own sake. It was a tool in the form of disconnected, simple rules which answered questions arising in the daily life of the people. Certainly nothing was done in mathematics that altered or affected the way of life. Although Babylonian mathematics was more advanced than the Egyptian, about the best one can say for both is that they showed some vigor, if not rigor, and more perseverance than brilliance.

All evaluation implies some sort of standard. It may be unfair but it is natural to compare the two civilizations with the Greek, which succeeded them. By this standard the Egyptians and Babylonians were crude carpenters, whereas the Greeks were magnificent architects. One does find more favorable, even laudatory, descriptions of the Babylonian and Egyptian achievements. But these are made by specialists who become, perhaps unconsciously, overimpressed by their own field of interest.

Bibliography

Boyer, Carl B.: *A History of Mathematics*, John Wiley and Sons, 1968, Chap. 2.

Cantor, Moritz: *Vorlesungen über Geschichte der Mathematik*, 2nd ed., B. G. Teubner, 1894, Vol. 1, Chap. 3.

Chace, A. B., *et al.*, eds.: *The Rhind Mathematical Papyrus*, 2 vols., Mathematical Association of America, 1927–29.

Childe, V. Gordon: *Man Makes Himself*, New American Library, 1951.

Karpinski, Louis C.: *The History of Arithmetic*, Rand McNally, 1925.

Neugebauer, O.: *The Exact Sciences in Antiquity*, Princeton University Press, 1952, Chap. 4.

————: *Vorgriechische Mathematik*, Julius Springer, 1934.

Sarton, George: *A History of Science*, Harvard University Press, 1952, Vol. 1, Chap. 2.

Smith, David Eugene: *History of Mathematics*, Dover (reprint), 1958, Vol. 1, Chap. 2; Vol. 2, Chaps. 2 and 4.

van der Waerden, B. L.: *Science Awakening*, P. Noordhoff, 1954, Chap. 1.

3
The Creation of Classical Greek Mathematics

> This, therefore, is mathematics: she reminds you of the invisible form of the soul; she gives life to her own discoveries; she awakens the mind and purifies the intellect; she brings light to our intrinsic ideas; she abolishes oblivion and ignorance which are ours by birth.   PROCLUS

1. *Background*

In the history of civilization the Greeks are preeminent, and in the history of mathematics the Greeks are the supreme event. Though they did borrow from the surrounding civilizations, the Greeks built a civilization and culture of their own which is the most impressive of all civilizations, the most influential in the development of modern Western culture, and decisive in founding mathematics as we understand the subject today. One of the great problems of the history of civilization is how to account for the brilliance and creativity of the ancient Greeks.

Though our knowledge of their early history is subject to correction and amplification as more archeological research is carried on, we now have reason to believe, on the basis of the *Iliad* and the *Odyssey* of Homer, the decipherment of ancient languages and scripts, and archeological investigations, that the Greek civilization dates back to 2800 B.C. The Greeks settled in Asia Minor, which may have been their original home, on the mainland of Europe in the area of modern Greece, and in southern Italy, Sicily, Crete, Rhodes, Delos, and North Africa. About 775 B.C. the Greeks replaced various hieroglyphic systems of writing with the Phoenician alphabet (which was also used by the Hebrews). With the adoption of an alphabet the Greeks became more literate, more capable of recording their history and ideas.

As the Greeks became established they visited and traded with the Egyptians and Babylonians. There are many references in classical Greek writings to the knowledge of the Egyptians, whom some Greeks erroneously considered the founders of science, particularly surveying, astronomy, and arithmetic. Many Greeks went to Egypt to travel and study. Others visited Babylonia and learned mathematics and science there.

The influence of the Egyptians and Babylonians was almost surely felt in Miletus, a city of Ionia in Asia Minor and the birthplace of Greek philosophy, mathematics, and science. Miletus was a great and wealthy trading city on the Mediterranean. Ships from the Greek mainland, Phoenicia, and Egypt came to its harbors; Babylonia was connected by caravan routes leading eastward. Ionia fell to Persia about 540 B.C., though Miletus was allowed some independence. After an Ionian revolt against Persia in 494 B.C. was crushed, Ionia declined in importance. It became Greek again in 479 B.C. when Greece defeated Persia, but by then cultural activity had shifted to the mainland of Greece with Athens as its center.

Though the ancient Greek civilization lasted until about A.D. 600, from the standpoint of the history of mathematics it is desirable to distinguish two periods, the classical, which lasted from 600 to 300 B.C., and the Alexandrian or Hellenistic, from 300 B.C. to A.D. 600. The adoption of the alphabet, already mentioned, and the fact that papyrus became available in Greece during the seventh century B.C. may account for the blossoming of cultural activity about 600 B.C. The availability of this writing paper undoubtedly helped the spread of ideas.

2. *The General Sources*

The sources of our knowledge of Greek mathematics are, peculiarly, less authentic and less reliable than our sources for the much older Babylonian and Egyptian mathematics, because no original manuscripts of the important Greek mathematicians are extant. One reason is that papyrus is perishable; though the Egyptians also used papyrus, by luck a few of their mathematical documents did survive. Some of the voluminous Greek writings might still be available to us if their great libraries had not been destroyed.

Our chief sources for the Greek mathematical works are Byzantine Greek codices (manuscript books) written from 500 to 1500 years after the Greek works were originally composed. These codices are not literal reproductions but critical editions, so that we cannot be sure what changes may have been made by the editors. We also have Arabic translations of the Greek works and Latin versions derived from Arabic works. Here again we do not know what changes the translators may have made or how well they understood the original texts. Moreover, even the Greek texts used by the Arabic and Byzantine authors were questionable. For example, though we do not have the Alexandrian Greek Heron's manuscript, we know that he made a number of changes in Euclid's *Elements*. He gave different proofs and added new cases of the theorems and converses. Likewise Theon of Alexandria (end of 4th cent. A.D.) tells us that he altered sections of the *Elements* in his edition.

The Greek and Arabic versions we have may come from such versions of the originals. However, in one or another of these forms we do have the works of Euclid, Apollonius, Archimedes, Ptolemy, Diophantus, and other Greek authors. Many Greek texts written during the classical and Alexandrian periods did not come down to us because even in Greek times they were superseded by the writings of these men.

The Greeks wrote some histories of mathematics and science. Eudemus (4th cent. B.C.), a member of Aristotle's school, wrote a history of arithmetic, a history of geometry, and a history of astronomy. Except for fragments quoted by later writers, these histories are lost. The history of geometry dealt with the period preceding Euclid's and would be invaluable were it available. Theophrastus (c. 372–c. 287 B.C.), another disciple of Aristotle, wrote a history of physics, and this, too, except for a few fragments, is lost.

In addition to the above, we have two important commentaries. Pappus (end of 3rd cent. A.D.) wrote the *Synagoge* or *Mathematical Collection*; almost the whole of it is extant in a twelfth-century copy. This is an account of much of the work of the classical and Alexandrian Greeks from Euclid to Ptolemy, supplemented by a number of lemmas and theorems that Pappus added as an aid to understanding. Pappus had also written the *Treasury of Analysis*, a collection of the Greek works themselves. This book is lost, but in Book VII of his *Mathematical Collection* he tells us what his *Treasury* contained.

The second important commentator is Proclus (A.D. 410–485), a prolific writer. Proclus drew material from the texts of the Greek mathematicians and from prior commentaries. Of his surviving works, the *Commentary*, which treats Book I of Euclid's *Elements*, is the most valuable. Proclus apparently intended to discuss more of the *Elements*, but there is no evidence that he ever did so. The *Commentary* contains one of the three quotations traditionally credited to Eudemus' history of geometry (see sec. 10) but probably taken from a later modification. This particular extract, the longest of the three, is referred to as the Eudemian summary. Proclus also tells us something about Pappus' work. Thus, besides the later editions and versions of some of the Greek classics themselves, Pappus' *Mathematical Collection* and Proclus' *Commentary* are the two main sources of the history of Greek mathematics.

Of original wordings (though not the manuscripts) we have only a fragment concerning the lunes of Hippocrates, quoted by Simplicius (first half of 6th cent. A.D.) and taken from Eudemus' lost *History of Geometry*, and a fragment of Archytas on the duplication of the cube. And of original manuscripts we have some papyri written in Alexandrian Greek times. Related sources on Greek mathematics are also immensely valuable. For example, the Greek philosophers, especially Plato and Aristotle, had much to say about mathematics and their writings have survived somewhat in the same way as have the mathematical works.

The reconstruction of the history of Greek mathematics, based on sources such as we have described, has been an enormous and complicated task. Despite the extensive efforts of scholars, there are gaps in our knowledge and some conclusions are arguable. Nevertheless the basic facts are clear.

3. The Major Schools of the Classical Period

The cream of the classical period's contributions are Euclid's *Elements* and Apollonius' *Conic Sections*. Appreciation of these works requires some knowledge of the great changes made in the very nature of mathematics and of the problems the Greeks faced and solved. Moreover, these polished works give little indication of the three hundred years of creative activity preceding them or of the issues which became vital in the subsequent history.

Classical Greek mathematics developed in several centers that succeeded one another, each building on the work of its predecessors. At each center an informal group of scholars carried on its activities under one or more great leaders. This kind of organization is common in modern times also and its reason for being is understandable. Today, when one great man locates at a particular place—generally a university—other scholars follow, to learn from the master.

The first of the schools, the Ionian, was founded by Thales (*c.* 640–*c.* 546 B.C.) in Miletus. We do not know the full extent to which Thales may have educated others, but we do know that the philosophers Anaximander (*c.* 610–*c.* 547 B.C.) and Anaximenes (*c.* 550–480 B.C.) were his pupils. Anaxagoras (*c.* 500–*c.* 428 B.C.) belonged to this school, and Pythagoras (*c.* 585–*c.* 500 B.C.) is supposed to have learned mathematics from Thales. Pythagoras then formed his own large school in southern Italy. Toward the end of the sixth century, Xenophanes of Colophon in Ionia migrated to Sicily and founded a center to which the philosophers Parmenides (5th cent. B.C.) and Zeno (5th cent. B.C.) belonged. The latter two resided in Elea in southern Italy, to which the school had moved, and so the group became known as the Eleatic school. The Sophists, active from the latter half of the fifth century onward, were concentrated mainly in Athens. The most celebrated school is the Academy of Plato in Athens, where Aristotle was a student. The Academy had unparalleled importance for Greek thought. Its pupils and associates were the greatest philosophers, mathematicians, and astronomers of their age; the school retained its pre-eminence in philosophy even after the leadership in mathematics passed to Alexandria. Eudoxus, who learned mathematics chiefly from Archytas of Tarentum (Sicily), founded his own school in Cyzicus, a city of northern Asia Minor. When Aristotle left Plato's Academy he founded another school, the Lyceum, in Athens. The Lyceum is commonly referred to as the Peripatetic school. Not all of the great mathematicians of the classical period can be identified with a school, but for the sake of coherence we shall occasionally discuss the work of a man in connection with a particular school even though his association with it was not close.

4. *The Ionian School*

The leader and founder of this school was Thales. Though there is no sure knowledge about Thales' life and work, he probably was born and lived in Miletus. He traveled extensively and for a while resided in Egypt, where he carried on business activities and reportedly learned much about Egyptian mathematics. He is, incidentally, supposed to have been a shrewd business-man. During a good season for olive growing, he cornered all the olive presses in Miletus and Chios and rented them out at a high fee. Thales is said to have predicted an eclipse of the sun in 585 B.C., but this is disputed on the ground that astronomical knowledge was not adequate at that time.

He is reputed to have calculated the heights of pyramids by comparing their shadows with the shadow cast by a stick of known height at the same time. By some such use of similar triangles he is supposed to have calculated the distance of a ship from shore. He is also credited with having made mathematics abstract and with having given deductive proofs for some theorems. These last two claims, however, are dubious. Discovery of the attractive power of magnets and of static electricity is also attributed to Thales.

The Ionian school warrants only brief mention so far as contributions to mathematics proper are concerned, but its importance for philosophy and the philosophy of science in particular is unparalleled (see Chap. 7, sec. 2). The school declined in importance when the Persians conquered the area.

5. *The Pythagoreans*

The torch was picked up by Pythagoras who, supposedly having learned from Thales, founded his own school in Croton, a Greek settlement in southern Italy. There are no written works by the Pythagoreans; we know about them through the writings of others, including Plato and Herodotus. In particular we are hazy about the personal life of Pythagoras and his followers; nor can we be sure of what is to be credited to him personally or to his followers. Hence when one speaks of the work of Pythagoras one really refers to the work done by the group between 585 B.C., the reputed date of his birth, and roughly 400 B.C. Philolaus (5th cent. B.C.) and Archytas (428–347 B.C.) were prominent members of this school.

Pythagoras was born on the island of Samos, just off the coast of Asia Minor. After spending some time with Thales in Miletus, he traveled to other places, including Egypt and Babylon, where he may have picked up some mathematics and mystical doctrines. He then settled in Croton. There he founded a religious, scientific, and philosophical brotherhood. It was a formal school, in that membership was limited and members learned from leaders. The teachings of the group were kept secret by the members, though the secrecy as to mathematics and physics is denied by some historians. The Pythagoreans were supposed to have mixed in politics; they allied themselves with the aristocratic faction and were driven out by the popular or democratic party. Pythagoras fled to nearby Metapontum and was murdered there about 497 B.C. His followers spread to other Greek centers and continued his teachings.

One of the great Greek contributions to the very concept of mathematics was the conscious recognition and emphasis of the fact that mathematical entities, numbers, and geometrical figures are abstractions, ideas entertained by the mind and sharply distinguished from physical objects or pictures. It is true that even some primitive civilizations and certainly the Egyptians and Babylonians had learned to think about numbers as divorced from physical objects. Yet there is some question as to how much they were consciously aware of the abstract nature of such thinking. Moreover, geometrical thinking in all pre-Greek civilizations was definitely tied to matter. To the Egyptians, for example, a line was no more than either a stretched rope or the edge of a field and a rectangle was the boundary of a field.

The recognition that mathematics deals with abstractions may with some confidence be attributed to the Pythagoreans. However, this may not have been true at the outset of their work. Aristotle declared that the Pythagoreans regarded numbers as the ultimate components of real, material objects.[1] Numbers did not have a detached existence apart from objects of sense. When the early Pythagoreans said that all objects were composed of (whole) numbers or that numbers were the essence of the universe, they meant it literally, because numbers to them were like atoms are to us. It is also believed that the sixth- and fifth-century Pythagoreans did not really distinguish numbers from geometrical dots. Geometrically, then, a number was an extended point or a very small sphere. However, Eudemus, as reported by Proclus, says that Pythagoras rose to higher principles (than had the Egyptians and Babylonians) and considered abstract problems for the pure intelligence. Eudemus adds that Pythagoras was the creator of pure mathematics, which he made into a liberal art.

The Pythagoreans usually depicted numbers as dots in sand or as pebbles. They classified the numbers according to the shapes made by the arrangements of the dots or pebbles. Thus the numbers 1, 3, 6, and 10 were called triangular because the corresponding dots could be arranged as triangles (Fig. 3.1). The fourth triangular number, 10, especially fascinated the Pythagoreans because it was a prized number for them, and had 4 dots on

1. *Metaphys.* I, v, 986a and 986a 21, Loeb Classical Library ed.

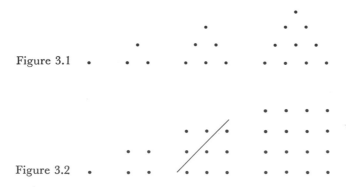

Figure 3.1

Figure 3.2

each side, 4 being another favorite number. They realized that the sums 1, 1 + 2, 1 + 2 + 3, and so forth gave the triangular numbers and that 1 + 2 + ⋯ + n = (n/2)(n + 1).

The numbers 1, 4, 9, 16, ... were called square numbers because as dots they could be arranged as squares (Fig. 3.2). Composite (nonprime) numbers which were not perfect squares were called oblong.

From the geometrical arrangements certain properties of the whole numbers became evident. Introducing the slash, as in the third illustration of Figure 3.2, shows that the sum of two consecutive triangular numbers is a square number. This is true generally, for as we can see, in modern notation,

$$\frac{n}{2}(n + 1) + \frac{n + 1}{2}(n + 2) = (n + 1)^2.$$

That the Pythagoreans could prove this general conclusion, however, is doubtful.

To pass from one square number to the next one, the Pythagoreans had the scheme shown in Figure 3.3. The dots to the right of and below the lines in the figure formed what they called a gnomon. Symbolically, what they saw here was that $n^2 + (2n + 1) = (n + 1)^2$. Further, if we start with 1 and

Figure 3.3

Figure 3.4. The shaded area is the gnomon.

add the gnomon 3 and then the gnomon 5, and so forth, what we have in our symbolism is

$$1 + 3 + 5 + \cdots + (2n - 1) = n^2.$$

As to the word "gnomon," originally in Babylonia it probably meant an upright stick whose shadow was used to tell time. In Pythagoras' time it meant a carpenter's square, and this is the shape of the above gnomon. It also meant what was left over from a square when a smaller square was cut out of one corner. Later, with Euclid, it meant what was left from a parallelogram when a smaller one was cut out of one corner provided that the parallelogram in the lower right-hand corner was similar to the one cut out (Fig. 3.4).

The Pythagoreans also worked with polygonal numbers such as pentagonal, hexagonal, and higher ones. As we can see from Figure 3.5, where each dot represents a unit, the first pentagonal number is 1, the second, whose dots form the vertices of a pentagon, is 5; the third is $1 + 4 + 7$, or 12, and so forth. The nth pentagonal number, in our notation, is $(3n^2 - n)/2$. Likewise the hexagonal numbers (Fig. 3.6) are 1, 6, 15, 28, . . . and generally $2n^2 - n$.

A number that equaled the sum of its divisors including 1 but not the number itself was called perfect; for example, 6, 28, and 496. Those exceeding the sum of the divisors were called excessive and those which were less were called defective. Two numbers were called amicable if each was the sum of the divisors of the other, for example, 284 and 220.

The Pythagoreans devised a rule for finding triples of integers which could be the sides of a right triangle. This rule implies knowledge of the Pythagorean theorem, about which we shall say more later. They found that when m is odd, then m, $(m^2 - 1)/2$, and $(m^2 + 1)/2$ are such a triple. However,

Figure 3.5. Pentagonal numbers

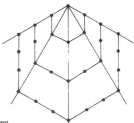

Figure 3.6. Hexagonal numbers

this rule gives only some sets of such triples. Any set of three integers which can be the sides of a right triangle is now called a Pythagorean triple.

The Pythagoreans studied prime numbers, progressions, and those ratios and proportions they regarded as beautiful. Thus if p and q are two numbers, the arithmetic mean A is $(p + q)/2$, the geometric mean G is $\sqrt{pq}$, and the harmonic mean H, which is the reciprocal of the arithmetic mean of $1/p$ and $1/q$, is $2pq/(p + q)$. Now G is seen to be the geometric mean of A and H. The proportion $A/G = G/H$ was called the perfect proportion and the proportion $p:(p + q)/2 = 2pq/(p + q):q$ was called the musical proportion.

Numbers to the Pythagoreans meant whole numbers only. A ratio of two whole numbers was not a fraction and therefore another kind of number, as it is in modern times. Actual fractions, expressing parts of a monetary unit or a measure, were employed in commerce, but such commercial uses of arithmetic were outside the pale of Greek mathematics proper. Hence the Pythagoreans were startled and disturbed by the discovery that some ratios—for example, the ratio of the hypotenuse of an isosceles right triangle to an arm or the ratio of a diagonal to a side of a square—cannot be expressed by whole numbers. Since the Pythagoreans had concerned themselves with whole-number triples that could be the sides of a right triangle, it is most likely that they discovered these new ratios in this work. They called ratios expressed by whole numbers commensurable ratios, which means that the two quantities are measured by a common unit, and they called ratios not so expressible, incommensurable ratios. Thus what we express as $\sqrt{2}/2$ is an incommensurable ratio. The ratio of incommensurable magnitudes was called αλογος (*alogos*, inexpressible). The term αρρητος (*arratos*, not having a ratio) was also used. The discovery of incommensurable ratios is attributed to Hippasus of Metapontum (5th cent. B.C.). The Pythagoreans were supposed to have been at sea at the time and to have thrown Hippasus overboard for having produced an element in the universe which denied the Pythagorean doctrine that all phenomena in the universe can be reduced to whole numbers or their ratios.

The proof that $\sqrt{2}$ is incommensurable with 1 was given by the Pythagoreans. According to Aristotle, their method was a *reductio ad absurdum*—that is, the indirect method. The proof showed that if the hypotenuse were commensurable with an arm then the same number would be both odd and even. It runs as follows: Let the ratio of hypotenuse to arm of an isosceles right triangle be $\alpha:\beta$ and let this ratio be expressed in the smallest numbers. Then $\alpha^2 = 2\beta^2$ by the Pythagorean theorem. Since α^2 is even, α must be even, for the square of any odd number is odd.[2] Now the ratio $\alpha:\beta$ is in its lowest terms. Hence β must be odd. Since α is even, let $\alpha = 2\gamma$. Then $\alpha^2 = 4\gamma^2 = 2\beta^2$. Hence $\beta^2 = 2\gamma^2$ and so β^2 is even. Then β is even. But β is also odd and so there is a contradiction.

This proof, which is of course the same as the modern one that $\sqrt{2}$ is irrational, was included in older editions of Euclid's *Elements* as Proposition 117 of Book X. However, it was most likely not in Euclid's original text and so is omitted in modern editions.

Incommensurable ratios are expressed in modern mathematics by irrational numbers. But the Pythagoreans would not accept such numbers. The Babylonians did work with such numbers by approximating them, though they probably did not know that their sexagesimal fractional approximations could never be made exact. Nor did the Egyptians recognize the distinctive nature of irrationals. The Pythagoreans did at least recognize that incommensurable ratios are entirely different in character from commensurable ones.

This discovery posed a problem that was central in Greek mathematics. The Pythagoreans had, up to this point, identified number with geometry. But the existence of incommensurable ratios shattered this identification. They did not cease to consider all kinds of lengths, areas, and ratios in geometry, but they restricted the consideration of numerical ratios to commensurable ones. The theory of proportions for incommensurable ratios and all kinds of magnitudes was provided by Eudoxus, whose work we shall consider shortly.

Some geometrical results are also credited to the Pythagoreans. The most famous is the Pythagorean theorem itself, a key theorem of Euclidean geometry. The Pythagoreans are also supposed to have discovered what we learn as theorems about triangles, parallel lines, polygons, circles, spheres, and the regular polyhedra. They knew in particular that the sum of the angles of a triangle is 180°. A limited theory of similar figures and the fact that a plane can be filled out with equilateral triangles, squares, and regular hexagons are included among their results.

The Pythagoreans started work on a class of problems known as

---

2. Any odd whole number can be expressed as $2n + 1$ for some n. Then $(2n + 1)^2 = 4n^2 + 4n + 1$, and this is necessarily odd.

application of areas. The simplest of these was to construct a polygon equal in area to a given polygon and similar to another given one. Another was to construct a specified figure with an area exceeding or falling short of another by a given area. The most important form of the problem of application of areas is: Given a line segment, construct on part of it or on the line segment extended a parallelogram equal to a given rectilinear figure in area and falling short (in the first case) or exceeding (in the second case) by a parallelogram similar to a given parallelogram. We shall discuss application of areas when we study Euclid's work.

The most vital contribution of the Greeks to mathematics is the insistence that all mathematical results be established deductively on the basis of explicit axioms. Hence the question arises as to whether the Pythagoreans proved their geometric results. No unequivocal answer can be given, but it is very doubtful that deductive proof on any kind of axiomatic basis, explicit or implicit, was a requirement in the early or middle period of Pythagorean mathematics. Proclus does affirm that they proved the angle sum theorem; this may have been done by the late Pythagoreans. The question of whether they proved the Pythagorean theorem has been extensively pursued, and the answer is that they probably did not. It is relatively easy to prove it by using facts about similar triangles, but the Pythagoreans did not have a complete theory of similar figures. The proof given in Proposition 47 of Book I of Euclid's *Elements* (Chap. 4, sec. 4) is a difficult one because it does not use the theory of similar figures, and this proof was credited by Proclus to Euclid himself. The most likely conclusion about proof in Pythagorean geometry is that during most of the life of the school the members affirmed results on the basis of special cases, much as they did in their arithmetic. However, by the time of the late Pythagoreans, that is, about 400 B.C., the status of proof had changed because of other developments; so these latter-day members of the brotherhood may have given legitimate proofs.

6. *The Eleatic School*

The Pythagorean discovery of incommensurable ratios brought to the fore a difficulty that preoccupied all the Greeks, namely, the relation of the discrete to the continuous. Whole numbers represent discrete objects, and a commensurable ratio represents a relation between two collections of discrete objects, or two lengths that have a common unit measure so that each length is a discrete collection of units. However, lengths in general are not discrete collections of units; this is why ratios of incommensurable lengths appear. Lengths, areas, volumes, time, and other quantities are, in other words, continuous. We would say that line segments, for example, can have irrational as well as rational lengths in terms of some unit. But the Greeks had not attained this view.

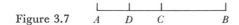

Figure 3.7   A   D   C   B

The problem of the relation of the discrete to the continuous was brought into the limelight by Zeno, who lived in the southern Italian city of Elea. Born some time between 495 and 480 B.C., Zeno was a philosopher rather than a mathematician, and like his master Parmenides was said to have been a Pythagorean originally. He proposed a number of paradoxes, of which four deal with motion. His purpose in posing these paradoxes is not clear because not enough of the history of Greek philosophy is known. He was said to be defending Parmenides, who had argued that motion or change is impossible. He was also attacking the Pythagoreans, who believed in extended but indivisible units, the points of geometry. We do not know precisely what Zeno said but must rely upon quotations from Aristotle, who cites Zeno in order to criticize him, and from Simplicius, who lived in the sixth century A.D. and based his statements on Aristotle's writings.

The four paradoxes on motion are distinct, but the import of all four taken together was probably intended to be the significant argument. Two opposing views of space and time were held in Zeno's day: one, that space and time are infinitely divisible, in which case motion is continuous and smooth; and the other, that space and time are made up of indivisible small intervals (like a movie), in which case motion is a succession of minute jerks. Zeno's arguments are directed against both theories, the first two paradoxes being against the first theory and the latter two against the second theory. The first paradox of each pair considers the motion of a single body and the second considers the relative motion of bodies.

Aristotle in his *Physics* states the first paradox, called the Dichotomy, as follows: "The first asserts the nonexistence of motion on the ground that that which is in motion must arrive at the half-way stage before it arrives at the goal." This means that to traverse *AB* (Fig. 3.7), one must first arrive at *C*; to arrive at *C* one must first arrive at *D*; and so forth. In other words, on the assumption that space is infinitely divisible and therefore that a finite length contains an infinite number of points, it is impossible to cover even a finite length in a finite time.

Aristotle, refuting Zeno, says there are two senses in which a thing may be infinite: in divisibility or in extent. In a finite time one can come into contact with things infinite in respect to divisibility, for in this sense time is also infinite; and so a finite extent of time can suffice to cover a finite length. Zeno's argument has been construed by others to mean that to go a finite length one must cover an infinite number of points and so must get to the end of something that has no end.

The second paradox is called Achilles and the Tortoise. According to

Figure 3.8

Aristotle: "It says that the slowest moving object cannot be overtaken by the fastest since the pursuer must first arrive at the point from which the pursued started so that necessarily the slower one is always ahead. The argument is similar to that of the Dichotomy, but the difference is that we are not dividing in halves the distances which have to be passed over." Aristotle then says that if the slowly moving object covers a finite distance, it can be overtaken for the same reason he gives in answering the first paradox.

The next two paradoxes are directed against "cinematographic" motion. The third paradox, called the Arrow, is given by Aristotle as follows: "The third paradox he [Zeno] spoke about, is that a moving arrow is at a standstill. This he concludes from the assumption that time is made up of instants. If it would not be for this supposition, there would be no such conclusion." According to Aristotle, Zeno means that at any instant during its motion the arrow occupies a definite position and so is at rest. Hence it cannot be in motion. Aristotle says that this paradox fails if we do not grant indivisible units of time.

The fourth paradox, called the Stadium or the Moving Rows, is put by Aristotle in these words: "The fourth is the argument about a set of bodies moving on a race-course and passing another set of bodies equal in number and moving in the opposite direction, the one starting from the end, the other from the middle and both moving at equal speed; he [Zeno] concluded that it follows that half the time is equal to double the time. The mistake is to assume that two bodies moving at equal speeds take equal times in passing, the one a body which is in motion, and the other a body of equal size which is at rest, an assumption which is false."

The probable point of Zeno's fourth paradox can be stated as follows: Suppose that there are three rows of soldiers, A, B, and C (Fig. 3.8), and that in the smallest unit of time B moves one position to the left, while in that time C moves one position to the right. Then relative to B, C has moved two positions. Hence there must have been a smaller unit of time in which C was one position to the right of B or else half the unit of time equals the unit of time.

It is possible that Zeno merely intended to point out that speed is relative. C's speed relative to B is not C's speed relative to A. Or he may have meant there is no absolute space to which to refer speeds. Aristotle says that Zeno's fallacy consists in supposing that things that move with the same speed past a moving object and past a fixed object take the same time. Neither Zeno's argument nor Aristotle's answer is clear. But if we think of

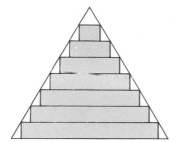

Figure 3.9

this paradox as attacking indivisible smallest intervals of time and indivisible smallest segments of space, which Zeno was attacking, then his argument makes sense.

We may include with the Eleatics Democritus (*c.* 460–*c.* 370 B.C.) of Abdera in Thrace. He is reputed to have been a man of great wisdom who worked in many fields, including astronomy. Since Democritus belonged to the school of Leucippus and the latter was a pupil of Zeno, many of the mathematical questions Democritus considered must have been suggested by Zeno's ideas. He wrote works on geometry, on number, and on continuous lines and solids. The works on geometry could very well have been significant predecessors of Euclid's *Elements*.

Archimedes says Democritus discovered that the volumes of a cone and a pyramid are 1/3 of the volumes of the cylinder and prism having the same base and height, but that the proofs were made by Eudoxus. Democritus regarded the cone as a series of thin indivisible layers (Fig. 3.9), but was troubled by the fact that if the layers were equal they should yield a cylinder and if unequal the cone could not be smooth.

7. *The Sophist School*

After the final defeat of the Persians at Mycale in 479 B.C., Athens became the major city in a league of Greek cities and a commercial center. The wealth acquired through trading, which made Athens the richest city of its time, was used by the famous leader Pericles to build up and adorn the city. Ionians, Pythagoreans, and intellectuals generally were attracted to Athens. Here emphasis was given to abstract reasoning and the goal of extending the domain of reason over the whole of nature and man was set.

The first Athenian school, the Sophist, embraced learned teachers of grammar, rhetoric, dialectics, eloquence, morals, and—what is of interest to us—geometry, astronomy, and philosophy. One of their chief pursuits was the use of mathematics to understand the functioning of the universe.

Many of the mathematical results obtained were by-products of efforts to solve the three famous construction problems: to construct a square equal in area to a given circle; to construct the side of a cube whose volume is double that of a cube of given edge; and to trisect any angle—all to be performed with straightedge and compass only.

The origin of these famous construction problems is accounted for in various ways. For example, one version of the origin of the problem of doubling the cube, found in a work of Eratosthenes (c. 284–192 B.C.), relates that the Delians, suffering from a pestilence, consulted the oracle, who advised constructing an altar double the size of the existing one. The Delians realized that doubling the side would not double the volume and turned to Plato, who told them that the god of the oracle had not so answered because he wanted or needed a doubled altar, but in order to censure the Greeks for their indifference to mathematics and their lack of respect for geometry. Plutarch also gives this story.

Actually, these construction problems are extensions of problems already solved by the Greeks. Since any angle could be bisected, it was natural to consider trisection. Since the diagonal of a square is the side of a square double in area to that of the original square, the corresponding problem for the cube becomes relevant. The problem of squaring the circle is typical of many Greek problems of constructing a figure of prescribed shape equal in area to a given figure. Another problem not quite so famous was to construct regular polygons of seven and more sides. Here, too, the construction of the square, regular pentagon, and regular hexagon suggested the next step.

Various explanations of the restriction to straightedge and compass have been given. The straight line and the circle were, in the Greek view, the basic figures, and the straightedge and compass are their physical analogues. Hence constructions with these tools were preferable. The reason is also given that Plato objected to other mechanical instruments because they involved too much of the world of the senses rather than the world of ideas, which he regarded as primary. It is very likely, however, that in the fifth century the restriction to straightedge and compass was not rigid. But, as we shall see, constructions played a vital role in Greek geometry and Euclid's axioms did limit constructions to those made with straightedge and compass. Hence from his time on, this restriction may have been taken more seriously. Pappus, for example, says that if a construction can be carried out with straightedge and compass, a solution using other means is not satisfactory.

The earliest known attempt to solve any of the three famous problems was made by the Ionian Anaxagoras, who is supposed to have worked on squaring the circle while in prison. We do not know any more about this work. One of the most famous attempts is due to Hippias of Elis, a city in the Peloponnesus. Hippias, a leading Sophist, was born about 460 B.C. and was a contemporary of Socrates.

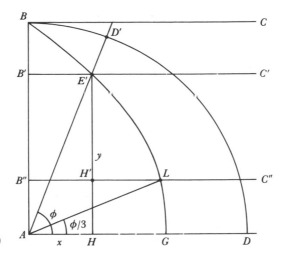

Figure 3.10

In his attempts to trisect an angle, Hippias invented a new curve, which, unfortunately, is not itself constructible with straightedge and compass. His curve is called the quadratrix and is generated as follows: Let AB (Fig. 3.10) rotate clockwise about A at a constant speed to the position AD. In the *same time* let BC move downward parallel to itself at a uniform speed to AD. Suppose AB reaches AD' as BC reaches $B'C'$. Let E' be the intersection of AD' and $B'C'$. Then E' is a typical point on the quadratrix $BE'G$. G is the final point on the quadratrix.[3]

The equation of the quadratrix in terms of rectangular Cartesian coordinates can be obtained as follows: Let AD' reach AD in some fraction t/T of the total time T that AB takes to reach AD. Since AD' and $B'C'$ move at constant speeds, $B'C'$ covers that part $E'H$ of BA in the same fraction of the total time. Hence

$$\frac{\phi}{\pi/2} = \frac{E'H}{BA}.$$

If we denote $E'H$ by y and BA by a, then

(1)
$$\frac{\phi}{\pi/2} = \frac{y}{a}$$

3. The point G cannot be obtained directly from the definition of the curve because AB reaches AD at the same instant as BC reaches AD and so there is no point of intersection of the rotating line and the horizontal line. G can be obtained only as the limit of preceding points of the quadratrix. By using the calculus we can show that $AG = 2a/\pi$ where $a = AB$.

or

$$y = a \cdot \phi \cdot \frac{2}{\pi}.$$

But if $AH = x$, then

$$\phi = \text{arc tan} \frac{y}{x}.$$

Hence

$$y = \frac{2a}{\pi} \text{arc tan} \frac{y}{x}$$

or

$$y = x \tan \frac{\pi y}{2a}.$$

The curve, if constructible, could be used to trisect any acute angle. Let ϕ be such an angle. Then trisect y so that $E'H' = 2H'H$. Draw $B''C''$ through H' and let it cut the quadratrix in L. Draw AL. Then $\sphericalangle LAD = \phi/3$, for, by the argument which led to (1),

$$\frac{\sphericalangle LAD}{\pi/2} = \frac{H'H}{a}$$

or

$$\frac{\sphericalangle LAD}{\pi/2} = \frac{y/3}{a}.$$

But by (1)

$$\frac{\phi}{\pi/2} = \frac{y}{a}.$$

Hence

$$\sphericalangle LAD = \frac{\phi}{3}.$$

Another famous discovery that resulted from the work on the construction problems was made by Hippocrates of Chios (5th cent. B.C.), the most famous mathematician of his century, who is to be distinguished from his contemporary Hippocrates of Cos, the father of Greek medicine. The mathematician Hippocrates flourished in Athens during the second half of the century; he was not a Sophist, but most likely a Pythagorean. He is credited with the idea of arranging theorems so that later ones can be proven on the

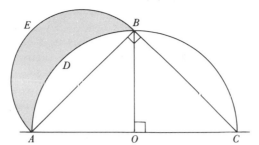

Figure 3.11

basis of earlier ones, in the manner familiar to us from the study of Euclid. He is also credited with introducing the indirect method of proof into mathematics. His text on geometry, called the *Elements*, is lost.

Hippocrates did not, of course, solve the problem of squaring the circle, but he did solve a related one. Let ABC be an isosceles right triangle (Fig. 3.11) and let it be inscribed in the semicircle with center at O. Let AEB be the semicircle with AB as diamter. Then

$$\frac{\text{Area semicircle } ABC}{\text{Area semicircle } AEB} = \frac{AC^2}{AB^2} = \frac{2}{1}.$$

Hence the area $OADB$ equals the area of the semicircle AEB. Now subtract the area ADB common to both. Then the area of the lune (shaded area) equals the area of triangle AOB. Thus the area of the lune, an area bounded by arcs, equals the area of a rectilinear figure; or, a curvilinear figure is reduced to a rectilinear one. This result is called a quadrature; that is, a curvilinear area has been computed in effect because it is equal to an area bounded by straight lines and the latter can be computed.

In this proof Hippocrates had to use the fact that the areas of two circles are to each other as the squares of their diameters. It is doubtful that Hippocrates really had a proof of this fact because the proof depends upon the method of exhaustion invented later by Eudoxus.

Hippocrates also squared three other lunes. This work on the lunes is known to us through the writings of Simplicius and is the only sizable fragment of *classical* Greek mathematics that we have as originally written.

Hippocrates also showed that the problem of doubling the cube can be reduced to finding two mean proportionals between the given side and one twice as long. In our algebraic notation, let x and y be such that

$$\frac{a}{x} = \frac{x}{y} = \frac{y}{2a}.$$

Then

$$x^2 = ay \quad \text{and} \quad y^2 = 2ax.$$

Since $y = x^2/a$ we obtain from the second equation that

$$x^3 = 2a^3.$$

This x is the desired answer. It cannot be constructed by straightedge and compass. Of course Hippocrates must have reasoned geometrically, in a manner that will be clearer when we examine Apollonius' *Conic Sections*.

One more very important idea was hit upon by the Sophists Antiphon (5th cent. B.C.) and Bryson (*c.* 450 B.C.). While trying to square the circle, Antiphon conceived the idea of approaching the area of a circle by inscribing polygons of more and more sides. Bryson added to this the idea of using circumscribed polygons. Antiphon further suggested that the circle be considered a polygon of an infinite number of sides. We shall see how these ideas were taken up by Eudoxus in the method of exhaustion (Chap. 4, sec. 9).

8. *The Platonic School*

The Platonic school succeeded the Sophists in the leadership of mathematical activity. Its forerunners, Theodorus of Cyrene in North Africa (born *c.* 470 B.C.) and Archytas of Tarentum in southern Italy (428–347 B.C.), were Pythagoreans and both taught Plato. Their teachings may have produced the strong Pythagorean influence in the entire Platonic school.

Theodorus is noted for having proved that the ratios that we represent as $\sqrt{3}, \sqrt{5}, \sqrt{7}, \ldots, \sqrt{17}$ are incommensurable with a unit. Archytas introduced the idea of regarding a curve as generated by a moving point and a surface as generated by a moving curve. He solved the duplication of the cube problem by finding two mean proportionals between two given quantities. These mean proportionals were constructed geometrically by finding the intersection of three surfaces: a circle rotated about a tangent, a cone, and a cylinder. The construction is quite detailed and does not warrant space here. Archytas also wrote on mathematical mechanics, designed machines, studied sound, and contributed inventions and some theory on musical scales.

The Platonic school was headed by Plato and included Menaechmus (4th cent. B.C.), his brother Dinostratus (4th cent. B.C.), and Theaetetus (*c.* 415–*c.* 369 B.C.). Many other members are known to us only by name.

Plato (427–347 B.C.) was born of a distinguished family and early in life had political ambitions. But the fate of Socrates convinced him there was no place in politics for a man of conscience. He traveled in Egypt and among the Pythagoreans in lower Italy; the Pythagorean influence may have been generated by these contacts. About 387 B.C. Plato founded in Athens his Academy, which in most respects was like a modern university. The Academy had grounds, buildings, students, and formal courses taught by Plato and his aides. During the classical period the study of mathematics and philosophy was favored there. Though the main center for mathematics shifted to Alexandria about 300 B.C., the Academy remained preeminent in philosophy throughout the Alexandrian period. It lasted nine hundred years until it was closed by the Christian emperor Justinian in A.D. 529 because it taught " pagan and perverse learning."

Plato, one of the most informed men of his day, was not a mathematician; but his enthusiasm for the subject and his belief in its importance for philosophy and for the understanding of the universe encouraged mathematicians to pursue it. It is noteworthy that almost all of the important mathematical work of the fourth century was done by the friends and pupils of Plato. Plato himself seems to have been more concerned to improve and perfect what was known.

Though we may not be sure to what extent the concepts of mathematics were treated as abstractions prior to Plato's time, there is no question that Plato and his successors did so regard them. Plato says that numbers and geometrical concepts have nothing material in them and are distinct from physical things. The concepts of mathematics are independent of experience and have a reality of their own. They are discovered, not invented or fashioned. This distinction between abstractions and material objects may have come from Socrates.

A quotation from Plato's *Republic* may serve to illustrate the contemporary view of the mathematical concepts. Socrates addresses Glaucon:

> And all arithmetic and calculation have to do with number.
> Yes....
> Then this is knowledge of the kind for which we are seeking, having a double use, military and philosophical; for the man of war must learn the art of number or he will not know how to array his troops, and the philosopher also, because he has to rise out of the sea of change and lay hold of true being, and therefore he must be an arithmetician.... Then this is a kind of knowledge which legislation may fitly prescribe; and we must endeavour to persuade those who are to be the principal men of our State to go and learn arithmetic, not as amateurs, but they must carry on the study until they see the nature of numbers with the mind only; nor again, like merchants or retail-traders, with a view to buying or selling, but for the sake of their military use, and of the soul herself; and because this will be the easiest way for her to pass from becoming to truth and being.... I mean, as I was saying, that arithmetic has a very great and elevating effect, compelling the soul to reason about abstract number, and rebelling against the introduction of visible or tangible objects into the argument....[4]

4. Book VII, sec. 525; in B. Jowett, *The Dialogues of Plato*, Clarendon Press, 1953, Vol. 2.

Another quotation[5] discusses the concepts of geometry. Speaking about mathematicians, Plato says: "And do you not know also that although they further make use of the visible forms and reason about them, they are thinking not of these, but of the ideals which they resemble . . . but they are really seeking to behold the things themselves, which can be seen only with the eye of the mind."

It is clear from these quotations that Plato and other Greeks for whom he speaks valued abstract ideas and preferred mathematical ideas as a preparation for philosophy. The abstract ideas with which mathematics deals are akin to others such as goodness and justice, the understanding of which is the goal of Plato's philosophy. Mathematics is the preparation for knowledge about the ideal universe.

Why did the Greeks prefer and stress the abstract concepts of mathematics? We cannot answer the question, but we should note that the early Greek mathematicians were philosophers and philosophers generally exerted a formative influence in the development of Greek mathematics. Philosophers are interested in ideas and typically show their propensity for abstractions in many domains. Thus the Greek philosophers thought about truth, goodness, charity, and intelligence. They speculated about the ideal society and the perfect state. The later Pythagoreans and the Platonists distinguished sharply between the world of ideas and the world of things. Relationships in the material world were subject to change and hence did not represent ultimate truth, but relationships in the ideal world were unchanging and absolute truths; these were the proper concern of the philosopher.

Plato in particular believed that the perfect ideals of physical objects are the reality. The world of ideals and relationships among them is permanent, ageless, incorruptible, and universal. The physical world is an imperfect realization of the ideal world and is subject also to decay. Hence the ideal world alone is worthy of study. Infallible knowledge can be obtained only about pure intelligible forms. About the physical world we can have only opinions; and physical science is sunk in the dregs of a sensuous world.

We are not sure whether the Platonists contributed the deductive structure of mathematics. They were concerned with proof and with the methodology of reasoning. Proclus and Diogenes Laertius (3rd cent. A.D.) credit the Platonists with two types of methodology. The first is the method of analysis, where what is to be established is regarded as known and the consequences deduced until a known truth or a contradiction is reached. If a contradiction is reached then the desired conclusion is false. If a known truth is reached then the steps are reversed, if possible, and the proof is made. The second is the method of *reductio ad absurdum* or the indirect method. The first method was probably not new with Plato, but perhaps he emphasized

5. *Republic*, Book VI, sec. 510.

the necessity for the subsequent synthesis. The indirect method, as already noted, is also attributed to Hippocrates.

The status of deductive structure with Plato is best indicated by a passage in *The Republic*.[6] He says

> You are aware that students of geometry, arithmetic and the kindred sciences assume the odd and the even and the figures and three kinds of angles and the like in their several branches of science; these are their hypotheses, which they and everybody are supposed to know, and therefore they do not deign to give any account of them either to themselves or others; but they begin with them, and go on until they arrive at last, and in a consistent manner, at their conclusion.

If this passage is indeed descriptive of the mathematics of the time, then proofs were certainly made, but the axiomatic basis was implicit or may have varied somewhat from one mathematician to another.

Plato did affirm the desirability of a deductive organization of knowledge. The task of science was to discover the structure of (ideal) nature and to give it an articulation in a deductive system. Plato was the first to systematize the rules of rigorous demonstration, and his followers are supposed to have arranged theorems in logical order. Also, we know that in Plato's Academy the question arose whether a given problem can be solved at all on the basis of the known truths and the hypotheses given in the problem. Whether or not mathematics was actually deductively organized on the basis of explicit axioms by the Platonists, there is no question that deductive proof from some accepted principles was required from at least Plato's time onward. By insisting on this form of proof the Greeks were discarding all rules, procedures, and facts that had been accepted in the body of mathematics for thousands of years preceding the Greek period.

Why did the Greeks insist on deductive proof? Since induction, observation, and experimentation were and still are vital sources of knowledge heavily and advantageously employed by the sciences, why did the Greeks prefer in mathematics deductive reasoning to the exclusion of all other methods? We know that the Greeks, the philosophical geometers as they were called, liked reasoning and speculation, as evidenced by their great contributions to philosophy, logic, and theoretical science. Moreover, philosophers are interested in obtaining truths. Whereas induction, experimentation, and generalizations based on experience yield only probable knowledge, deduction gives absolutely certain results if the premises are correct. Mathematics in the classical Greek world was part of the body of truths philosophers sought and accordingly had to be deductive.

Still another reason for the Greek preference for deduction may be found in the contempt shown by the educated class of the classical Greek

6. Book VI, sec. 510.

period toward practical affairs. Though Athens was a commercial center, business as well as such professions as medicine were carried on by the slave class. Plato contended that the engagement of freemen in trade should be punished as a crime, and Aristotle said that in the perfect state no citizen (as opposed to slaves) should practice any mechanical art. To thinkers in such a society, experimentation and observation would be alien. Hence no results, scientific or mathematical, would be derived from such sources.

There is, incidentally, evidence that in the sixth and fifth centuries B.C. the Greek attitude toward work, trade, and technical skills had been quite different and that mathematics had been applied to the practical arts. Thales used his mathematics to improve navigation. Solon, a ruler of the sixth century, invested the crafts with honor and inventors were esteemed. *Sophia*, the Greek word usually taken to mean wisdom and abstract thought, at that time meant technical skill. It was the Pythagoreans, Proclus says, who "transformed mathematics into a free education," that is, an education for free men rather than a skill for slaves.

Plutarch in his life of Marcellus substantiates the change in attitude toward such devices as mechanical instruments:

> Eudoxus and Archytas had been the first originators of this far-famed and highly-prized art of mechanics, which they employed as an elegant illustration of geometrical truths, and as means of sustaining experimentally, to the satisfaction of the senses, conclusions too intricate for proof by words and diagrams. As, for example, to solve the problem, so often required in constructing geometrical figures, given the two extremes, to find the two mean lines of a proportion, both of these mathematicians had recourse to the aid of instruments, adapting to their purpose certain curves and sections of lines. But what with Plato's indignation at it, and his invectives against it as the mere corruption and annihilation of the one good of geometry, which was thus shamefully turning its back upon the unembodied objects of pure intelligence to recur to sensation, and to ask help (not to be obtained without base supervisions and deprivation) from matter, so it was that mechanics came to be separated from geometry, and, repudiated and neglected by philosophers, took its place as a military art.

This accounts for the poor development of experimental science and the science of mechanics in the classical Greek period.

Whether or not historical research has isolated the relevant factors to explain the Greek preference for deductive reasoning, we do know that they were the first to insist on deductive reasoning as the sole method of *proof* in mathematics. This requirement has been characteristic of mathematics ever since and has distinguished mathematics from all other fields of knowledge or investigation. However, we have yet to see to what extent later mathematicians remained true to this principle.

Figure 3.12

As far as the content of mathematics is concerned, Plato and his school improved the definitions and are also supposed to have proved new theorems of plane geometry. Further, they gave impetus to solid geometry. In Book VII, section 528 of *The Republic*, Plato says that before one can consider astronomy, which treats solids in motion, one needs a science of such solids. But this science, he says, has been neglected. He complains that the investigators of solid figures have not received due support from the state. Plato and his associates proceeded to study solid geometry and are supposed to have proved new theorems. They studied the properties of the prism, pyramid, cylinder, and cone; and they knew that there can be at most five regular polyhedra. The Pythagoreans undoubtedly knew that one can form three of these solids with 4, 8, and 20 equilateral triangles, the cube with squares, and the dodecahedron with 12 pentagons, but the proof that there can be no more than five is probably due to Theaetetus.

The Platonic school's most significant discovery was the conic sections. The discovery is attributed by the Alexandrian Eratosthenes to Menaechmus, a geometer and astronomer, who was a pupil of Eudoxus but a member of Plato's Academy. While it is not known for certain what led to the discovery of the conic sections, a common belief is that it resulted from the work on the famous construction problems. We know that Hippocrates of Chios solved the problem of doubling the cube by finding an x and y such that

$$a:x = x:y = y:2a.$$

But these equations say that

$$x^2 = ay, \qquad y^2 = 2ax, \quad \text{and} \quad xy = 2a^2.$$

Hence we can see through coordinate geometry that x and y are the coordinates of the point of intersection of two parabolas or a parabola and a hyperbola. Menaechmus worked on the problem and saw both ways of solving it through pure geometry. According to the mathematical historian Otto Neugebauer (1899–), the conic sections might have originated in work on the construction of sundials.

Menaechmus introduced the conic sections by using three types of cones (Fig. 3.12), right-angled, acute-angled, and obtuse-angled, and cutting each by a plane perpendicular to an element. Only one branch of the hyperbola was recognized at this time.

Among other mathematical studies made by the Platonists was Theaetetus' work on incommensurables. Previously Theodorus of Cyrene had proved that (in our notation and language) $\sqrt{3}$, $\sqrt{5}$, $\sqrt{7}$, and other square roots are irrational. Theaetetus investigated other and higher types of irrationals and classified them. We shall note these types when we study Book X of Euclid's *Elements*. In this work of Theaetetus we see how the number system was being extended to more irrationals, but only those incommensurable ratios were studied which arose from geometrical thinking and could be constructed geometrically as lengths. Dinostratus, another Platonist, showed how to use the quadratrix of Hippias to square the circle. Aristaeus the Elder (*c.* 320 B.C.) is said by Pappus to have written a work in five books called *Elements of Conic Sections*.

9. *The School of Eudoxus*

The greatest of the classical Greek mathematicians and second only to Archimedes in all antiquity was Eudoxus. Eratosthenes called him "godlike." He was born in Cnidos in Asia Minor about 408 B.C., studied under Archytas in Tarentum, traveled in Egypt, where he learned some astronomy, and then founded a school at Cyzicus in northern Asia Minor. About 368 B.C. he and his followers joined Plato. Some years later he returned to Cnidos and died there about 355 B.C. An astronomer, physician, geometer, legislator, and geographer, he is most noted for the creation of the first astronomical theory of the heavenly motions (Chap. 7).

His first great contribution to mathematics was a new theory of proportion. The discovery of more and more irrationals (incommensurable ratios) made it necessary for the Greeks to face these numbers. Were they in fact numbers? They occurred in geometrical arguments, whereas the whole numbers and ratios of whole numbers occurred both in geometry and in the general study of quantity. Moreover, how could proofs of geometry which had been made for commensurable lengths, areas, and volumes be extended to incommensurable ones?

Eudoxus introduced the notion of a magnitude (Chap. 4, sec. 5). It was not a number but stood for entities such as line segments, angles, areas, volumes, and time which could vary, as we would say, continuously. Magnitudes were opposed to numbers, which jumped from one value to another, as from 4 to 5. No quantitative values were assigned to magnitudes. Eudoxus then defined a ratio of magnitudes and a proportion, that is, an equality of two ratios, to cover commensurable and incommensurable ratios. However, again, no numbers were used to express such ratios. The concepts of ratio and proportion were tied to geometry, as we shall see when we study Book V of Euclid.

What Eudoxus accomplished was to avoid irrational numbers as numbers. In effect, he avoided giving numerical values to lengths of line segments, sizes of angles, and other magnitudes, and to ratios of magnitudes. While Eudoxus' theory enabled the Greek mathematicians to make tremendous progress in geometry by supplying the necessary logical foundation for incommensurable ratios, it had several unfortunate consequences.

For one thing, it forced a sharp separation between number and geometry, for only geometry could handle incommensurable ratios. It also drove mathematicians into the ranks of the geometers, and geometry became the basis for almost all rigorous mathematics for the next two thousand years. We still speak of x^2 as x square and x^3 as x cube instead of x second or x third, say, because the magnitudes x^2 and x^3 had only geometric meaning to the Greeks.

The Eudoxian solution to the problem of treating incommensurable lengths or the irrational number actually reversed the emphasis of previous Greek mathematics. The early Pythagoreans had certainly emphasized number as the fundamental concept, and Archytas of Tarentum, Eudoxus' teacher, stated that arithmetic alone, not geometry, could supply satisfactory proofs. However, in turning to geometry to handle irrational numbers, the classical Greeks abandoned algebra and irrational numbers as such. What did they do about solving quadratic equations, where the solutions can indeed be irrational numbers? And what did they do about the simple problem of finding the area of a rectangle whose sides are incommensurable? The answer is that they converted most of algebra to geometry, in a manner we shall examine in the next chapter. The geometric representation of irrationals and of operations with irrationals was, of course, not practical. It might be logically satisfactory to think of $\sqrt{2} \cdot \sqrt{3}$ as an area of a rectangle, but if one needed to know the product in order to buy floor covering, he would not have it.

Though the Greeks devoted their deepest efforts in mathematics to geometry, we must keep in mind that whole numbers and ratios of whole numbers were still acceptable concepts. This area of mathematics, as we shall see, was built up deductively in Books VII, VIII, and IX of Euclid's *Elements*. The material is essentially what we call the theory of numbers or the properties of integers.

The question also arises: What did the classical Greeks do about the need for numbers in scientific work and in commerce and other practical affairs? Classical Greek science, as we shall see, was qualitative. As for the practical uses of numbers, we mentioned earlier that the intellectuals of that period confined themselves to philosophical and scientific activities and took no hand in commerce or the trades; educated people did not concern themselves with practical problems. But one could think about all rectangles in geometry without concerning himself in the least with the actual dimensions of even one rectangle. Mathematical thought was thus separated from practical needs, and there was no compulsion for the mathematicians to improve arithmetical and algebraic techniques. When the barrier between the cultured and slave classes was breached in the Alexandrian period (300 B.C. to about A.D. 600) and educated men interested themselves in practical affairs, the emphasis shifted to quantitative knowledge and the development of arithmetic and algebra.

To return to the contributions of Eudoxus, the powerful Greek method of establishing the areas and volumes of curved figures, now called the method of exhaustion, is also due to him. We shall examine the method and its use, as given by Euclid, later. It is really the first step in the calculus but does not use an explicit theory of limits. With it Eudoxus proved, for example, that the areas of two circles are to each other as the squares of their radii, the volumes of two spheres are to each other as the cubes of their radii, the volume of a pyramid is one-third the volume of a prism of the same base and altitude, and the volume of a cone is one-third the volume of the corresponding cylinder.

Some authority can be found to credit every school from Thales' onward with having introduced the deductive organization of mathematics. There is no question, however, that the work of Eudoxus established the deductive organization on *the basis of explicit axioms*. The necessity for understanding and operating with incommensurable ratios is undoubtedly the reason for this step. Since Eudoxus undertook to provide the precise logical basis for these ratios, he most likely saw the need to formulate axioms and deduce consequences one by one so that no mistakes would be made with these unfamiliar and troublesome magnitudes. This need to work with incommensurable ratios also undoubtedly reinforced the earlier decision to rely only on deductive reasoning for proof.

Because the Greeks sought truths and had decided on deductive proof, they had to obtain axioms that were themselves truths. They did find statements whose truth was self-evident to them, though the justifications given for accepting the axioms as indisputable truths varied. Almost all Greeks believed that the mind was capable of recognizing truths. Plato applied his theory of anamnesis, that we have had direct experience of truths in a period of existence as souls in another world before coming to earth, and we have but to recall this experience to know that these truths included the axioms of geometry. No experience on earth is necessary. Some historians read into statements by Plato and Proclus the idea that there can be some arbitrariness in the axioms, provided only that they are clear and true in the mind of the individual. The important thing is to reason deductively on the basis of the ones chosen. Aristotle had a good deal to say about axioms and we shall note his views in a moment.

10. *Aristotle and His School*

Aristotle (384–322 B.C.) was born in Stageira, a city in Macedonia. For twenty years he was a pupil and colleague of Plato, and for three years, from 343 to 340 B.C., he was the tutor of Alexander the Great. In 335 B.C. he founded his own school, the Lyceum. It had a garden, a lecture room, and an altar to the Muses.

Aristotle wrote on mechanics, physics, mathematics, logic, meteorology, botany, psychology, zoology, ethics, literature, metaphysics, economics, and many other fields. There is no one book on mathematics but discussions of the subject occur in a variety of places, and he uses it to illustrate a number of points.

He viewed the sciences as falling into three types—theoretical, productive, and practical. The theoretical ones, which seek truth, are mathematics, physics (optics, harmonics, and astronomy), and metaphysics; of these mathematics is the most exact. The productive sciences are the arts; and the practical ones, for example ethics and politics, seek to regulate human actions. In the theoretical sciences, logic is preliminary to the several subjects included there, and the metaphysician discusses and explains what the mathematician and natural philosopher (scientist) take for granted, for example, the being or reality of the subject matter and the nature of axioms.

Though Aristotle did not contribute significant new mathematical results (a few theorems in Euclid are his), his views on the nature of mathematics and its relation to the physical world were highly influential. Whereas Plato believed that there was an independent, eternally existing world of ideas which constituted the reality of the universe and that mathematical concepts were part of this world, Aristotle favored concrete matter or substance. However, he too arrived at an *emphasis* on ideas, namely, the universal essences of physical objects, such as hardness, softness, heaviness, lightness, sphericity, coldness, and warmth. Numbers and geometrical forms, too, were properties of real objects; they were recognized by abstraction but belonged to the objects. Thus mathematics deals with abstract concepts, which are derived from properties of physical bodies.

Aristotle discusses definition. His notion of definition is modern; he calls it a name for a collection of words. He also points out that definition must be in terms of something prior to the thing defined. Thus he criticizes the definition, "a point is that which has no part," because the words "that which" do not say what they refer to, except possibly "point" and so the definition is not proper. He grants the need for undefined terms, since there must be a starting point for the series of definitions, but later mathematicians lost sight of this need until the end of the nineteenth century.

He also notes (as Plato, according to Plutarch, did earlier) that a definition tells us what a thing is but not that the thing exists. The existence of defined things has to be proved except in the case of a few primary things such as point and line, whose existence is assumed along with the first principles or axioms. Thus one can define a square, but such a figure may not exist; that is, the properties demanded in the definition may be incompatible. Leibniz gave the example of a regular polyhedron with ten faces; one can define such a figure but it does not exist. If one did not realize that this figure did not exist, and proceeded to prove theorems about it, his results would be nonsensical. The method of proving existence that Aristotle and Euclid adopted was construction. The first three axioms in Euclid's *Elements* grant the construction of straight lines and circles; all other mathematical concepts must be constructed to establish their existence. Thus angle trisectors, though definable, are not constructible with straight lines and circles and so could not be considered in Greek geometry.

Aristotle also treats the basic principles of mathematics. He distinguishes between the axioms or common notions, which are truths common to all sciences, and the postulates, which are acceptable first principles for any one science. Among axioms he includes logical principles, such as the law of contradiction, the law of excluded middle, the axiom that if equals are added or subtracted from equals the results are equal, and other such principles. The postulates need not be self-evident but their truth must then be attested to by the consequences derived from them. The collection of axioms and postulates should be of the fewest possible, provided they enable all the results to be proved. Though, as we shall see, Euclid uses Aristotle's distinction between common notions and postulates, all mathematicians up to the late nineteenth century overlooked this distinction and treated axioms and postulates as equally self-evident. According to Aristotle, the axioms are obtained from the observation of physical objects. They are immediately apprehended generalizations. He and his followers gave many definitions and axioms or improved earlier ones. Some of the Aristotelian versions were taken up by Euclid.

Aristotle discusses the fundamental problem of how points and lines can be related. A point, he says, is indivisible and has position. But then no accumulation of points, however far it may be carried, can give us anything divisible, whereas of course a line is a divisible magnitude. Hence points cannot make up anything continuous like a line, for point cannot be continuous with point. A point, he says, is like the now in time; now is indivisible and not a part of time. A point may be an extremity, beginning, or divider of a line but is not a part of it or of magnitude. It is only by *motion* that a point can generate a line and thus be the origin of magnitude. He also argues that a point has no length and so if a line were composed of points, it would have no length. Similarly, if time were composed of instants there would be no interval of time. His definition of continuity, which a line possesses, is: A thing is continuous when the limits at which any two successive parts touch are one and the same and are, as the word continuous implies, held together. Actually Aristotle makes many statements about continuous magnitudes which are not in agreement with each other. The substance of his doctrine, nevertheless, is that points and numbers are discrete quantities and must be distinguished from the continuous magnitudes of geometry. There is no continuum in arithmetic. As to the relation of the two fields, he considers arithmetic—that is, the theory of numbers—more accurate, because numbers lend themselves to abstraction more readily than the geometric concepts. He also considers arithmetic to be prior to geometry because the number three is needed to consider a triangle.

In discussing infinity he makes a distinction, which is important today, between the potentially infinite and the actually infinite. The age of the earth, if it had a sudden beginning, is potentially infinite but is never at any time actually infinite. According to Aristotle, only the potentially infinite exists. The positive integers, he grants, are potentially infinite in that we can always add 1 to any number and get a new one, but the infinite set as such does not exist. Further, most magnitudes cannot be even potentially infinite, because if they were continually added to they could exceed the bounds of the universe. Space, however, is potentially infinite, in that it can be repeatedly subdivided, and time is potentially infinite in both ways.

A major achievement of Aristotle was the founding of the science of logic. In producing correct laws of mathematical reasoning the Greeks had laid the groundwork for logic, but it took Aristotle to codify and systematize these laws into a separate discipline. Aristotle's writings make it abundantly clear that he derived logic from mathematics. His basic principles of logic— the law of contradiction, which asserts that a proposition cannot be both true and false, and the law of excluded middle, which maintains that a proposition must be either true or false—are the heart of the indirect method of mathematical proof. Further, Aristotle used mathematical examples taken from contemporary texts to illustrate his principles of reasoning. Aristotelian logic remained unchallenged until the nineteenth century.

Though the science of logic was derived from mathematics, logic eventually came to be considered independent of and prior to mathematics and applicable to all reasoning. Even Aristotle, as already noted, regarded logic as preliminary to science and philosophy. In mathematics he emphasized deductive proof as the sole basis for establishing facts. For Plato, who believed that mathematical truths preexist or exist in a world independent of man, reasoning was not the guarantee of the correctness of theorems; the logical powers played only a secondary role. They made explicit, so to speak, what was already known to be true.

One member of Aristotle's school especially worthy of note is Eudemus of Rhodes, who lived in the last part of the fourth century B.C. and was the author of the Eudemian summary quoted by Proclus and Simplicius. As we noted earlier, Eudemus wrote histories of arithmetic, geometry, and astronomy. He is the first historian of science on record. But what is more significant is that the knowledge already existing in his time should have been sufficiently extensive to warrant histories.

The last of the men of the classical period we shall mention here is Autolycus of Pitane, an astronomer and geometer, who flourished about 310 B.C. He was not a member of Plato's or Aristotle's schools, though he did teach one of the leaders who succeeded Plato. Of three books he wrote, two have come down to us; they are the earliest Greek books we have intact, though only through manuscripts that presumably are copies of Autolycus' work. These books, *On the Moving Sphere* and *On Risings and Settings*, were eventually included in a collection called the *Little Astronomy* (as distinguished from Ptolemy's later *Great Collection*, or the *Almagest*). *On the Moving Sphere* treats meridian circles, great circles generally, and what we would call parallels of latitude, as well as the visible and invisible areas produced by a distant light source shining on a rotating sphere, as the sun does on the earth. The book presupposes theorems of spherical geometry which must, therefore, have been known to the Greeks of that time. Autolycus' second book, on the rising and setting of stars, belongs to observational astronomy.

The form of the book on moving spheres is significant. Letters denote points on diagrams. The propositions are logically ordered. Each proposition is first stated generally, then repeated, but with explicit reference to the figure; finally, the proof is given. This is the style Euclid uses.

Bibliography

Apostle, H. G.: *Aristotle's Philosophy of Mathematics*, University of Chicago Press, 1952.

Ball, W. W. Rouse: *A Short Account of the History of Mathematics*, Dover (reprint), 1960, Chaps. 2–3.

Boyer, Carl B.: *A History of Mathematics*, John Wiley and Sons, 1968, Chaps. 4–6.

Brumbaugh, Robert S.: *Plato's Mathematical Imagination*, Indiana University Press, 1954.

Gomperz, Theodor: *Greek Thinkers*, 4 vols., John Murray, 1920.

Guthrie, W. K. C.: *A History of Greek Philosophy*, Cambridge University Press, 1962 and 1965, Vols. 1 and 2.

Hamilton, Edith: *The Greek Way to Western Civilization*, New American Library, 1948.

Heath, Thomas L.: *A History of Greek Mathematics*, Oxford University Press, 1921, Vol. 1.

———: *A Manual of Greek Mathematics*, Dover (reprint), 1963, Chaps. 4–9.

———: *Mathematics in Aristotle*, Oxford University Press, 1949.

Jaeger, Werner: *Paideia*, 3 vols., Oxford University Press, 1939–44.

Lasserre, François: *The Birth of Mathematics in the Age of Plato*, American Research Council, 1964.

Maziarz, Edward A., and Thomas Greenwood: *Greek Mathematical Philosophy*, F. Unger, 1968.

Sarton, George: *A History of Science*, Harvard University Press, 1952, Vol. 1, Chaps. 7, 11, 16, 17, and 20.

Scott, J. F.: *A History of Mathematics*, Taylor and Francis, 1958, Chap. 2.

Smith, David Eugene: *History of Mathematics*, Dover (reprint), 1958, Vol. 1, Chap. 3.

van der Waerden, B. L.: *Science Awakening*, P. Noordhoff, 1954, Chaps. 4–6.

Wedberg, Anders: *Plato's Philosophy of Mathematics*, Almqvist and Wiksell, 1955.

4
Euclid and Apollonius

We have learned from the very pioneers of this science not to
have any regard to mere plausible imaginings when it is a
question of the reasonings to be included in our geometrical
doctrine.
                                                        PROCLUS

1. *Introduction*

The cream of the mathematical work created by the men of the classical
period has fortunately come down to us in the writings of two men, Euclid
and Apollonius. Chronologically, both belong to the second great period of
Greek history, the Hellenistic or Alexandrian. It is quite certain that Euclid
lived in Alexandria about 300 B.C. and trained students there, though his
own education was probably acquired in Plato's Academy. This information, incidentally, is about all we have on Euclid's personal life and even this
comes from a one-paragraph passage in Proclus' *Commentary*. Apollonius died
in 190 B.C., so his life too falls within the Alexandrian period. It is customary,
however, to identify Euclid's work with the classical period, because his
books are accounts of what was developed in that age. Euclid's work is
actually an organization of the separate discoveries of the classical Greeks;
this is clear from a comparison of its contents with what is known of the
earlier work. The *Elements* in particular is as much a mathematical history
of the age just brought to a close as it is the logical development of a subject.
Apollonius' work is generally classed with that of the Alexandrian period but
the content and spirit of his major work, *Conic Sections*, is of the classical
period. In fact Apollonius acknowledged that the first four of the work's
eight books were a revision of Euclid's lost work on the same subject. Pappus
mentions that Apollonius spent a long time with the pupils of Euclid at
Alexandria, which readily explains the kinship with Euclid. The justification
for identifying Apollonius with the work of the classical period will be more
apparent when we have learned the characteristics of the work of the
Alexandrian period.

2. *The Background of Euclid's* Elements

Euclid's most famous work is the *Elements*. The major sources of the material in this work can generally be identified despite our slim knowledge of the classical period. Euclid undoubtedly owes much of his material to the Platonists with whom he studied. Moreover, Proclus says that Euclid put into his *Elements* many of Eudoxus' theorems, perfected theorems of Theaetetus, and made irrefragable demonstrations of results loosely proved by his predecessors.

The particular choice of axioms, the arrangement of the theorems, and some proofs are his, as are the polish and the rigor of the demonstrations. The form of presentation of proof has, however, already been noted in Autolycus and was pretty surely used by others who preceded Euclid. Despite all he may have taken from earlier texts and other sources, Euclid was unquestionably a great mathematician. His other writings support this judgment despite any question as to how much of the *Elements* is original with Euclid. Proclus makes it clear that the *Elements* was highly valued in Greece and refers by way of substantiation to the numerous commentaries on it. Among the most important must have been those by Heron (*c.* 100 B.C.–*c.* A.D. 100), Porphyry (3rd cent. A.D.), and Pappus (end of 3rd cent. A.D.). Presumably Euclid's book was so good that it superseded the ones supposed to have been written by Hippocrates of Chios and by the Platonists Leon and Theudius.

There are no extant manuscripts written by Euclid himself. Hence his writings have had to be reconstructed from numerous recensions, commentaries, and remarks by other writers. All of the English and Latin editions of Euclid's *Elements* stem originally from Greek manuscripts. These were Theon of Alexandria's recension of Euclid's *Elements* (end of the 4th cent. A.D.), copies of Theon's recension, written versions of lectures by Theon, and one Greek manuscript found by François Peyrard (1760–1822) in the Vatican Library. This tenth-century manuscript is a copy of an edition of Euclid that precedes Theon's. Hence the historians J. L. Heiberg and Thomas L. Heath have used principally this manuscript for their study of Euclid, comparing it of course with the other available manuscripts and commentaries. There are also Arabic translations of Greek works and Arabic commentaries, presumably based on Greek manuscripts no longer available. These, too, have been used to decide what was in Euclid's *Elements*. But the Arabic translations and revisions are on the whole inferior to the Greek manuscripts. Of course, the reconstruction, since it is based on so many sources, leaves some matters in doubt. The purpose of Euclid's *Elements* is in question. It is considered by some as a treatise for learned mathematicians and by others as a text for students. Proclus gives some weight to the latter belief.

In view of the length and incomparable historical importance of this work, we shall devote several sections of this chapter to a review of and comment on the contents. Since we still learn Euclidean geometry, we may be somewhat surprised by the contents of the *Elements*. The high school versions most widely used during our century are patterned on Legendre's modification of Euclid's work. Some algebra used by Legendre is not in the *Elements*, though, as we shall see, the equivalent geometrical material is.

3. *The Definitions and Axioms of the* Elements

The *Elements* contains thirteen books. Two others, surely by later writers, were included in some editions. Book I begins with the definitions of the concepts to be used in the first part of the work. We shall note only the most important ones; these are numbered as in Heath's edition.[1]

DEFINITIONS

1. A point is that which has no part.

2. A line is breadthless length.
   The word line means curve.

3. The extremities of a line are points.
   This definition makes clear that a line or curve is always finite in length. A curve extending to infinity does not occur in the *Elements*.

4. A straight line is a line which lies evenly with the points on itself.
   In keeping with definition 3, the straight line of Euclid is our line segment. The definition is believed to be suggested by the mason's level or an eye looking along a line.

5. A surface is that which has length and breadth only.

6. The extremities of a surface are lines.
   Hence a surface, too, is a bounded figure.

7. A plane surface is a surface which lies evenly with the straight lines on itself.

15. A circle is a plane figure contained by one line such that all the straight lines falling upon it from one point among those lying within the figure are equal to one another.

16. And the point is called the center of the circle.

17. A diameter of the circle is any straight line drawn through the center and terminated in both directions by the circumference [not defined explicitly] of the circle, and such a straight line also bisects the circle.

1. T. L. Heath: *The Thirteen Books of Euclid's* Elements, Dover (reprint), 1956, 3 vols.

23. Parallel straight lines are straight lines which, being in the same plane and being produced indefinitely in both directions, do not meet one another in either direction.

The opening definitions are framed in terms of concepts that are not defined, and hence serve no logical purpose. Euclid might not have realized that the initial concepts must be undefined and was naïvely explaining their meaning in terms of physical concepts. Some commentators say he appreciated that the definitions were not logically helpful but wanted to explain what his terms represented intuitively so that his readers would be convinced that the axioms and postulates were applicable to these concepts.

Euclid next lays down five postulates and five common notions (Proclus uses the term "axiom" for "common notion"). He adopts the distinction already made by Aristotle, namely, that the common notions are truths applicable to all sciences whereas the postulates apply only to geometry. As we have noted, Aristotle said that the postulates need not be known to be true but that their truth would be tested by whether the results deduced from them agreed with reality. Proclus even speaks of all of mathematics as hypothetical; that is, it merely deduces what must follow from the assumptions, whether or not the latter are true. Presumably Euclid accepted Aristotle's views concerning the truth of the postulates. However, in the subsequent history of mathematics, both the postulates and the common notions were accepted as unquestionable truths, at least until the advent of non-Euclidean geometry.

Euclid postulates the following:

POSTULATES

1. [It is possible] to draw a straight line from any point to any point.

2. [It is possible] to extend a finite straight line continuously in a straight line.

3. [It is possible] to describe a circle with any center and distance [radius].

4. That all right angles are equal to one another.

5. That if a straight line falling on two straight lines makes the interior angles on the same side less than two right angles, the two straight lines, if produced indefinitely, meet on that side on which the angles are less than the two right angles.

COMMON NOTIONS

1. Things which are equal to the same thing are also equal to one another.

2. If equals be added to equals, the wholes are equal.

3. If equals be subtracted from equals, the remainders are equal.

4. Things which coincide with one another are equal to one another.

5. The whole is greater than the part.

Euclid does not naïvely assume that the defined concepts exist or are consistent; as Aristotle had pointed out, one might define something that had incompatible properties. The first three postulates, since they declare the possibility of constructing lines and circles, are existence assertions for these two entities. In the development of Book I, Euclid proves the existence of the other entities by constructing them. An exception is the plane.

Euclid presupposes that the line in Postulate 1 is unique; this assumption is implicit in Book I, Proposition 4. It would have been better, however, to make it explicit. Likewise, in Postulate 2 Euclid assumes the extension is unique. He uses the uniqueness explicitly in Book XI, Proposition 1, but has actually already used it unconsciously at the very beginning of Book I.

Postulate 5 is Euclid's own; it is a mark of his genius that he recognized its necessity. Many Greeks objected to this postulate because it was not clearly self-evident and hence lacked the appeal of the others. The attempts to prove it from the other axioms and postulates—which, according to Proclus, commenced even in Euclid's own time—all failed. The full history of these efforts will be related in the discussion of non-Euclidean geometry.

As to the common notions, there are differences of opinion over which ones were in Euclid's original version. Common Notion 4, which is the basis for proof by superposition, is geometrical in character and should be a postulate. Euclid uses superposition in Book I, Propositions 4 and 8, though apparently he was unhappy about the method; he could have used it to prove Proposition 26 ($a.s.a. = a.s.a.$ and $s.a.a. = s.a.a.$), but instead uses a longer proof. He probably found the method in works of older geometers and did not know how to avoid it. More axioms were added to Euclid's by Pappus and others who found Euclid's set inadequate.

4. *Books I to IV of the* Elements

Books I to IV treat the basic properties of rectilinear figures and circles. Book I contains familiar theorems on congruence, parallel lines, the Pythagorean theorem, elementary constructions, equivalent figures (figures with equal area), and parallelograms. All figures are rectilinear, that is, composed of line segments. Of special note are the following theorems (the wording is not verbatim):

Proposition 1. On a given straight line to construct an equilateral triangle.
   The proof is simple. With A as center (Fig. 4.1) and AB as radius con-

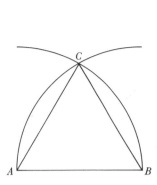

Figure 4.1

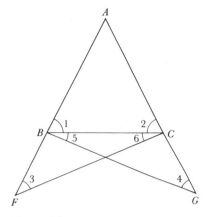

Figure 4.2

struct a circle. With B as center and BA as radius construct a circle. Let C be a point of intersection. Then ABC is the required triangle.

Proposition 2. To place at a given point (as an extremity) a straight line equal to a given straight line.

One would think this can be done immediately by using Postulate 3. But to do so means that the compass must keep its spread of the given length while being transferred to the point at which the equal length is to be constructed. Euclid, however, assumes a collapsible compass and gives a more complicated construction. Of course, he does assume that the compass remains fixed while describing a circle with given center and radius, that is, as long as the compass remains on the paper.

Proposition 4. Two triangles are congruent if two sides and the included angle of one triangle are equal to corresponding parts of the other.

The proof is made by placing one triangle on the other and showing that they must coincide.

Proposition 5. The base angles of an isosceles triangle are equal.

The proof is better than in many current elementary books, which assume at this stage the existence of the bisector of angle A. But the proof of its existence depends upon Proposition 5. Euclid extends AB to F (Fig. 4.2) and AC to G so that $BF = CG$. Then $\triangle AFC \cong \triangle AGB$. Hence $FC = GB$, $\angle ACF = \angle ABG$ and $\angle 3 = \angle 4$. Now $\triangle CBF \cong \triangle BCG$. Hence $\angle 5 = \angle 6$. Therefore $\angle 1 = \angle 2$. Pappus proves the theorem by regarding the given triangle as $\triangle ABC$ and $\triangle ACB$. Then Proposition 4 applies and the base angles are equal.

Proposition 16. An exterior angle of a triangle is greater than either remote interior angle.

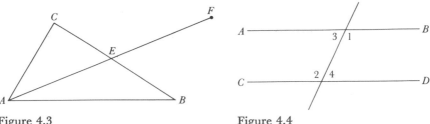

Figure 4.3   Figure 4.4

The proof (Fig. 4.3) requires an infinitely extensible straight line, because in it AE is extended its own length to F, and it must be possible to do this.

Proposition 20. The sum of any two sides of a triangle is greater than the third side.

This theorem is as close as one comes in Euclidean geometry to the fact that the straight line is the shortest distance between two points.

Proposition 27. If a straight line falling on two straight lines makes the alternate interior angles equal to one another, the straight lines will be parallel to one another.

The proof is made by contradiction using the proposition on the exterior angle of a triangle. The theorem establishes the existence of at least one parallel line to a given line and through a given point.

Proposition 29. A straight line falling on parallel straight lines makes the alternate interior angles equal to one another, the exterior angle equal to the interior and opposite angle [corresponding angles are equal], and the interior angles on the same side equal to two right angles.

The proof (Fig. 4.4) supposes $\sphericalangle 1 \neq \sphericalangle 2$. If $\sphericalangle 2$ is greater, add $\sphericalangle 4$ to each. Then $\sphericalangle 2 + \sphericalangle 4 > \sphericalangle 1 + \sphericalangle 4$. This implies that $\sphericalangle 1 + \sphericalangle 4$ is less than two right angles. By the parallel postulate, which is used for the first time, the two given lines AB and CD would have to meet whereas they are parallel by hypothesis.

Proposition 44. To a given straight line to apply, in a given rectilinear angle, a parallelogram equal to a given triangle.

This proposition (Fig. 4.5) says that we are given a triangle C, an angle D, and a line segment AB. We are to construct on AB as one side a parallelogram equal in area to C and containing angle D as one angle. We shall not give Euclid's proof, which depends upon preceding propositions. The major point of note is that this is the first of the problems included under the theory of application of areas, the theory ascribed to the Pythagoreans by Eudemus (as reported by Proclus). In this case we apply (exactly) an area to AB.

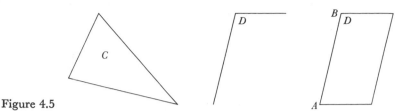

Figure 4.5

Secondly, this is an example of transformation from one area to another. Thirdly, in the special case when D is a right angle the parallelogram must be a rectangle. Then the area of the given triangle and AB may be regarded as given quantities. The other side of the rectangle is then the quotient of the given area C and AB. Hence we have division performed geometrically; this theorem is an example of geometrical algebra.

Proposition 47. In right-angled triangles the square on the side subtending the right angle is equal to the sum of the squares on the sides containing the right angle.

This is of course the Pythagorean theorem. The proof is made by means of areas, as in many high school texts. One shows (Fig. 4.6) that $\triangle ABD \cong \triangle FBC$, that rectangle $BL = 2\triangle ABD$, and rectangle $GB = 2\triangle FBC$. Then rectangle $BL = $ square GB. Likewise rectangle $CL = $ square AK.

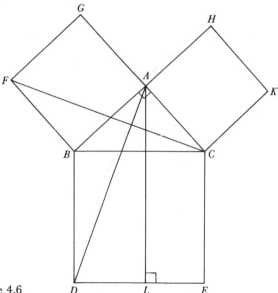

Figure 4.6

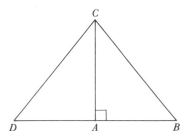

Figure 4.7

The theorem also shows how to obtain a given square whose area is a sum of two given squares, that is, how to find x such that $x^2 = a^2 + b^2$. Hence it is another example of geometrical algebra.

Proposition 48. If in a triangle the square on one side be equal to the sum of the squares on the remaining two sides of the triangle, the angle contained by the remaining two sides is a right angle.

This proposition is the converse of the Pythagorean theorem. The proof in Euclid (Fig. 4.7) constructs AD perpendicular to AC and equal to AB. We have by hypothesis that

$$AB^2 + AC^2 = BC^2$$

and we have from the right triangle ADC that

$$AD^2 + AC^2 = DC^2.$$

Since $AB = AD$ then $BC^2 = DC^2$ and so $DC = BC$. Hence the two triangles are congruent by s.s.s., so that angle CAB must be a right angle.

The outstanding material in Book II is the contribution to geometrical algebra. We have already pointed out that the Greeks did not recognize the existence of irrational numbers and so could not handle all lengths, areas, angles, and volumes numerically. In Book II all quantities are represented geometrically, and thereby the problem of assigning numerical values is avoided. Thus numbers are replaced by line segments. The product of two numbers becomes the area of a rectangle with sides whose lengths are the two numbers. The product of three numbers is a volume. Addition of two numbers is translated into extending one line by an amount equal to the length of the other and subtraction into cutting off from one line the length of a second. Division of two numbers, which are treated as lengths, is merely indicated by a statement that expresses a ratio of the two lines; this is in accord with the principles introduced later in Books V and VI.

Division of a product (an area) by a third number is performed by finding a rectangle with the third number (length) as one side and equal in

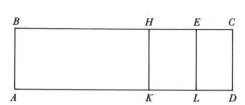

Figure 4.8

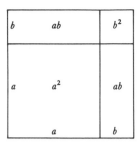

Figure 4.9

area to the given product. The other side of the rectangle is, of course, the quotient. The construction uses the theory of application of areas already touched upon in Proposition 44 of Book I. The addition and subtraction of products are the addition and subtraction of rectangles. The sum or difference is transformed into a single rectangle by means of the method of application of areas. The extraction of a square root is, in this geometrical algebra, the finding of a square equal in area to a rectangle whose area is the given quantity; this is done in Proposition 14 (see below).

The first ten propositions of Book II deal geometrically with the following equivalent algebraic propositions. Stated in our notation some of these are:

1. $a(b + c + d + \cdots) = ab + ac + ad + \cdots$;
2. $(a + b)a + (a + b)b = (a + b)^2$;
3. $(a + b)a = ab + a^2$;
4. $(a + b)^2 = a^2 + 2ab + b^2$;

5. $ab + \left\{\frac{1}{2}(a + b) - b\right\}^2 = \left\{\frac{1}{2}(a + b)\right\}^2$;

6. $(2a + b)b + a^2 = (a + b)^2$.

The geometric statement of (1) is contained in:

Proposition 1. If there be two straight lines (Fig. 4.8) and one of them be cut into any number of segments whatever, the rectangle contained by the two straight lines is equal to the rectangles contained by the uncut straight line and each of the segments.

Propositions 2 and 3 are really special cases of Proposition 1 but are separately stated and proved by Euclid. The geometric form of (4) above is well known. Euclid's statement is:

Proposition 4. If a straight line be cut at random (Fig. 4.9), the square on the whole is equal to the squares on the segments and twice the rectangle contained by the segments.

The proof gives the obvious geometrical facts shown in the figure.

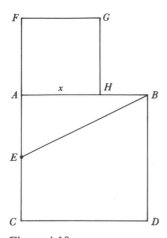

Figure 4.10

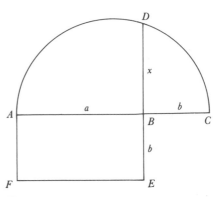

Figure 4.11

Proposition 11. To cut a given straight line so that the rectangle contained by the whole and one of the segments is equal to the square on the remaining segment.

This requires that we divide AB (Fig. 4.10) at some point H so that $AB \cdot BH = AH \cdot AH$. Euclid's construction is as follows: Let AB be the given line. Construct the square $ABDC$. Let E be the midpoint of AC. Draw BE. Let F on CA produced be such that $EF = EB$. Construct the square $AFGH$. Then H is the desired point on AB, that is,

$$AB \cdot BH = AH \cdot AH.$$

The proof is made by means of areas, using preceding theorems, including the Pythagorean theorem—the crucial theorem being Proposition 6.

The importance of the theorem is that AB of length a is divided into two segments of lengths x and $a - x$ so that

$$(a - x)a = x^2$$

or

$$x^2 + ax = a^2.$$

Hence we have a geometric way of solving this quadratic equation. Also, AB is divided in extreme and mean ratio, that is, from $AB \cdot BH = AH \cdot AH$ we have $AB:AH = AH:BH$. Other propositions in Book II amount to solving the quadratics $ax - x^2 = b^2$ and $ax + x^2 = b^2$.

Proposition 14. To construct a square equal to a given rectilinear figure.

The given rectilinear figure can be any polygon, but if the given figure is a *rectangle, ABEF* (Fig. 4.11), then Euclid's method amounts to the follow-

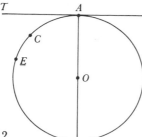

Figure 4.12

ing: Extend AB to C so that $BC = BE$. Construct the circle on AC as diameter. Erect the perpendicular DB at B. The desired square is the square on DB. Euclid gives a proof in terms of areas. This theorem solves $x^2 = ab$ or yields the square root of ab. As we shall see, in Book VI more complicated quadratic equations are solved geometrically.

Book III, which contains 37 propositions, begins with some definitions appropriate to the geometry of circles and then proceeds to discuss properties of chords, tangents, secants, central and inscribed angles, and so on. These theorems are in the main familiar to us from high school geometry. The following theorems are worthy of special note.

Proposition 16. The straight line drawn at right angles to the diameter of a circle from its extremity will fall outside the circle, and into the space between the straight line and the circumference another straight line cannot be interposed; further the angle of the semicircle is greater, and the remaining angle less, than any acute rectilinear angle.

The novelty in the theorem is that Euclid considers the space (Fig. 4.12) between the tangent TA and the arc ACE; he not only says that no line can be drawn in this space and through A, which falls entirely outside the circle, but also considers the angle formed by TA and the arc ACE. Whether this angle, which the Greeks called the hornlike angle, had a definite magnitude was a subject of controversy. Proposition 16 says this angle is smaller than any acute angle formed by straight lines but it does not say the angle is of zero magnitude.

Proclus speaks of horn angles as true angles. The subject of the size of horn angles was debated also in the late Middle Ages and in the Renaissance by Cardan, Peletier, Vieta, Galileo, Wallis, and others. What made the horn angle especially troublesome to later commentators on Euclid is that one can draw circles of smaller and smaller diameters passing through A and tangent to TA and it seems intuitively clear that the horn angle should increase in size but, according to the above proposition, it does not. On the other hand, if any two horn angles are of zero size and therefore equal, they should be superposable. But they are not. Some commentators concluded that a horn angle was not an angle.[2]

Book IV, in its 16 propositions, deals with figures inscribed in and circumscribed about circles, for example, triangles, squares, regular pentagons, and regular hexagons. The last proposition, which shows how to inscribe a 15-sided regular polygon in a given circle, is said to have been used in astronomy; the angle of the ecliptic (the angle made by the plane of the earth's equator and the plane of the earth's orbit around the sun) up to the time of Eratosthenes was taken to be 24°, or 1/15 of 360°.

5. Book V: The Theory of Proportion

Book V, based on Eudoxus' work, is considered to be the greatest achievement of Euclidean geometry; its contents have been more extensively discussed and its meaning more heavily debated than those of any other portion of the *Elements*. The Pythagoreans are supposed to have had a theory of proportion, that is, the equality of two ratios, for commensurable magnitudes, or magnitudes whose ratio could be expressed by a ratio of whole numbers. Though we do not know the details of this theory, it is presumed to be what we shall encounter in Book VII and is supposed to have been applied to statements about similar triangles. The mathematicians before Eudoxus who used proportions in general did not have a secure foundation for incommensurable magnitudes. Book V extends the theory of proportion to incommensurable ratios but avoids irrational numbers.

The notion of magnitude is intended to cover quantities or entities that are commensurable or incommensurable with each other. Thus lengths, areas, volumes, angles, weights, and time are magnitudes. The magnitudes length and area have already appeared, for example, in Book II. But thus far Euclid has had no occasion to treat other kinds of magnitudes or to treat ratios of magnitudes and proportions. Hence he does not introduce the general notion of magnitude until this point. His particular emphasis now is on proportions for all kinds of magnitudes.

Despite the fact that the definitions are important in this book, there is no definition of magnitude as such. Euclid starts with:

Definition 1. A magnitude is part of a magnitude, the less of the greater, when it measures the greater.

Part is used here to mean submultiple, as 2 is of 6, whereas 4 is not a submultiple of 6.

---

2. By the usual current definition of an angle between two curves, the horn angle is of zero size.

Definition 2. The greater is a multiple of the less when it is measured by the less.

Thus multiple means integral multiple.

Definition 3. A ratio is a sort of relation in respect to size between two magnitudes of the same kind.

It is difficult to separate the significance of this third definition from that of the next.

Definition 4. Magnitudes are said to have a ratio to one another which are capable, when multiplied, of exceeding one another.

The meaning of this definition is that the magnitudes a and b have a ratio if some integral multiple n (including 1) of a exceeds b and some integral multiple (including 1) of b exceeds a. The definition excludes the concept that appeared later, namely, the infinitely small quantity which is not 0, called the infinitesimal. Euclid's definition does not allow a ratio between two magnitudes if one is so small that some finite multiple of it does not exceed the other. The definition also excludes infinitely large magnitudes because then no finite multiple of the smaller one will exceed the larger. The next definition is the key one.

Definition 5. Magnitudes are said to be in the same ratio, the first to the second and the third to the fourth, when, if any equimultiples whatever be taken of the first and third, and any equimultiples whatever of the second and fourth, the former equimultiples alike exceed, are alike equal to, or alike fall short of, the latter equimultiples respectively taken in corresponding order.

The definition says that

$$\frac{a}{b} = \frac{c}{d}$$

if when we multiply a and c by *any* whole number m, say, and b and d by *any* whole number n, then for all such choices of m and n,

$$ma < nb \quad \text{implies} \quad mc < nd,$$
$$ma = nb \quad \text{implies} \quad mc = nd,$$

and

$$ma > nb \quad \text{implies} \quad mc > nd.$$

Just to see the meaning of this definition, let us use modern numbers. To test whether

$$\frac{\sqrt{2}}{1} = \frac{\sqrt{6}}{\sqrt{3}}$$

we should, at least theoretically, know that if m is *any* whole number and n *any* other, then

$$m\sqrt{2} < n \cdot 1 \quad \text{implies} \quad m\sqrt{6} < n\sqrt{3}$$

and

$$m\sqrt{2} = n \cdot 1 \quad \text{implies} \quad m\sqrt{6} = n\sqrt{3}$$

and

$$m\sqrt{2} > n \cdot 1 \quad \text{implies} \quad m\sqrt{6} > n\sqrt{3}.$$

Of course, in this present example the equality $m\sqrt{2} = n \cdot 1$ will not occur because m and n are whole numbers whereas $\sqrt{2}$ is irrational, but this means that $m\sqrt{6} = n\sqrt{3}$ need not occur. The definition merely says that if the left side of any one of the three possibilities occurs then the right side must occur. An alternative statement of Definition 5 is that the integral m and n for which $ma \lessdot nb$ are the same as those m' and n' for which $m'c < n'd$.

It may be desirable to point out at once what Euclid does with the above definitions. When we wish to prove that if $a/b = c/d$, then $(a + b)/b = (c + d)/d$, we regard the ratios and the proportion as numbers even if the ratios are incommensurable, and we use algebra to prove the result. We know that we can operate with irrationals by the laws of algebra. Euclid cannot and does not. The Greeks had not thus far justified operations with ratios of incommensurable magnitudes; hence Euclid proves this theorem by using the definitions he has given and, in particular, Definition 5. In effect, he is laying the basis for an algebra of magnitudes.

Definition 6. Let magnitudes which have the same ratio be called proportional.

Definition 7. When, of the equimultiples, the multiple of the first magnitude exceeds the multiple of the second, but the multiple of the third does not exceed the multiple of the fourth, then the first is said to have a greater ratio to the second than the third has to the fourth.

The definition states that if for even one m and one n, $ma > nb$ but mc is not greater than nd, then $a/b > c/d$. Hence, given an incommensurable ratio a/b, it is possible to place it among all other such ratios, namely those less and those greater.

Definition 8. A proportion in three terms is the least possible.

In this case $a/b = b/c$.

Definition 9. When three magnitudes are proportional, the first is said to have to the third the duplicate ratio of that which it has to the second.

Thus if $A/B = B/C$, then A has the duplicate ratio to C that it has to B. This means that $A/C = A^2/B^2$, for $A = B^2/C$ so that $A/C = B^2/C^2 = A^2/B^2$.

Definition 10. When four magnitudes are continuously proportional the first is said to have to the fourth the triplicate ratio of that which it has to the second, and so on continually, whatever the proportion.

Thus if $A/B = B/C = C/D$, then A has the triplicate ratio to D that it has to B. That is, $A/D = A^3/B^3$, for $A = B^2/C$ so that $A/D = B^2/CD = (B^2/C^2)(C/D) = A^3/B^3$.

Definitions 11 to 18 define corresponding magnitudes, alternation, inversion, composition, separation, conversion, etc. These refer to forming $(a + b)/b$, $(a - b)/b$, and other ratios from a/b.

Book V now proceeds to prove twenty-five theorems about magnitudes and ratios of magnitudes. The proofs are verbal and depend only on the definitions and on the common notions or axioms, such as that equals subtracted from equals give equals. The postulates are not used. Euclid uses line segments as illustrations of magnitudes to help his readers perceive the meaning of the theorems and proofs, but the theorems apply to all kinds of magnitudes.

We shall state some of the propositions of Book V in modern algebraic language, using m, n, and p for integers and a, b, and c for magnitudes. However, to illustrate Euclid's language, let us note his wording of the first proposition.

Proposition 1. If there be any number of magnitudes whatever which are, respectively, equimultiples of any magnitudes equal in multitude, then, whatever multiple one of the magnitudes is of one, that multiple also will be of all.

In algebraic language this means that $ma + mb + mc + \cdots = m(a + b + c + \cdots)$.

Proposition 4. If $a/b = c/d$, then $ma/nb = mc/nd$.

Proposition 11. If $a/b = c/d$ and $c/d = e/f$ then $a/b = e/f$.

Note the equality of ratios depends on the definition of proportion and Euclid is careful to prove that equality is transitive.

Proposition 12. If $a/b = c/d = e/f$ then $a/b = (a + c + e)/(b + d + f)$.

Proposition 17. If $a/b = c/d$ then $(a - b)/b = (c - d)/d$.

Proposition 18. If $a/b = c/d$ then $(a + b)/b = (c + d)/d$.

Some of the propositions seem to duplicate propositions in Book II. However, the propositions in the latter book are asserted and proved only for line segments, whereas Book V provides the theory for all kinds of magnitudes.

Book V was crucial for the subsequent history of mathematics. The classical Greeks did not introduce irrational numbers and sought to avoid them in part by working geometrically, as we have already noted in our review of Books I to IV. However, this use of geometry did not take care of ratios and proportions of incommensurable magnitudes of all sorts, and this lack was filled by Book V, which started anew with a general theory of magnitude. It thereby placed all of Greek geometry that dealt with magnitudes on a sound basis. The critical question, however, has been whether the theory of magnitudes provided a logical basis for a theory of real numbers, including, of course, the irrational numbers.

There is no question about how succeeding generations of mathematicians interpreted Euclid's theory of magnitudes. They regarded it as applicable only to geometry and therefore took the attitude that only geometry was rigorous. Hence, when in the Renaissance and in the following centuries irrational numbers were reintroduced and used, many mathematicians objected because these numbers had no logical foundation.

A critical examination of Book V seems to establish that they were right. It is true that the definitions and proofs as presented by Euclid in Book V make no use of geometry. As already noted his use of line segments in his presentation of the propositions and proofs is pedagogical only. However, if Euclid had really offered a theory of irrationals in his theory of magnitudes, it would have had to come from either one of two possible interpretations. The first is that magnitudes themselves could be taken to be the irrational numbers, and the second is that the ratios of two magnitudes could be the irrational numbers.

Let us suppose the magnitudes themselves could be the irrational numbers. Then, even if we leave aside any criticisms of Euclid's rigor when judged by modern standards, the following difficulties enter. Euclid never defines what he means by a magnitude, or the equality or equivalence of magnitudes. Moreover, Euclid works not with the magnitudes themselves but with proportions. A product of two magnitudes a and b occurs only when a and b are lengths, thereby enabling Euclid to consider the product as an area. The product ab could not then be a number because the product has no general meaning in Euclid. Further, Euclid proves in Book V a number of theorems on proportion which in themselves can readily be restated, as in fact we did above, as algebraic theorems. However, to prove Proposition 18 of Book V he needs to establish the fourth proportional to three given magnitudes, which he is able to do only for magnitudes that are line segments (Book VI, Proposition 12). Hence, not only is his theory of general magnitudes incomplete (even for proofs he himself makes in Book XII), but what he does establish for lengths is dependent on geometry. Moreover, Euclid insists in Definition 3 that a ratio can be formed only of magnitudes of the same kind. Clearly if magnitudes were numbers this limitation would be meaningless. His concept of magnitude as used later adheres to the definition and so is tied to geometry. Another difficulty is that there is no system of rational numbers to which a theory of irrationals could be added. Ratios of whole numbers occur, but only as members of a proportion, and even these ratios are not regarded as fractions. Finally, there is no product of a/b and c/d even when all four quantities are whole numbers, let alone magnitudes.

Now let us consider interpreting Euclid's ratios of magnitudes as numbers, so that the incommensurable ratios would be the irrational numbers and the commensurable ratios, the rational numbers. If these ratios are numbers it should be possible at least to add and multiply them. But nowhere in Euclid does one find what $(a/b) + (c/d)$ means when a, b, c, and d are magnitudes. In Euclid's usage the ratios occur only as elements of a proportion and hence have no general significance. Finally, as noted above, Euclid does not have the concept of a rational number on which to build a theory of irrationals.

Thus the course that the history of mathematics actually took until about 1800, namely, treating continuous quantities rigorously solely on a geometric basis, was necessary. As far as Euclid's *Elements* is concerned, there was no foundation for irrational numbers.

6. *Book VI: Similar Figures*

Book VI, which treats similar figures and uses the theory of proportion of Book V, opens with some definitions. We note just a few:

Definition 1. Similar rectilinear figures are those having corresponding angles equal and the sides about the equal angles proportional.

Definition 3. A straight line is cut in extreme and mean ratio when the whole line is to the greater segment as the greater to the less.

Definition 4. The height of any figure is the perpendicular drawn from the vertex to the base.
   This definition is certainly vague but Euclid does employ it.
   In the proofs of the theorems in this book, Euclid, using his theory of proportion, does not have to treat separately the commensurable and incommensurable cases, a separation introduced by Legendre, who used an algebraic definition of proportion limited to commensurable quantities and so had to treat the incommensurable cases by another argument such as a *reductio ad absurdum*.
   We shall note only some of the thirty-three theorems. Again we shall find some basic results of modern algebra treated geometrically.

Proposition 1. Triangles and parallelograms [i.e. the areas] which are under the same height [have the same altitudes] are to one another as their bases.

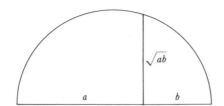

Figure 4.13

Here Euclid uses a proportion among four magnitudes two of which are areas.

Proposition 4. In equiangular triangles the sides about the equal angles are proportional, and those are the corresponding sides which subtend the equal angles.

Proposition 5. If two triangles have their sides proportional, the triangles will be equiangular and will have those angles equal which the corresponding sides subtend.

Proposition 12. To three given straight lines to find a fourth proportional.

Proposition 13. To two given straight lines to find a mean proportional.

The method is the familiar one (Fig. 4.13). It means from an algebraic standpoint that, given a and b, we can find $\sqrt{ab}$.

Proposition 19. [The areas of] similar triangles are to one another in the duplicate ratio of the corresponding sides.

We express this theorem today by the statement that the ratio of the areas of two similar triangles equals the ratio of the squares of two corresponding sides.

Proposition 27. Of all the parallelograms applied to the same straight line [constructed on part of the straight line] and deficient [from the parallelogram on the entire straight line] by parallelogramic figures similar and similarly situated to that [given parallelogram] described on the half of the straight line, [the area of] that parallelogram is greatest which is applied to the half of the straight line and is similar to the defect.

The meaning of this proposition is as follows: We start (Fig. 4.14) with a given parallelogram AD constructed on AC, which is one-half of a given line segment AB. Now we consider a parallelogram AF on AK which is part of AB subject to the condition that the defect of AF, which is FB, is a parallelogram similar to AD. We can of course obtain many parallelograms meeting the conditions that AF does. Euclid's theorem states that of all such parallelograms, the one constructed on AC, which is half of AB, has the largest area.

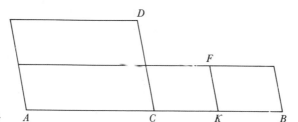

Figure 4.14

The proposition has a vital algebraic meaning. Suppose the given parallelogram AD is a rectangle (Fig. 4.15) and that the ratio of its sides is c to b, where b is AC. Now consider the rectangle AF which is required to meet the condition that its defect, rectangle FB, is similar to AD. If we denote FK by x, then KB is bx/c. Let the length of AB be a; then $AK = a - (bx/c)$. Hence the areas S of AF is

(1)
$$S = x\left(a - \frac{bx}{c}\right).$$

Proposition 27 says that S is a maximum when AF is AD. But $AC = a/2$ and $CD = ac/2b$. Hence

$$S \leq \frac{a^2 c}{4b}.$$

On the other hand the condition that equation (1) regarded as a quadratic in x have a real root is that its discriminant be greater than or equal to 0. That is,

$$a^2 - 4\frac{b}{c}S \geq 0$$

or

$$S \leq \frac{a^2 c}{4b}.$$

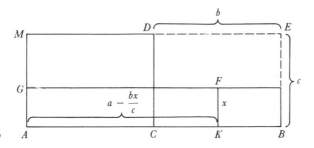

Figure 4.15

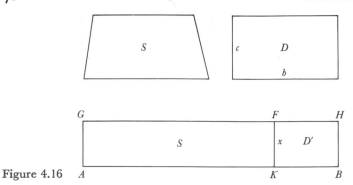

Figure 4.16

Thus the proposition tells us not only what the largest possible value of S is but that for all possible values there is an x which satisfies (1) and which geometrically furnishes one side, KF, of the rectangle AF. This result will be used in the next proposition.

Before considering it let us note an interesting special case of Proposition 27. Suppose the given parallelogram AD (Fig. 4.15) is a square. Then of all rectangles on AB deficient by a square similar to AD, the square on AC is the greatest. But the rectangle AF on (part of) AB has area $AK \cdot KF$ and since $KF = KB$, the perimeter of this rectangle is the same as that of square DB or square AD. But AD has larger area than AF. Thus of all rectangles with the same perimeter the square has the greatest area.

Proposition 28. To a given straight line to apply [on part of the line as side] a parallelogram equal to a given rectilinear figure [S] and deficient [from the parallelogram on the entire line] by a parallelogramic figure similar to a given one [D]. Thus [by Proposition 27] the given rectilinear figure [S] must not be greater than the parallelogram described on half of the straight line and similar to the defect.

This theorem is the geometrical equivalent of the solution of the quadratic equation $ax - (b/c)x^2 = S$, where S is the area of the given rectilinear figure and is subject to the condition for a real solution, namely, S is not greater than $a^2 c/4b$. To see this, suppose (for convenience) that the parallelograms are rectangles (Fig. 4.16), S is the given rectilinear figure, D is the other given rectangular figure with sides c and b, a is AB, and x is one side of the desired rectangle. What Euclid constructs is the rectangle $AKFG$ of area S and such that the defect D' is similar to D. But $AKFG = ABHG - D'$. Since D' is similar to D its area is bx^2/c. Hence

$$(2) \qquad\qquad S = ax - \frac{b}{c}x^2.$$

Thus to construct $AKFG$ is to find AK and x such that x satisfies equation (2).

Proposition 29. To a given straight line to apply a parallelogram equal to a given rectilinear figure [S] and exceeding by a parallelogramic figure similar to a given one [D].

In algebraic terms this theorem solves

$$ax - \frac{b}{c}x^2 = S$$

where a, b, c, and S are given. S is not limited because for all positive S the equation has a real solution. With Propositions 28 and 29, Euclid has shown how, in modern language, one can solve any quadratic equation when one or both roots are positive. His constructions furnish the root as a length.

In Proposition 28 the parallelogram constructed falls short of the parallelogram on the entire line AB and in Proposition 29 the parallelogram constructed exceeds the one on the given line AB. The respective parallelograms were called in Greek *elleipsis* and *hyperbolē*. A parallelogram of specified area constructed on the entire given line as base, as in Book I, Proposition 44, was called *parabolē*. These terms were carried over to the conic sections for a reason which will be obvious when we study Apollonius' work.

Proposition 31. In right-angled triangles, the figure on the side subtending the right angle is equal to the similar and similarly described figures on the sides containing the right angle.

This is a generalization of the Pythagorean theorem.

7. *Books VII, VIII, and IX: The Theory of Numbers*

Books VII, VIII, and IX treat the theory of numbers, that is, the properties of whole numbers and ratios of whole numbers. These three books are the only ones in the *Elements* that treat arithmetic as such. In them Euclid represents numbers as line segments and the product of two numbers as a rectangle, but the arguments do not depend on the geometry. The statements and proofs are verbal as opposed to the modern symbolic form.

Many of the definitions and theorems, particularly those on proportion, duplicate what was done in Book V. Hence historians have considered the question of why Euclid proves all over again propositions for numbers instead of referring to propositions already proven in Book V.

The answers vary. Aristotle did include number as one kind of magnitude, but he also emphasized the cleavage between the discrete and the continuous, and we do not know whether Euclid was influenced by either of Aristotle's views on this matter. Nor can one decide on the basis of the vague definitions in Book V whether he meant his notion of magnitude to include whole numbers. If one judged by the fact that he treated number independently one would conclude that his magnitudes do not include numbers.

Another explanation of this independent treatment of numbers is that the theory of numbers and of commensurable ratios existed before Eudoxus' work and Euclid followed tradition in presenting what were two independent developments, the pre-Eudoxian and largely Pythagorean theory and the Eudoxian theory. He may also have believed that since the theory of numbers can be built up on simpler foundations than the theory of magnitudes, it was wise to do so. One finds alternative approaches in modern contributions to mathematics too, and for the same reason—that they are simpler. Although Euclid separates number and magnitude, he does have a few theorems that relate them. For example, Proposition 5 of Book X states that commensurable magnitudes have to one another the ratio which a number has to a number.

In the three books under discussion, as in other books, Euclid assumes facts that he does not state explicitly. Thus he assumes without mention that if A divides (evenly into) B and B divides C, then A divides C. Also, if A divides B and divides C, it divides $B + C$ and $B - C$.

Book VII begins with some definitions:

Definition 3. A number is a part of a number, the less of the greater, when it measures the greater. [The number which is part of another divides evenly into the latter.]

Definition 5. The greater number is a multiple of the less when it is measured by the less.

Definition 11. A prime number is that which is measured by a unit alone.

Definition 12. Numbers prime to one another [relatively prime] are those which are measured by unit alone as a common measure.

Definition 13. A composite number is that which is measured by some number [other than 1].

Definition 16. And when two numbers having multiplied one another make some number, the number so produced is called plane, and its sides are the numbers which have multiplied one another.

Definition 17. And when three numbers having multiplied one another make some number, the number so produced is solid, and its sides are the numbers which have multiplied one another.

Definition 20. Numbers are proportional when the first is the same multiple, or the same part, or the same parts of the second that the third is of the fourth.

Definition 22. A perfect number is that which is equal to [the sum of] its own parts.

Propositions 1 and 2 give the process of finding the greatest common measure (divisor) of two numbers. Euclid describes the process by saying that if A and B are the numbers and $B < A$, then subtract B from A enough times until a number C less than B is left. Then subtract C from B enough times until a number less than C is left. And so on. If A and B are relatively prime we arrive at 1 as the last remainder. Then 1 is the greatest common divisor. If A and B are not relatively prime we arrive at some stage where the last number measures the one before it. This last number is the greatest common divisor of A and B. This process is referred to today as the Euclidean algorithm.

Simple theorems about numbers follow. For example, if $a = b/n$ and $c = d/n$, then $a \pm c = (b \pm d)/n$. Some are just the theorems on proportion previously proved for magnitudes and now proved all over again for numbers. Thus if $a/b = c/d$, then $(a - c)/(b - d) = a/b$. Also, in Definition 15 $a \cdot b$ is defined as b added to itself a times. Hence Euclid proves that $ab = ba$.

Proposition 30. If two numbers by multiplying one another make some number, and any prime number measures the product, it will also measure one of the original numbers.

This result is fundamental in the modern theory of numbers. We say that if a prime p divides a product of two whole numbers it must divide at least one of the factors.

Proposition 31. Any composite number is measured by some prime number.

Euclid's proof says that if A is composite it is, by definition, measured by some number B. If B is not prime and hence composite, B is measured by C. Then C measures A. If C is not prime, etc. Then he says, "If the investigation be continued in this way, some prime number will be found which will measure the number before it, which will also measure A. For if it is not found, an infinite series of numbers will measure the number A, each of which is less than the other: which is impossible in numbers." He assumes here that any set of whole numbers has a least number.

Book VIII continues with the theory of numbers; no new definitions are needed. In essence the book treats geometrical progressions. A geometrical progression is to Euclid a set of numbers in continued proportion, that is, $a/b = b/c = c/d = d/e = \cdots$. This continued proportion satisfies our definition of geometric progression, for if $a, b, c, d, e, \ldots$ are in geometrical progression the ratio of any term to the next one is a constant.

Book IX concludes the work on the theory of numbers. There are theorems on square and cube numbers, plane and solid numbers, and more theorems on continued proportions. Of note are the following:

Proposition 14. If a number be the least that is measured by prime numbers, it will not be measured by any other prime number except those originally measuring it.

This means that if a is the product of the primes p, q, $\ldots$, then this decomposition of a into primes is unique.

Proposition 20. Prime numbers are more than any assigned multitude of prime numbers.

In other words, the number of primes is infinite. Euclid's proof of this proposition is a classic. He supposes that there is just a finite number of primes, $p_1, p_2, \ldots, p_n$. He then forms $p_1 \cdot p_2 \cdots \cdot p_n + 1$ and argues that if this new number is a prime, we have a contradiction, because this prime is larger than any of the n primes and so we would have more than n primes. On the other hand, if this new number is composite it must be divisible (exactly) by a prime. But this prime divisor is not $p_1, p_2, \ldots,$ or p_n because these leave a remainder of 1. Hence there must be some other prime; and again we have a contradiction of the assumption that there are just the n primes $p_1, p_2, \ldots, p_n$.

Proposition 35 of Book IX gives an elegant proof for the sum of a geometric progression. Proposition 36 gives a famous theorem on perfect numbers, namely, if the sum of the terms (starting with 1) of the geometric progression

$$1 + 2 + 2^2 + \cdots + 2^{n-1}$$

is prime, the product of that sum and the last term, that is,

$$(1 + 2 + \cdots + 2^{n-1})2^{n-1} \quad \text{or} \quad (2^n - 1)2^{n-1},$$

is a perfect number. The first four perfect numbers, 6, 28, 496, and 8128, and perhaps the fifth, were known to the Greeks.

8. Book X: The Classification of Incommensurables

Book X of the *Elements* undertakes to classify types of irrationals, that is, magnitudes incommensurable with given magnitudes. Augustus De Morgan describes the general contents of this book by saying, "Euclid investigates every possible variety of line which can be represented [in modern algebra] by $\sqrt{\sqrt{a} \pm \sqrt{b}}$, a and b representing two commensurable lines." Of course not all irrationals are so representable, and Euclid covers only those that arise in his geometrical algebra.

The first proposition in Book X is important for developments in later books of the *Elements*.

Proposition 1. Two unequal magnitudes being set out, if from the greater there be subtracted a magnitude greater than its half, and from that which is left a magnitude greater than its half, and if this process be repeated continually, then there will be left some magnitude which will be less than the lesser magnitude set out.

At the conclusion of the proof Euclid says the theorem can be proven if the parts subtracted be halves. One step in the proof utilizes an axiom, not consciously recognized as such by Euclid, to the effect that of two unequal magnitudes the smaller can be added to itself a finite number of times, so as to have the sum exceed the larger. Euclid bases the questionable step on the definition of a ratio between two magnitudes (Definition 4 of Book V). But this definition does not justify the step. It says that two magnitudes have a ratio when either can be added to itself enough times to have the sum exceed the other; hence Euclid should prove that this can be done for the magnitudes he deals with. Instead he assumes that his magnitudes have a ratio and uses the fact that the smaller can be added to itself enough times to exceed the greater. According to Archimedes, the axiom in question (strictly an equivalent statement) was used by Eudoxus, who had established it as a lemma. Archimedes uses this lemma without proof and so, in effect, he too uses it as an axiom. It is called today the axiom of Archimedes–Eudoxus.

There are 115 propositions in Book X, though Propositions 116 and 117 are found in some editions of Euclid. The latter gives the proof already described in Chapter 3 of the irrationality of $\sqrt{2}$.

9. *Books XI, XII, and XIII: Solid Geometry and the Method of Exhaustion*

Book XI begins the treatment of solid geometry, though some important theorems on plane geometry are yet to come. The book opens with definitions.

Definition 1. A solid is that which has length, breadth, and depth.

Definition 2. An extremity of a solid is a surface.

Definition 3. A straight line is at right angles to a plane when it makes right angles with all the straight lines which meet it and are in the plane.

Definition 4. A plane is at right angles to a plane when the straight lines drawn, in one of the planes, at right angles to the common section of the planes, are at right angles to the remaining plane.

Definition 6. The inclination of a plane to a plane is the acute angle contained by the straight lines drawn at right angles to the common section, at the same point, one in each of the planes.

We call this acute angle the plane angle of the dihedral angle.

Also defined are parallel planes, similar solid figures, solid angle, pyramid, prism, sphere, cone, cylinder, cube, the regular octahedron, the

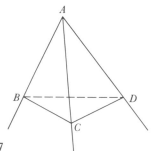

Figure 4.17

regular icosahedron, and other figures. The sphere is defined as the figure comprehended by a semicircle rotated around a diameter. The cone is defined as the figure comprehended by rotating a right-angled triangle about one of the arms. Then if the fixed arm or axis is less, equal to, or greater than the other arm, the cone is obtuse-angled, right-angled, or acute-angled, respectively. The cylinder is the figure comprehended by rotating a rectangle around one side. The significance of these last three definitions is that solid figures, except for the regular polyhedra, arise from rotating plane figures about an axis.

The definitions are loose, unclear, and often assume theorems. For example, in Definition 6 it is assumed that the acute angle is the same no matter where it is formed on the common section of the planes. Also Euclid intends to consider convex solids only, but does not specify this in his definitions of the regular polyhedra.

The book considers only figures formed by plane elements. The first 19 of the 39 theorems of this book treat properties of lines and planes, for example, theorems on lines perpendicular and parallel to planes. The proofs of the early theorems in this book are not adequate, and many general theorems about polyhedra are proved only for special cases.

Proposition 20. If a solid angle be contained by three plane angles, any two, taken together in any manner, are greater than the remaining one.

That is, of the three plane angles (Fig. 4.17) *CAB*, *CAD*, and *BAD*, the sum of any two is greater than the third one.

Proposition 21. Any solid angle is contained by plane angles [the sum of which is] less than four right angles.

Proposition 31. Parallelepipedal solids which are on equal bases and of the same height are equal [equivalent] to one another.

Proposition 32. Parallelepipedal solids which are of the same height are to one another as their bases.

Book XII contains 18 theorems on areas and volumes, particularly of curvilinear figures and figures bounded by surfaces. The dominant idea of the book is the method of exhaustion, which comes from Eudoxus. To prove, for example, that the areas of two circles are to each other as the squares on their diameters, the method approximates the areas of the two circles more and more closely by inscribed regular polygons and, since the theorem in question is true for the polygons, it is proved to be true for the circles. The term exhaustion comes from the fact that the successive inscribed polygons "exhaust" the circle. This term was not used by the Greeks; it was introduced in the seventeenth century. The term, as well as this loose description, may suggest that the method is approximate and just a step in the direction of the rigorous limit concept. But, as we shall see, the method is rigorous. There is no explicit limiting process in it; it rests on the indirect method of proof and in this way avoids the use of a limit. Actually Euclid's work on areas and volumes is sounder than that of Newton and Leibniz, who tried to build on algebra and the number system and sought to use the limit concept.

For a better understanding of the method of exhaustion, let us consider one example in some detail. (In the next chapter we shall consider some examples in Archimedes' work.) Book XII opens with:

Proposition 1. Similar polygons inscribed in circles are to one another as the squares on the diameters of the circles.

We shall not give the proof because no special feature is involved. We come now to the crucial proposition.

Proposition 2. Circles are to one another as the squares on the diameters.

The following describes the essence of Euclid's proof: He first proves that the circle can be "exhausted" by polygons. Inscribe a square in the circle (Fig. 4.18). The area of the square is more than 1/2 of the area of the circle because the former area is 1/2 of that of the circumscribed square and this area is larger than the area of the circle. Now let AB be any side of the inscribed square. Bisect arc AB at the point C and join AC and CB. Draw the tangent at C and then draw AD and BE perpendicular to the tangent. $\sphericalangle 1 = \sphericalangle 2$ because each is 1/2 of arc CB. It follows that DE is parallel to AB, and so $ABED$ is a rectangle whose area is greater than that of segment $ABFCG$. Hence triangle ABC, which is half the rectangle, is greater than 1/2 of segment $ABFCG$. By repeating the process at each side of the square, we obtain a regular octagon, which encloses not only the square but more than half of the difference between the area of the circle and the area of the square. On each side of the octagon we may construct a triangle, just as triangle ACB was constructed on AB. We then obtain a sixteen-sided regular polygon, which encloses the octagon and more than half of the difference between the area of the circle and the area of the octagon. The process may

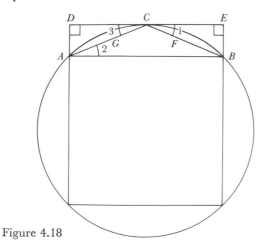

Figure 4.18

be repeated as often as one wishes. Euclid next employs Proposition 1 of Book **X** to affirm that the difference between the area of the circle and the area of some regular polygon of a sufficiently large number of sides can be made less than any given magnitude.

Now let S and S' be the areas of two circles (Fig. 4.19) and let d and d' be their diameters. Euclid wishes to prove that

(3) $S:S' = d^2:d'^2$.

Suppose that this equality does not hold but that

(4) $S:S'' = d^2:d'^2$,

where S'' is some area greater or less than S'. (The existence of the fourth proportional as an area is assumed here and elsewhere in Book **XII**.) Suppose $S'' < S'$. We construct regular polygons of more and more sides in

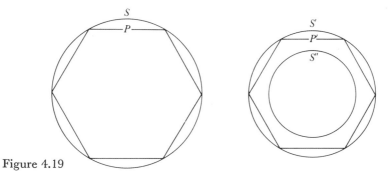

Figure 4.19

S' until we arrive at one, P' say, which is such that its area differs from S' by less than $S' - S''$. This polygon can be constructed because we proved above that the difference between the circle S' and inscribed regular polygons can be made less than any given magnitude and so less than $S' - S''$. Then

(5) $$S' > P' > S''.$$

Inscribe in S a polygon P similar to P'. Then, by Proposition 1,

$$P:P' = d^2:d'^2.$$

By reason of (4) we also have

$$P:P' = S:S''$$

or

$$P:S = P':S''.$$

However, since $P < S$ then

$$P' < S''.$$

But by (5) this is a contradiction.

Similarly, one can show that S'' cannot be greater than S'. Hence $S'' = S'$, and in view of (4), the proportion (3) is established.

This method is used to prove such critical and difficult theorems as:

Proposition 5. Pyramids which are of the same height and have triangular bases are to one another as the bases.

Proposition 10. Any cone is a third part of the cylinder which has the same base with it and equal height.

Proposition 11. Cones and cylinders which are of the same height are to one another as their bases.

Proposition 12. Similar cones and cylinders are to one another in the triplicate ratio [ratio of the cubes] of the diameters in their bases.

Proposition 18. Spheres are to one other in the triplicate ratio of their respective diameters.

Book XIII considers properties of regular polygons as such and when inscribed in a circle, and the problem of how to inscribe the five regular solids in a sphere. It also proves that no more than the five types of (convex) regular solids (polyhedra) exist. This last result is a corollary to Proposition 18, the last in the book.

The proof that no more than five regular solids can exist depends upon an earlier theorem, Book XI, Proposition 21, that the faces of a solid angle must contain less than 360°. Hence if we put together equilateral triangles we can have three meet at each vertex of the regular solid to form a tetrahedron; we can use four at a time to form an octahedron; and we can use five to form the icosahedron. Six equilateral triangles at one vertex would add up to 360° and so cannot be used. We can use three squares at any one vertex and thus form the cube. Then we can use three regular pentagons at each vertex to form the dodecahedron. No other regular polygons can be used, for even three coming together at one point will form an angle of 360° or more. Note that Euclid assumes convex regular solids. There are other, nonconvex regular solids.

The thirteen books of the *Elements* contain 467 propositions. In some of the old editions there are two more books, both of which contain more results on regular solids, though Book XV is unclear and inaccurate. However, both postdate Euclid. Book XIV is due to Hypsicles (*c.* 150 B.C.), and parts of Book XV were probably written as late as the sixth century A.D.

10. *The Merits and Defects of the* Elements

Since the *Elements* is the first substantial source of mathematical knowledge and one that was used by all succeeding generations, it influenced the course of mathematics as no other book has. The very concept of mathematics, the notion of proof, and the logical ordering of theorems were learned by studying it, and its contents determined the course of subsequent thinking. Hence we should note the characteristics that influenced so strongly the future of mathematics.

Though, as mentioned before, the form of presentation of the individual propositions is not original with Euclid, the form of presentation of the entire work—the statement of all the axioms at the outset, the explicit statement of all definitions, and the orderly chain of theorems—is his own. Moreover, the theorems are arranged to go from the simple to more and more complex ones.

Euclid also selected the theorems that he regarded as prior in importance. Thus he does not give, for example, the theorem that the altitudes of a triangle meet in a point. There are theorems in other works by Euclid, which we shall discuss shortly, that he did not deem worth including in the *Elements*.

Though the requirement that the existence of figures must be demonstrated before the figures can be received into the logical structure antedates Euclid, he carries out this prerequisite with skill and sophistication. In accordance with Postulates 1, 2, and 3, the constructions permitted involve only the drawing of straight lines and circles. This means, in effect, straightedge and compass constructions. It is because Euclid could not establish the existence of angle trisectors that he proved no theorems involving them.

Despite some omissions and errors of proof that we shall point out shortly, Euclid's choice of axioms is remarkable. From a small set he was able to prove hundreds of theorems, many of them deep ones. Moreover, his choice was sophisticated. His handling of the parallel axiom is especially clever. Euclid undoubtedly knew that any such axiom states explicitly or implicitly what must happen in the infinite reaches of space and that any pronouncement on what must be true of infinite space is physically dubious because man's experiences are limited. Nevertheless, he also realized that some such axiom is indispensable. He therefore chose a version that states conditions under which two lines will meet at a finitely distant point. Moreover, he proved all the theorems he could before calling upon this axiom.

Though Euclid used superposition of figures to establish congruence, a method which rests on Common Notion 4, he was evidently concerned about the soundness of the method. There are two objections to it: First, the concept of motion is utilized and there is no logical basis for this concept; and second, the method of superposition assumes that a figure retains all its properties when moved from one position to another. The displaced figure may indeed be proved congruent to a second one but the first figure in its original position may not be congruent to the second one. To assume that moving a figure does not change its properties is to make a strong assumption about physical space. Indeed the very purpose of Euclidean geometry is to compare figures in different positions. The evidence for Euclid's concern about the method's soundness is that he did not use it for proofs that he could make by other means, even though superposition would have permitted a simpler proof.

Though mathematicians generally did regard Euclid's work as a model of rigor until well into the nineteenth century, there are serious defects that a few mathematicians recognized and struggled with. The first is the use of superposition. The second is the vagueness of some of his definitions and the pointlessness of others. The initial definitions of point, line, and surface have no precise mathematical meanings and, as we now recognize, could not have been given any because any independent mathematical development must have undefined terms (see sec. 3). As to the vagueness of many definitions, we have but to refer back to those in Book V as an example. An additional objection to the definitions is that several, such as Definition 17 of Book I, presuppose an axiom.

A critical study of Euclid, with, of course, the advantage of present insights, shows that he uses dozens of assumptions that he never states and undoubtedly did not recognize. A few have been mentioned in our survey. What Euclid and hundreds of the best mathematicians of later generations did was to use facts either evident from the figures or intuitively so evident that they did not realize they were using them. In a few instances the unconscious assumptions could be obviated by proofs based on the explicit assumptions, but this is not true generally.

Among the assumptions made unconsciously are those concerning the continuity of lines and circles. The proof of Proposition 1 of Book I assumes that the two circles have a point in common. Each circle is a collection of points, and it could be that though the circles cross each other there is no common point on the two circles at the supposed point or points of intersection. The same criticism applies to the straight line. Two lines may cross each other and yet not have a common point as far as the logical basis in the *Elements* is concerned.

There are also defects in the proofs actually given. Some of these are mistakes made by Euclid that can be remedied, though new proofs would be needed in a few instances. Another kind of defect that runs throughout the *Elements* is the statement of a general theorem that is proved only for special cases or for special positions of the given data.

Though we have praised Euclid for the overall organization of the contents of the *Elements*, the thirteen books are not a unity, but are to an extent compilations of previous works. For example, we have already noted that Books VII, VIII, and IX repeat for whole numbers many results given for magnitudes. The first part of Book XIII repeats results of Books II and IV. Books X and XIII probably were a unit before Euclid and were due to Theaetetus.

Despite these defects, many of which were pointed out by later commentators (Chap. 42, sec. 1) and very likely also by immediate successors of Euclid, the *Elements* was so successful that it displaced all previous texts on geometry. In the third century B.C., when others were still extant, even Apollonius and Archimedes referred to the *Elements* for prior results.

11. *Other Mathematical Works by Euclid*

Euclid wrote a number of other mathematical and physical works, many significant in the development of mathematics. We shall reserve discussion of his chief physical works, the *Optics* and the *Catoptrica*, for a later chapter.

Euclid's *Data* was included by Pappus in his *Treasury of Analysis*. Pappus describes it as consisting of supplementary geometrical material concerned with "algebraic problems." When certain magnitudes are given or determined, the theorems determine other magnitudes. The material is not different in nature from what appears in the *Elements* but the specific theorems are different. The *Data* may have been intended as a set of exercises to review the *Elements*. It is known in full.

Of the works of Euclid, next to the *Elements*, his *Conics* played the most vital role in the history of mathematics. According to Pappus, the contents of this lost work of four books became substantially the first three books of Apollonius' *Conic Sections*. Euclid treated the conics as sections of the three different types of cones (right-angled, acute-angled, and obtuse-angled). The ellipse was also obtained as a section of any cone and of a circular cylinder. As we shall see, Apollonius changed the approach to the conic sections.

The *Pseudaria* of Euclid contained correct and false geometric proofs and was intended for the training of students. The work is lost.

On Divisions [of figures], mentioned by Proclus, treats the subdivision of a given figure into other figures, as a triangle into smaller triangles or a triangle into triangles and quadrilaterals. A Latin translation, probably due to Gerard of Cremona (1114–87), of an incorrect and incomplete Arabic version exists. In 1851 Franz Woepcke found and translated another Arabic version that seems to be correct. There is an English translation by R. C. Archibald.

Another lost work is the *Porisms*. The contents and even the nature of the work are largely unknown. Pappus in his *Mathematical Collection* says that the *Porisms* consisted of three books. It is believed from the remarks of Pappus and Proclus that the *Porisms* dealt essentially with constructions of geometric objects whose existence was already assured. Thus these problems were intermediate between pure theorems and constructions establishing existence. To find the center of a circle under some given conditions would be typical of the problems in the *Porisms*.

The work *Surface-Loci*, composed of two books, is mentioned by Pappus in his *Collection*. This work, which is not extant, probably dealt with loci that are surfaces.

Euclid's *Phaenomena*, though a text on astronomy, contains 18 propositions on spherical geometry and others on uniformly rotating spheres. The earth is treated as a sphere. Some versions are extant.

12. *The Mathematical Work of Apollonius*

The other great Greek who belongs to the classical period, in the two senses of summarizing and adding to the kind of mathematics the classical period produced, is Apollonius (*c.* 262–190 B.C.). Apollonius was born in Perga, a city in the northwestern part of Asia Minor, which was under the rule of Pergamum during his lifetime. He came to Alexandria in his youth and learned mathematics from Euclid's successors. As far as we know, he remained in Alexandria and became an associate of the great mathematicians who worked there. His chief work was on the conic sections but he also wrote on other subjects. His mathematical powers were so extraordinary that he became known in his time and thereafter as "the Great Geometer." His reputation as an astronomer was almost as great.

The conic sections, as we know, were studied long before Apollonius' time. In particular, Aristaeus the Elder and Euclid had written works on them. Also Archimedes' work, which we shall study later, contains some results on this subject. Apollonius, however, stripped the knowledge of all irrelevancies and fashioned it systematically. Besides being comprehensive, his *Conic Sections* contains highly original material and is ingenious, extremely adroit, and excellently organized. As an achievement it is so monumental that it practically closed the subject to later thinkers, at least from the purely geometrical standpoint. It may truly be regarded as the culmination of classical Greek geometry.

The *Conic Sections* was written in eight books and contained 487 propositions. Of these books we have the first four reproduced in Greek manuscripts of the twelfth and thirteenth centuries and the next three in an Arabic translation written in A.D. 1290. The eighth book is lost, though a restoration based on indications by Pappus was made by Halley in the seventeenth century.

Euclid's predecessors, Euclid himself, and Archimedes treated the conic sections as arising from the three kinds of right circular cones—as they had been introduced by the Platonist Menaechmus. Euclid and Archimedes were aware that the ellipse can also be obtained as a section of the two other types of right circular cones, and Archimedes knew in addition that sections of *oblique* circular cones made by planes cutting all the generators are ellipses. He probably realized that the other conic sections can be obtained from oblique circular cones.

Apollonius, however, was the first to base the theory of all three conics on sections of one circular cone, right or oblique. He was also the first to recognize both branches of the hyperbola. One presumed reason that Menaechmus and the other predecessors of Apollonius used sections perpendicular to one of the elements of the three types of right circular cones is not that they did not see that other sections can be made of these cones, but rather that they wanted to treat the converse problem. Given curves whose geometric properties are those of the conic sections, the proof that these curves are obtainable as sections of a cone is more readily made when the plane of the section is perpendicular to an element of the cone.

We consider first the definitions and basic properties of the conics, which are in Book I. Given a circle BC and any point A (Fig. 4.20) outside the plane of the circle, a double cone is generated by a line through A and moving around the circumference of the circle. The circle is called the base of the cone. The axis of the cone is the line from A to the center of the circle (not shown in the figure). If this line is perpendicular to the base, the cone is right circular; otherwise it is scalene or oblique. A section of the cone by a plane cuts the plane of the base in a line DE. Take the diameter BC of the base circle perpendicular to DE. Then ABC is a triangle in whose interior

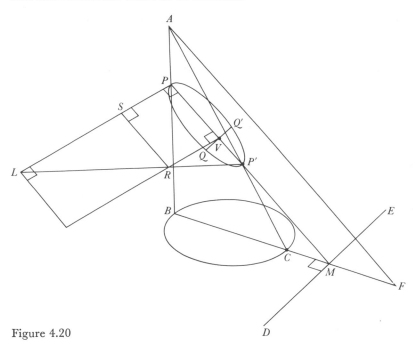

Figure 4.20

the axis of the cone lies; ABC is called an axial triangle. Let this triangle cut the conic in PP'. (PP' need not be an axis of the conic section.) $PP'M$ is the line determined by the intersection of the cutting plane and the axial triangle.[3] In the conic section let $Q'Q$ be any chord which is parallel to DE. Hence QQ' need not be perpendicular to PP'. Apollonius then proves that $Q'Q$ is bisected by PP' so that VQ is half of $Q'Q$.

Now draw AF parallel to PM to meet BM in F, say. Next draw PL perpendicular to PM and in the plane of the section. For the ellipse and hyperbola L is chosen to satisfy the condition

$$\frac{PL}{PP'} = \frac{BF \cdot FC}{AF^2},$$

and for the parabola L is chosen to satisfy

$$\frac{PL}{PA} = \frac{BC^2}{BA \cdot AC}.$$

3. Apollonius points out that if the cone is scalene then PM is not necessarily perpendicular to DE. Perpendicularity holds only for right circular cones or when the plane of ABC is perpendicular to the base of a scalene cone.

In the cases of ellipse and hyperbola we now draw $P'L$. From V draw VR parallel to PL to meet $P'L$ in R. (In the case of the hyperbola the location of P' is on the other branch and $P'L$ has to be extended to locate R.)

After some subordinate constructions which we do not give Apollonius proves that for the ellipse and hyperbola

(6) $$QV^2 = PV \cdot VR.$$

Now Apollonius refers to QV as an ordinate and the result (6) says that the square of the ordinate equals a rectangle applied to PL, namely $PV \cdot VR$. Moreover, he proves that in the case of the ellipse this rectangle falls short of the entire rectangle $PV \cdot PL$ by the rectangle LR which is similar to the entire rectangle formed by PL and PP'. Hence the term "ellipse" (sec. 6).

In the case of the hyperbola, (6) still holds but the construction would show that VR is longer than PL so that the rectangle $PV \cdot VR$ exceeds the rectangle applied to PL, that is, $PL \cdot PV$, by the rectangle LR which is similar to the rectangle formed by PL and PP'. Hence the term hyperbola. In the case of the parabola Apollonius shows that in place of (6),

(7) $$QV^2 = PV \cdot PL$$

so that the rectangle which equals QV^2 is exactly the rectangle applied to PL with width PV. Hence the term "parabola."

Apollonius introduced the terminology parabola, ellipse, and hyperbola for the conics in place of Menaechmus' sections of the right-angled, acute-angled, and obtuse-angled cone. Where the words parabola and ellipse occur in Archimedes, as in his *Quadrature of the Parabola* (Chap. 5, sec. 3) they were introduced by later transcribers.

Equations (6) and (7) are the basic plane properties of the conic sections. Having derived them, Apollonius disregards the cone and derives further properties from these equations. In effect, where we now use abscissa, ordinate, and the equation of a conic to derive properties, Apollonius uses PV, the ordinate or semichord QV, and a geometric equality, namely (6) or (7). Of course, no algebra appears in Apollonius' treatment.

We can readily transcribe Apollonius' basic properties into modern coordinate geometry. If we denote PL, which Apollonius calls the latus rectum or the parameter of the ordinates, by $2p$ and denote the length of the diameter PP' by d, if x is the distance PV measured from P, and if y is the distance QV (which means we are using oblique coordinates), then one sees immediately from (7) that the equation of the parabola is

$$y^2 = 2px.$$

For the ellipse, we note that we first get from the defining equation (6) that

$$y^2 = PV \cdot VR.$$

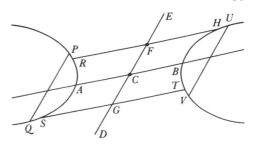

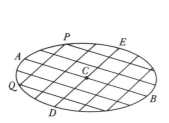

Figure 4.21   Figure 4.22

But $PV \cdot VR = x(2p - LS)$. Also, because the rectangle LR is similar to the rectangle determined by PL and PP',

$$\frac{LS}{PL} = \frac{x}{d}.$$

Hence $LS = 2px/d$. Then

$$y^2 = x\left(2p - \frac{2px}{d}\right) = 2px - \frac{2px^2}{d}.$$

For the hyperbola we get

$$y^2 = 2px + \frac{2px^2}{d}.$$

In the Apollonian construction, d is infinite for the parabola and so we see how the parabola appears as a limiting case of ellipse or hyperbola.

To pursue further Apollonius' treatment of the conics we need some definitions of concepts that are still important in modern geometry. Consider a set of parallel chords in an ellipse, say the set parallel to PQ in Figure 4.21. Apollonius proves that the centers of these chords lie on one line AB, which is called a diameter of the conic. (The line PP' in the basic Figure 4.20 is a diameter.) He then proves that if through C, the midpoint of AB, a line DE be drawn parallel to the original family of chords, this line will bisect all the chords parallel to AB. The line DE is called the conjugate diameter to AB. In the case of the hyperbola (Fig. 4.22), the chords may be inside the branches, e.g. PQ, and the length of the diameter is that cut off (if it is cut off) between the two branches, AB in the figure. The chords parallel to AB, for example RH, then lie between the branches. The conjugate diameter to AB, namely DE, which is defined to be the mean proportional between AB and the latus rectum of the hyperbola, does not cut the curve. In the parabola any diameter, that is, a line passing through the midpoints of a family of parallel chords, is always parallel to the axis of symmetry, but there is no diameter conjugate to a given diameter because each chord of the family of chords parallel to the given diameter is infinite in length. The axes of an

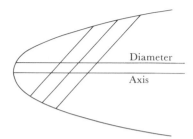

Figure 4.23

ellipse or a hyperbola are two diameters that are perpendicular to each other. For the parabola (Fig. 4.23), the axis is the diameter whose corresponding chords are perpendicular to the diameter.

After introducing the basic properties of the conic sections, Apollonius proves simple facts about conjugate diameters. Book I also treats tangents to conics. Apollonius conceives of a tangent as a line which has just one point in common with a conic section but which everywhere else lies outside of it. He then shows that a straight line drawn through an extremity of a diameter (point P in the basic Figure 4.20) and parallel to the corresponding chords of that diameter (parallel to QQ' in that figure) will fall outside the conic and no other straight line can fall between the said straight line and the conic (see *Elements*, Book III, Proposition 16). Therefore the said straight line touches the conic, that is, is the tangent at P.

Another theorem on tangents asserts the following: Suppose PP' (Fig. 4.24) is a diameter of a parabola and QV is one of the chords corresponding to that diameter. Then if a point T be taken on the diameter but outside the curve and such that $TP = PV$, where V is the foot of the ordinate (chord) from Q to the diameter PP', then the line TQ will touch the parabola at Q. There are analogous theorems for the ellipse and hyperbola.

Apollonius proves next that if any diameter of the conic other than PP' in the basic figure (4.20) be taken, the definitive property of the conic, equations (6) and (7), remains the same; of course QV then refers to the

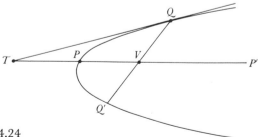

Figure 4.24

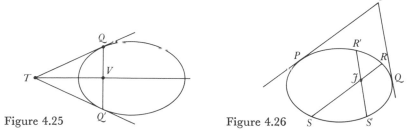

Figure 4.25   Figure 4.26

chords of that diameter. What he has done amounts in our language to transformation from one oblique coordinate system to another. In this connection, he also proves that from any diameter and the ordinates to it, one can transform to a diameter (axis) to which the ordinates are perpendicular. Then, in our language, the coordinate system is rectangular. Apollonius also shows how to construct conics given certain data—for example, a diameter, the latus rectum, and the inclination of the ordinates to the diameter. He does this by first constructing the cone of which the desired conic is then a section.

Book II starts with the construction and properties of the asymptotes to a hyperbola. He shows, for example, not only the existence of the asymptotes but also that the distance between a point on the curve and the asymptote becomes smaller than any given length by going far enough out on the curve. Then the conjugate hyperbola to a given one is introduced. It is shown to have the same asymptotes.

Additional theorems of Book II show how to find a diameter of a conic, the center of a central conic, the axis of a parabola, and the axes of a central conic. For example, if T (Fig. 4.25) is external to a conic, TQ and TQ' are tangents at the points Q and Q' of the conic, and V is the midpoint of the chord QQ', then TV is a diameter. Another method of finding the diameter of a conic is to draw two parallel chords; the line joining the midpoints of the chords is a diameter. The point of intersection of any two diameters is the center of a central conic. The book concludes with methods of constructing tangents to conics satisfying given conditions, as, for example, passing through a given point.

Book III begins with theorems about areas of figures formed by tangents and diameters. One of the chief results here (Fig. 4.26) is that if OP and OQ are tangents to a conic, if RS is any chord parallel to OP, and $R'S'$ any chord parallel to OQ, and if RS and $R'S'$ intersect in J (internally or externally), then

$$\frac{RJ \cdot JS}{R'J \cdot JS'} = \frac{OP^2}{OQ^2}.$$

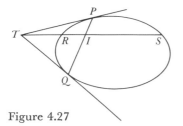

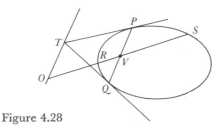

Figure 4.27   Figure 4.28

The theorem is a generalization of a well-known theorem in elementary geometry, namely, that if two chords intersect in a circle the product of the segments of one equals the product of the segments of the other, for in this case $OP^2/OQ^2 = 1$.

Book III then treats what we call the harmonic properties of pole and polar. Thus if TP and TQ are tangents to a conic (Fig. 4.27) and if TRS is any line meeting the conic in R and S and meeting PQ in I, then

$$\frac{TR}{TS} = \frac{IR}{IS}.$$

That is, T divides RS externally in the same ratio as I divides it internally. The line PQ is called the polar of the point T, and T, R, I, and S are said to form a harmonic set of points. Also if any line through V (Fig. 4.28), the midpoint of PQ, meets the conic in R and S and meets the parallel to PQ through T in O, then

$$\frac{OR}{OS} = \frac{VR}{VS}.$$

The line through T is the polar of V and O, R, V, and S are a harmonic set of points.

The book continues with the subject of the focal properties of central conics; the focus of a parabola is not mentioned here. The foci (the word is not used by Apollonius) are defined for the ellipse and the hyperbola (Fig. 4.29) as the points F and F' on the (major) axis AA' such that $AF \cdot FA' = AF' \cdot F'A' = 2p \cdot AA'/4$. Apollonius proves for the ellipse and the hyperbola that the lines PF and PF' from a point P on the conic make equal angles with the tangent at P and that the sum (for the ellipse) of the focal distances PF and PF' equals AA' and the difference of the focal distances (for the hyperbola) equals AA'.

No concept of directrix appears in this work, but the fact that a conic is a locus of points the ratio of whose distances from a fixed point (focus) and a fixed line (directrix) is constant was known to Euclid and is stated and proved by Pappus (Chap. 5, sec. 7).

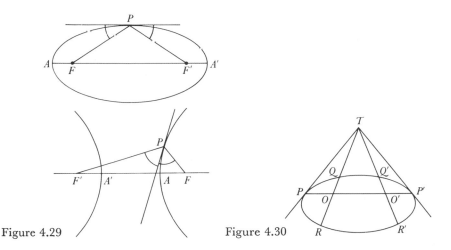

Figure 4.29   Figure 4.30

There is a famous problem, partly solved by Euclid, of determining the locus of points for each of which the distances p, q, r, and s to four given lines satisfy the condition $pq = \alpha rs$, where α is a given number. Apollonius says in his preface to the *Conic Sections* that this problem can be solved by the propositions of Book III. This can indeed be done; and Pappus too knew that this locus is a conic.

Book IV treats further properties of pole and polar. For example, one proposition gives a method of drawing two tangents to a conic from an external point T (Fig. 4.30). Draw TQR and $TQ'R'$. Let O be the fourth harmonic point to T on QR; that is, $TQ:TR = OQ:OR$, and let O' be the fourth harmonic point to T on $Q'R'$. Draw OO'. Then P and P' are the points of tangency.

The remainder of the book deals with the number of possible intersections of conics in various positions. Apollonius proves that two conics can intersect in at most four points.

Book V is the most remarkable for its novelty and originality. It deals with the maximum and minimum lengths that can be drawn from particular points to a conic. Apollonius starts with special points on the major axis of a central conic or on the axis of a parabola and finds the lines of maximum and minimum distances from such points to the curve. Then he takes points on the minor axis of an ellipse and does the same thing. He also proves that if O be any point within any conic and if OP is a maximum or minimum straight line from O to the conic, then the line perpendicular to OP at P is tangent at P; and if O' be any point on OP produced outside the conic, then $O'P$ is a minimum line from O' to the conic. The perpendicular to a tangent at a point of tangency we now call a normal, and so the maximum and minimum

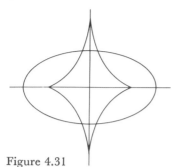

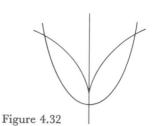

Figure 4.31   Figure 4.32

lines are normals. Apollonius next considers properties of normals to any conic. For example, in a parabola or an ellipse a normal at any point will meet the curve again. He then shows how from given points within or without a conic one can construct the normals to the conic.

In the course of his investigation of the (relative) maximum and minimum lines that can be drawn from a point to any conic, Apollonius determines the positions of points from which two, three, and four such lines can be drawn. For each of the conics he determines the locus of points such that from points on one side one number of normals can be drawn, and from points on the other, another number of normals can be drawn. The locus itself, which Apollonius does not discuss, is what we now call the evolute of the conic, or the locus of points of intersection of "nearby" normals to the conic, or the envelope of the family of normals to the conic. Thus from any point inside the evolute of the ellipse (Fig. 4.31) four normals to the ellipse can be drawn but from any point outside two normals can be drawn. (There are exceptional points.) In the case of a parabola, the evolute (Fig. 4.32) is the curve called a semicubical parabola (first studied by William Neile [1637–70]). From any point in the plane above the semicubical parabola three normals to the parabola can be drawn and from any point below, only one can be drawn. From a point on the semicubical parabola two can be drawn.

Book VI treats congruent and similar conics and segments of conics. Segments are regions cut off by a chord just as in the circle. Apollonius also shows how, given a right circular cone, one can construct on it a conic section equal to a given one.

Book VII has no outstanding propositions. It treats properties of conjugate diameters of a central conic. Apollonius compares these properties with the corresponding properties of the axes. Thus if a and b are the axes and a' and b' are two conjugate diameters of an ellipse or a hyperbola, $a + b < a' + b'$. Further, the sum of the squares on any two conjugate diameters of an ellipse equals the sum of the squares on the axes. For the hyperbola the corresponding proposition holds but with difference instead of sum. Also in an ellipse or a hyperbola the area of the parallelogram determined by any two conjugate diameters and the angle at which they intersect equals the area of the rectangle determined by the axes.

Book VIII is lost. It probably contained propositions on how to determine conjugate diameters of a (central) conic such that certain functions of their lengths have given values.

Pappus mentions six other mathematical works of Apollonius. One of these, *On Contacts*, whose contents were reconstructed by Vieta, contained the famous Apollonian problem: Given any three points, lines, or circles, or any combination of three of these, to construct a circle passing through the points and tangent to the given lines and circles. Many mathematicians, including Vieta and Newton, gave solutions to this problem.

The strict deductive mathematics of Euclid and Apollonius has given rise to the impression that mathematicians create by reasoning deductively. Our review of the three hundred years of activity preceding Euclid should show that conjectures preceded proofs and that analysis preceded synthesis. In fact, the Greeks did not think much of propositions obtained by simple deduction. Results that sprung readily from a theorem the Greeks called corollaries or porisms. Such results, obtained without additional labor, were regarded by Proclus as windfalls or bonuses.

We have not exhausted the contributions of the Greek genius to mathematics. What we have discussed thus far belongs to the classical Greek period; the significant epoch extending from about 300 B.C. to A.D. 600 still awaits us. Before turning the page let us recall that the classical period contributed more than content; it created mathematics in the sense in which we understand the word today. The insistence on deduction as a method of proof and the preference for the abstract as opposed to the concrete determined the character of mathematics, while the selection of a most fruitful and highly acceptable set of axioms and the divination and proof of hundreds of theorems sent the science well on its way.

Bibliography

Ball, W. W. Rouse: *A Short Account of the History of Mathematics*, Dover (reprint), 1960, Chaps. 2–3.

Boyer, Carl B.: *A History of Mathematics*, John Wiley and Sons, 1968, Chaps. 7 and 9.

Coolidge, Julian L.: *A History of Geometrical Methods*, Dover (reprint), 1963, Book 1, Chaps. 2–3.

Heath, Thomas L.: *A Manual of Greek Mathematics*, Dover (reprint), 1963, Chaps. 3–9 and 12.

———: *A History of Greek Mathematics*, Oxford University Press, 1921, Vol. 1, Chaps. 3–11; Vol. 2, Chap. 14.

———: *The Thirteen Books of Euclid's* Elements, 3 vols., Dover (reprint), 1956.

————: *Apollonius of Perga*, Barnes and Noble (reprint), 1961.

Neugebauer, Otto: *The Exact Sciences in Antiquity*, Princeton University Press, 1952, Chap. 6.

Proclus: *A Commentary on the First Book of Euclid's* Elements, Princeton University Press, 1970.

Sarton, George: *A History of Science*, Harvard University Press, 1952 and 1959, Vol. 1, Chaps. 8, 10, 11, 17, 20; Vol. 2, Chap. 3.

Scott, J. F.: *A History of Mathematics*, Taylor and Francis, 1958, Chap. 2.

Smith, David Eugene: *History of Mathematics*, Dover (reprint), 1958, Vol. 1, Chap. 3; Vol. 2, Chap. 5.

Struik, Dirk J.: *A Concise History of Mathematics*, 3rd ed., Dover, 1967, Chap. 3.

van der Waerden, B. L.: *Science Awakening*, P. Noordhoff, 1954, Chaps. 4–6.

5

The Alexandrian Greek Period: Geometry and Trigonometry

> Without the concepts, methods and results found and developed by previous generations right down to Greek antiquity one cannot understand either the aims or the achievements of mathematics in the last fifty years. HERMANN WEYL

1. *The Founding of Alexandria*

The course of mathematics has been very much dependent on the course of history. Conquests launched by the Macedonians, a Greek people living in the northern part of the mainland of Greece, led to the destruction of the classical Greek civilization and paved the way for another essentially Greek civilization of quite different character. The conquests were begun in 352 B.C. by Philip II of Macedonia. Athens was beaten in 338 B.C. In 336 B.C. Alexander the Great, son of Philip, took command and conquered Greece, Egypt, and the Near East as far east as India and as far south as the cataracts of the Nile. He constructed new cities everywhere, both as strongholds and as centers of commerce. The main one, Alexandria, centrally located in Alexander's empire and intended as his capital, was founded in Egypt in 332 B.C. Alexander chose the site and drew up the plans for the buildings and for colonizing the city, but the work was not completed for many years thereafter.

Alexander envisioned a cosmopolitan culture in his new empire. Because the only other leading civilization was Persian, Alexander deliberately sought to fuse the two. In 325 B.C. he himself married Statira, daughter of the Persian ruler Darius, and compelled a hundred of his generals and ten thousand of his soldiers to marry Persians. He incorporated twenty thousand Persians into his army and mixed them with Macedonians in the same phalanxes. He also brought colonists of all nations to the various cities he founded. After his death, written orders were found to transport large groups of Europeans to Asia and vice versa.

Alexander died in 323 B.C. before he could complete his capital and while still engaged in conquests. After his death, his generals fought each other for power. Following several decades of political instability, the empire was split into three independent parts. The European portion became the Antigonid empire (from the Greek general Antigonus); the Asian part, the Seleucid empire (after the Greek general Seleucus); and Egypt, ruled by the Greek Ptolemy dynasty, became the third empire. Antigonid Greece and Macedonia gradually fell under Roman domination and became unimportant as far as the development of mathematics is concerned. The mathematics generated in the Seleucid empire was largely a continuation of Babylonian mathematics, though influenced by the developments we are about to consider. The major creations following the classical Greek period were made in the Ptolemaic empire, primarily in Alexandria.

That the Ptolemaic empire became the mathematical heir of classical Greece was not accidental. The kings of the empire were wise Greeks and pursued Alexander's plan to build a cultural center at Alexandria. Ptolemy Soter, who ruled from 323 to 285 B.C., his immediate successors, Ptolemy II, called Philadelphus, who ruled from 285 to 247 B.C., and Ptolemy Euergetes, who reigned from 247 to 222 B.C., were well aware of the cultural importance of the great Greek schools such as those of Pythagoras, Plato, and Aristotle. These rulers therefore brought to Alexandria scholars from all the existing centers of civilization and supported them with state funds. About 290 B.C Ptolemy Soter built a center in which the scholars could study and teach. This building, dedicated to the muses, became known as the Museum, and it housed poets, philosophers, philologists, astronomers, geographers, physicians, historians, artists, and most of the famous mathematicians of the Alexandrian Greek civilization.

Adjacent to the Museum, Ptolemy built a library, not only for the preservation of important documents but for the use of the general public. This famous library was said at one time to contain 750,000 volumes, including the personal library of Aristotle and his successor Theophrastus. Books, incidentally, were more readily available in Alexandria than in classical Greece because Egyptian papyrus was at hand. In fact, Alexandria became the center of the book-copying trade of the ancient world.

The Ptolemies also pursued Alexander's plan of encouraging a mixture of peoples, so that Greeks, Persians, Jews, Ethiopians, Arabs, Romans, Indians, and Negroes came unhindered to Alexandria and mingled freely in the city. Aristocrat, citizen, and slave jostled each other and, in fact, the class distinctions of the older Greek civilization broke down. The civilization in Egypt was influenced further by the knowledge brought in by traders and by the special expeditions organized by the scholars to learn more about other parts of the world. Consequently, intellectual horizons were broadened. The long sea voyages of the Alexandrians called for far better knowledge of geography, methods of telling time, and navigational techniques, while commercial competition generated an interest in materials, in efficiency of production, and in improvement of skills. Arts that had been despised in the classical period were taken up with new zest and training schools were established. Pure science continued to be pursued but was also applied.

The mechanical devices created by the Alexandrians are astonishing even by modern standards. Pumps to bring up water from wells and cisterns, pulleys, wedges, tackles, systems of gears, and a mileage-measuring device no different from what may be found in the modern automobile were used commonly. Steam power was employed to drive a vehicle along the city streets in the annual religious parade. Water or air heated by fire in secret vessels of temple altars was used to make statues move. The awe-struck audience observed gods who raised their hands to bless the worshipers, gods shedding tears, and statues pouring out libations. Water power operated a musical organ and made figures on a fountain move automatically while compressed air was used to operate a gun. New mechanical instruments, including an improved sundial, were invented to refine astronomical measurements.

The Alexandrians had an advanced knowledge of such phenomena as sound and light. They knew the law of reflection of light and had an empirical grip on the law of refraction (Chap. 7, sec. 7), knowledge which they applied to the design of mirrors and lenses. In this period there appeared for the first time a work on metallurgy, which contained far more chemistry than the few empirical facts known to the earlier Egyptian and Greek scholars. Poisons were a specialty. Medicine flourished, partly because the dissection of human bodies, forbidden in classical Greece, was now permitted, and the art of healing reached its pinnacle in the work of Galen (129–c. 201), who, however, lived chiefly in Pergamum and Rome. Hydrostatics, the science of the equilibrium of bodies immersed in fluids, was investigated intensively and indeed founded in systematic form. The greatest of their scientific achievements was the first truly quantitative astronomical theory (Chap. 7, sec. 4).

2. *The Character of Alexandrian Greek Mathematics*

The work of the scholars at the Museum was divided into four departments—literature, mathematics, astronomy, and medicine. Since two of these were essentially mathematical and medicine, through astrology, involved some mathematics, we see that mathematics occupied a dominant place in the Alexandrian world. The character of the mathematics was very much affected by the new civilization and culture. No matter what mathematicians may say about the purity of their subject and their indifference to and elevation above their environment, the new Hellenistic civilization produced a mathematics entirely different in character from that of the classical period.

Of course Euclid and Apollonius were Alexandrians; but, as we have

already noted, Euclid organized the work of the classical period, and
Apollonius is exceptional in that he too organized and extended classical
Greek mathematics—though in his astronomy and his work on irrational
numbers (both of which will be presented in later chapters), he was somewhat affected by the Alexandrian culture. To be sure, the other great
Alexandrian mathematicians, Archimedes, Eratosthenes, Hipparchus, Nicomedes, Heron, Menelaus, Ptolemy, Diophantus, and Pappus, continued to
display the Greek genius for theoretical and abstract mathematics, but with
striking differences. Alexandrian geometry was devoted in the main to
results useful in the calculation of length, area, and volume. It is true that
some such theorems are also in Euclid's *Elements*. For example, Proposition 10
of Book XII asserts that any cone is a third part of the cylinder which has
the same base and height. Hence if one knows the volume of a cylinder he
can compute the volume of a cone. However, such theorems are relatively
scarce in Euclid, whereas they were the major concern of the Alexandrian
geometers. Thus, while Euclid was content to prove that the areas of two
circles are to each other as the squares on their diameters—which leaves us
with the knowledge that the area $A = kd^2$ but without a value for k—
Archimedes obtained a close approximation to π so that circular areas
could be computed.

Further, the classical Greeks, because they would not entertain irrationals as numbers, had produced a purely qualitative geometry. The
Alexandrians, following the practice of the Babylonians, did not hesitate
to use irrationals and in fact applied numbers freely to lengths, areas, and
volumes. The climax of this work was the development of trigonometry.

Even more significant is the fact that the Alexandrians revived and
extended arithmetic and algebra, which became subjects in their own right.
This development of the science of number was, of course, necessary if
quantitative knowledge was to be obtained either from geometrical results
or from the direct use of algebra.

The Alexandrian mathematicians took an active hand in the work on
mechanics. They calculated centers of gravity of bodies of various shapes;
they dealt with forces, inclined planes, pulleys, and gears; and they were
often inventors. They were also the chief contributors of their time to the
work on light, mathematical geography, and astronomy.

In the classical period mathematics had embraced arithmetic (of the
whole numbers only), geometry, music, and astronomy. The scope of mathematics was broadened immeasurably in the Alexandrian period. Proclus,
who drew material from Geminus of Rhodes (1st cent. B.C.), cites the latter
on the divisions of mathematics (presumably in Geminus' time): arithmetic
(our theory of numbers), geometry, mechanics, astronomy, optics, geodesy,
canonic (science of musical harmony), and logistic (applied arithmetic).
According to Proclus, Geminus says: "The entire mathematics was separated

into two main divisions with the following distinction: one part concerned itself with the intellectual concepts and the other with material concepts." Arithmetic and geometry were intellectual. The other division was material. However, this distinction was gradually lost sight of, if it was still significant as late as the first century B.C. One can say, as a broad generalization, that the mathematicians of the Alexandrian period severed their relation with philosophy and allied themselves with engineering.

We shall treat first the Alexandrian work in geometry and trigonometry. In the next chapter we shall discuss the arithmetic and algebra.

3. *Areas and Volumes in the Work of Archimedes*

There is no one individual whose work epitomizes the character of the Alexandrian age so well as Archimedes (287–212 B.C.), the greatest mathematician in antiquity. The son of an astronomer, he was born in Syracuse, a Greek settlement in Sicily. As a youth he came to Alexandria, where he received his education. Though he returned to Syracuse and spent the rest of his life there, he kept contact with Alexandria. He was well known in the Greek world and greatly admired and respected by his contemporaries.

Archimedes was possessed of a lofty intellect, great breadth of interests—both practical and theoretical—and excellent mechanical skill. His work in mathematics included the finding of areas and volumes by the method of exhaustion, the calculation of π (in the course of which he approximated square roots of small and large numbers), and a new scheme for representing large numbers in verbal language. In mechanics, he found the centers of gravity of many plane and solid figures and gave theorems on the lever. The area of hydrostatics that deals with the equilibrium of bodies floating in water was founded by him. He is also reputed to have been a good astronomer.

His inventions so far excelled the technique of his times that endless stories and legends grew up about him. Indeed in popular esteem his inventions overshadowed his mathematics, though he is ranked with Newton and Gauss as one of the three greatest in that field. In his youth he constructed a planetarium, a contrivance operated by water power that reproduced the motions of the sun, moon, and planets. He invented a pump (Archimedean screw) for raising water from a river; he showed how to use the lever to move great weights; he used compound pulleys to launch a galley for King Hieron of Syracuse; and he invented military engines and catapults to protect Syracuse when it was under attack by the Romans. Taking advantage of the focusing properties of a paraboloidal mirror, he concentrated the sun's rays on the Roman ships besieging Syracuse and burned them.

Perhaps the most famous of the stories about Archimedes is his discovery of the method of testing the debasement of a crown of gold. The king of

Syracuse had ordered the crown. When it was delivered, he suspected that it was filled with baser metals and sent it to Archimedes to devise some method of testing the contents without, of course, destroying the workmanship. Archimedes pondered the problem; one day while bathing he observed that his body was partly buoyed up by the water and suddenly grasped the principle that enabled him to solve the problem. He was so excited by this discovery that he ran out into the street naked shouting "Eureka!" ("I have found it!") He had discovered that a body immersed in water is buoyed up by a force equal to the weight of the water displaced, and by means of this principle was able to determine the contents of the crown (see Chap. 7, sec. 6).

Though Archimedes was a remarkably ingenious and successful inventor, Plutarch says that these inventions were merely "the diversions of geometry at play." According to Plutarch, Archimedes "possessed so high a spirit, so profound a soul, and such treasures of scientific knowledge that, though these inventions had obtained for him the renown of more than human sagacity, he yet would not deign to leave behind him any written work on such subjects, but, regarding as ignoble and sordid the business of mechanics and every sort of art which is directed to use and profit, he placed his whole ambition in those speculations in whose beauty and subtlety there is no admixture of the common needs for life." However Plutarch's status as a storyteller is far higher than his status as a historian. Archimedes did write books on mechanics and we have one entitled *On Floating Bodies* and another, *On the Equilibrium of Planes*; two others, *On Levers* and *On Centers of Gravity*, are lost. He also wrote a work on optics that is lost, and did deign to write about his inventions; though the work has vanished, we know definitely that he wrote *On Sphere-making*, which describes an invention displaying the motions of the sun, the moon, and the five planets about the (fixed) earth.

The death of Archimedes portended what was to happen to the entire Greek world. In 216 B.C. Syracuse allied itself with Carthage in the second Punic war between that city and Rome. The Romans attacked Syracuse in 212 B.C. While drawing mathematical figures in the sand, Archimedes was challenged by one of the Roman soldiers who had just taken the city. Story has it that Archimedes was so lost in thought that he did not hear the challenge of the Roman soldier. The soldier thereupon killed him, despite the order of the Roman commander, Marcellus, that Archimedes be unharmed. Archimedes was then seventy-five and still in full possession of his powers. By way of "compensation," the Romans built an elaborate tomb upon which they inscribed a famous Archimedean theorem.

Archimedes' writings took the form of small tracts rather than large books. Our knowledge of these works comes from extant Greek manuscripts and from Latin manuscripts translated from the Greek from the thirteenth century onward. Some of the Latin versions were made from Greek

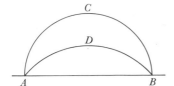

Figure 5.1

manuscripts available to those translators but not to us. In 1543 Tartaglia did a translation into Latin of some works of Archimedes.

Archimedes' geometrical work is the zenith of Alexandrian Greek mathematics. In his mathematical derivations Archimedes uses theorems of Euclid and Aristaeus, and still other results which he says are manifest, that is, can readily be proved from the known results. His proofs are therefore solidly established but not easy for us to follow because we are not familiar with many of the methods and results of the Greek geometers.

In his work *On the Sphere and Cylinder*, Archimedes starts with definitions and assumptions. The first assumption or axiom is that of all lines (curves) which have the same extremities the straight line is shortest. Other axioms involve the lengths of concave curves and surfaces. For example *ADB* (Fig. 5.1) is assumed to be less than *ACB*. These axioms enable Archimedes to compare perimeters of inscribed and circumscribed polygons with the perimeter of the circle.

After some preliminary propositions, he proves in Book I:

Proposition 13. The surface of any right circular cylinder excluding the bases is equal to [the area of] a circle whose radius is a mean proportional between the side [a generator] and the diameter of its base.

This is followed by many theorems about the volumes of cones. Of great interest are:

Proposition 33. The surface of any sphere is four times the [area of the] greatest circle on it.
Corollary to Prop. 34. Every cylinder whose base is the greatest circle in a sphere and whose height is equal to the diameter of the sphere is 3/2 of [the volume of] the sphere, and its surface together with its bases is 3/2 of the surface of the sphere.

That is, he compares the surface area and volume of a sphere and a cylinder circumscribed about the sphere. This is the famous theorem which in accordance with Archimedes' wishes, was inscribed on his tombstone.

He then proves in Propositions 42 and 43 that the surface of the segment *ALMNP* of a sphere is the area of a circle whose radius is *AL* (Fig. 5.2). The segment can be less or more than a hemisphere.

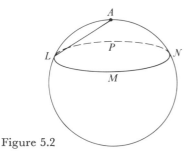

Figure 5.2

The theorems on surface area and volume are proved by the method of exhaustion. Archimedes uses inscribed and circumscribed rectilinear figures to "exhaust" the area or volume, and then, like Euclid, uses the indirect method of proof to complete the argument.

Some theorems in the second book of *On the Sphere and Cylinder*, which is concerned largely with segments of a sphere, are significant because they contain new geometrical algebra. For example, he gives:

Proposition 4. To cut a given sphere by a plane so that the volumes of the segments are to one another in a given ratio.

This problem amounts algebraically to the solution of the cubic equation

$$(a - x):c = b^2:x^2$$

and Archimedes solves it geometrically by finding the intersection of a parabola and a rectangular hyperbola.

The work *On Conoids and Spheroids* treats properties of figures of revolution generated by conics. Archimedes' right-angled conoid is a paraboloid of revolution. (In Archimedes' time the parabola was still regarded as a section of a right-angled cone.) The obtuse-angled conoid is one branch of a hyperboloid of revolution. Archimedes' spheroids are what we call oblate and prolate spheroids, which are figures of revolution generated by ellipses. The main object of the work is the determination of volumes of segments cut off from the three solids by planes. The book also contains some of Archimedes' work on the conic sections already alluded to in the discussion of Apollonius. As in other works, he presupposes theorems that he deems easily proved or that can be proved by methods he has previously used. Many of the proofs use the method of exhaustion. Some examples of the contents are furnished by the following propositions:

Proposition 5. If AA' and BB' be the major and minor axes of an ellipse and if d be the diameter of any circle, then the area of the ellipse is to the area of the circle as $AA' \cdot BB'$ is to d^2.

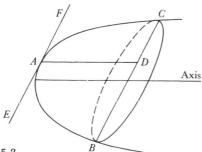

Figure 5.3

The theorem says that if $2a$ is the major axis and $2b$ is the minor axis and s and s' are the areas of the ellipse and circle, then $s/s' = 4ab/d^2$. Since $s' = (\pi/4)d^2$, $s = \pi ab$.

Proposition 7. Given an ellipse with center C and a line CO perpendicular to the plane of the ellipse, it is possible to find a circular cone with vertex O and such that the ellipse is a section of it.

Clearly Archimedes realized that some, at least, of the several conic sections can be obtained from the same cone, a fact that Apollonius utilized.

Proposition 11. If a paraboloid of revolution be cut by a plane through, or parallel to, the axis [of revolution] the section will be a parabola equal to the original parabola which generated the paraboloid . . . . If the paraboloid be cut by a plane at right angles to its axis, the section will be a circle whose center is on the axis.

There are similar results for the hyperboloid and spheroid.

Among principal results of the work are:

Proposition 21. Any segment [the volume] of a paraboloid of revolution is half as large again as the cone or segment of a cone which has the same base and the same axis.

The base is the area (Fig. 5.3) of the plane figure, ellipse or circle, which is cut out on the paraboloid by the plane determining the segment. The parabolic sections BAC and BC on the base are cut out by a plane through the axis of the paraboloid and perpendicular to the original cutting plane. EF is the tangent to the parabola that is parallel to BC, and A is the point of tangency. AD, drawn parallel to the axis of the paraboloid, is the axis of the segment. It can be shown that D is the midpoint of CB. Also, if the base is an ellipse then CB is its major axis; if the base is a circle then CB is its diameter. The cone has the same base as the segment, and has vertex A and axis AD.

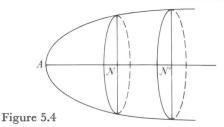

Figure 5.4

Proposition 24. If from a paraboloid of revolution two segments be cut off by planes drawn in *any* manner, the segments will be to one another as the squares on their axes.

To illustrate the theorem, suppose the planes are perpendicular to the axis of the paraboloid (Fig. 5.4); then the two volumes are to each other as AN^2 is to AN'^2. There are similar theorems for segments of hyperboloids and spheroids.

One of the very novel works of Archimedes is a short treatise known as *The Method*, in which he shows how he used ideas from mechanics to obtain correct mathematical theorems. This work was discovered as recently as 1906 in a library in Constantinople. The manuscript was written in the tenth century on a parchment that contains other works of Archimedes already known through other sources. Archimedes illustrates his method of discovery with the problem of finding the area of a parabolic segment CBA (Fig. 5.5). In this basically physical argument he uses theorems on centers of gravity established elsewhere by him.

ABC (Fig. 5.5) is any segment of a parabola bounded by the straight line AC and the arc ABC. Let CE be the tangent to the parabola at C; let D be the midpoint of CA; and let DBE be the diameter through D (line parallel to the axis of the parabola). Then Archimedes, referring to Euclid's *Conics*, states that

(1) $$EB = BD,$$

though Euclid's proof of this fact is not known. Now draw AF parallel to ED and let CB cut AF in K. Then, by (1) and the use of similar triangles, one proves that $FK = KA$. Produce CK to H so that $CK = KH$. Further, let $MNPO$ be any diameter of the parabola. Then $MN = NO$, because of (1) and the use of similar triangles.

Now Archimedes compares the area of the segment and the area of triangle CFA. He regards the first area as the sum of line segments such as PO and the area of the triangle as the sum of line segments such as MO. He then proves that

$$HK \cdot OP = KN \cdot MO.$$

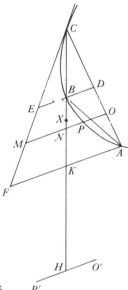

Figure 5.5

Physically this means that if we regard KH and KN as arms of a lever with a fulcrum at K, then OP regarded as a weight placed at H would balance the weight of MO placed at N. Consequently the sum of all the line segments such as PO placed at H (i.e., $P'O'$) will balance the sum of all the line segments such as MO each concentrated at its midpoint, which is the center of gravity of a line segment. But the collection of segments MO each placed at its center of gravity is "equivalent" to the triangle CAF placed at its center of gravity. In his book *On the Equilibrium of Planes*, Archimedes shows that this center is X on CK where $KX = (1/3)CK$. By the law of the lever, $KX \cdot$ the area of triangle $CFA = HK \cdot$ area of parabolic segment or

(2)
$$\frac{\triangle CFA}{\text{segment } CBA} = \frac{HK}{KX} = \frac{3}{1}.$$

Archimedes wishes to relate the area of the segment to triangle ABC. He points out that (the area of) this triangle is one half of triangle CKA because both have the same base CA and the altitude of one is readily shown to be half of the altitude of the other. Moreover, triangle CKA is one half of triangle CFA (because KA is half of FA). Hence triangle ABC is one fourth of triangle CFA and, from (2), he has that the area of segment ABC is to the area of triangle ABC as 4 is to 3.

In this mechanical method Archimedes regards the area of the parabolic segment and of triangle CFA as sums of an infinite number of line segments.

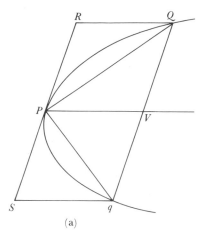
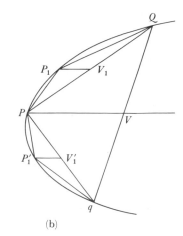

Figure 5.6, a and b

This method, he says, is one of discovery but not of rigorous geometrical proof. He shows in this treatise how effective the method is by using it to discover other theorems on segments of spheres, cylinders, spheroids, and paraboloids of revolution.

In his book *Quadrature of the Parabola* Archimedes gives two methods for finding the area of the parabolic segment. The first of these is similar to the mechanical argument just examined in that he again balances areas by means of the principle of the lever, but his choice of areas is different. His conclusion is of course the same as that in (2) above. It is given in Proposition 16. Now Archimedes knows the result he wishes to prove and he proceeds to do so by rigorous mathematics in a sequence of theorems (Propositions 18–24).

The first step is to prove that the parabolic segment can be "exhausted" by a series of triangles. Let QPq (Fig. 5.6a) be the parabolic segment and let PV be the diameter bisecting all chords parallel to the base Qq of the segment so that V is the midpoint of Qq. It is intuitively apparent and proved in Proposition 18 that the tangent at P is parallel to Qq. Next take QR and qS parallel to PV. Then triangle QPq is one half of parallelogram $QRSq$ and so triangle QPq is greater than one half of the parabolic segment.

As a corollary to this result, Archimedes shows that the parabolic segment can be approximated by a polygon as closely as one pleases, for by constructing a triangle in the segment cut off by PQ (Fig. 5.6b) wherein P_1V_1 is the diameter of that segment one can prove by simple geometry (Proposition 21) that (the area of) triangle $PP_1Q = (1/8)$ triangle PQq. Hence triangle PP_1Q and triangle PP'_1q, which is constructed on Pq and has the same properties as triangle PP_1Q, are together $1/4$ of triangle PQq;

moreover, by the result in the preceding paragraph, the two smaller triangles fill out more than half of the parabolic segments in which they lie. The process of constructing triangles on the new chords QP_1, P_1P, PP'_1, and P'_1q can be continued. This part of the proof is entirely analogous to the corresponding part of Euclid's theorem on the areas of two circles.

Hence we have the conditions sufficient to apply Proposition 1 of Book X of Euclid's *Elements*; that is, we can assert that the area of the polygonal figure obtained by adding triangles to the original triangle PQq, that is, the area

(3) $$\triangle PQq + (1/4)\triangle PQq + (1/16)\triangle PQq \cdots$$

to a *finite* number of terms can approximate the parabolic segment as closely as one pleases; that is, the difference between the area of the segment and the finite sum (3) can be made less than any preassigned quantity.

Now Archimedes applies the indirect method of proof that completes the proof by the method of exhaustion. He first proves that for n terms of a geometrical progression in which the common ratio is $1/4$,

(4) $$A_1 + A_2 + \cdots + A_n + (1/3)A_n = (4/3)A_1.$$

This is readily proven in many ways; we can do it by our formula for the sum of n terms of a geometrical progression. In the application of (4), A_1 is triangle PQq.

Then Archimedes shows that the area A of the parabolic segment cannot be greater or less than $(4/3)A_1$. His proof is simply that if the area A exceeded $(4/3)A_1$ then he could get a (finite) set of triangles whose sum S would differ from the area of the segment by less than any given magnitude, and hence the sum S would exceed $(4/3)A_1$. That is,

$$A > S > (4/3)A_1.$$

But by (4) if S contains m terms, say, then

$$S + (1/3)A_m = (4/3)A_1$$

or

$$S < (4/3)A_1.$$

Hence there is a contradiction.

Likewise, suppose one assumes that the area A of the parabolic segment is less than $(4/3)A_1$. Then $(4/3)A_1 - A$ is a definite number. Since the triangles Archimedes forms get smaller he can get a sequence of inscribed triangles such that

(5) $$(4/3)A_1 - A > A_m$$

where A_m is the mth term of the sequence and geometrically represents the sum of 2^{m-1} triangles. But since by (4)

$$(6) \qquad A_1 + A_2 + \cdots + A_m + \frac{1}{3} A_m = \frac{4}{3} A_1,$$

then

$$\frac{4}{3} A_1 - (A_1 + A_2 + \cdots + A_m) = \frac{1}{3} A_m,$$

or

$$(7) \qquad \frac{4}{3} A_1 - (A_1 + A_2 + \cdots + A_m) < A_m.$$

It follows from (5) and (7) that

$$(8) \qquad A_1 + A_2 + \cdots + A_m > A.$$

But a sum consisting of inscribed triangles is always less than the area of the segment. Hence (8) is impossible.

Of course, in effect Archimedes has summed an infinite geometric progression, because when n becomes infinite in (4), A_n approaches 0, and the sum of the infinite progression is $4A_1/3$.

The work of Archimedes on the mechanical and mathematical methods of obtaining the area of a parabolic segment shows how clearly he distinguishes physical from mathematical reasoning. His rigor is far superior to that which we shall find in the work of Newton and Leibniz.

In the work *On Spirals* Archimedes defines the spiral as follows. Suppose a line (ray) rotates at a constant rate about one end while remaining in one plane and a point starting from the fixed end moves out at a constant speed along the line; then the point will describe a spiral. In our polar coordinates the equation of the spiral is $\rho = a\theta$. As the curve is drawn in Figure 5.7, θ is a positive clockwise. The deepest result in the work is:

Proposition 24. The area bounded by the first turn of the spiral and the initial line [the shaded area in the figure] is equal to one third of the first circle.

The first circle is the circle with radius OA, which equals $2\pi a$, and so the shaded area is $\pi(2\pi a)^2/3$.

The proof is by the method of exhaustion. In preceding theorems, which prepare the ground, the area of a region bounded by an arc of a spiral, the arc $BPQRC$ in Figure 5.8, and by two radii vectors OB and OC, is enclosed between two sets of circular sectors. Thus $Bp', Pq', Qr', \ldots$ are arcs of circles with center at O and likewise $Pb, Qp, Rq, \ldots$ are arcs of circles. The circular sectors of the inscribed set are $OBp', OPq', OQr', \ldots$ and the circular sectors of the circumscribed set are $OPb, OQp, ORq, \ldots$. Thus circular

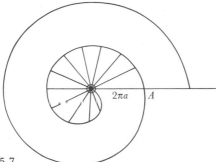

Figure 5.7

sectors replace inscribed and circumscribed polygons as the approximating figures in the method of exhaustion. (We use such approximating figures in the calculus when we determine areas in polar coordinates.) The novel feature in this application of the method of exhaustion is that Archimedes chooses smaller and smaller circular sectors so that the difference between the area under the arc of the spiral and the sum of the areas of the finite number of "inscribed" circular sectors (and the sum of the areas of the finite number of "circumscribed" circular sectors) can be made less than any given magnitude. This manner of approximating the desired area is not the same as "exhausting" it by adding more and more rectilinear figures. However, in the last part of the proof Archimedes uses the indirect method of proof as he does in the work on the parabola and as Euclid does in his proofs by the method of exhaustion. There is no explicit limit process.

Archimedes also gives the result for the area bounded by the arc of the spiral after the radius vector has rotated twice completely around O; and there are other related results on area. Incidentally, later mathematicians used the spiral to trisect an angle and in fact to divide an angle into any number of equal parts.

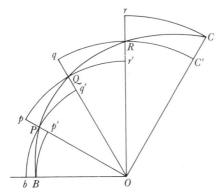

Figure 5.8

It is at once apparent from a study of the geometrical work of Archimedes that he is concerned to obtain useful results on area and volume. This work and his mathematical work in general are not spectacular as to conclusions nor especially new in method or subject matter, but he tackles very difficult and original problems. He often says that the suggestions for problems came from reading the works of his predecessors; for example, the work of Eudoxus on the pyramid, cone, and cylinder (given in Euclid's *Elements*) suggested to Archimedes his work on the sphere and cylinder, and the problem of squaring the circle suggested the quadrature of the parabolic segment. Archimedes' work on hydrostatics, however, is entirely novel; and his work on mechanics is new in that he gives mathematical demonstrations (Chap. 7, sec. 6). His writing is elegant, ordered, finished, and to the point.

4. *Areas and Volumes in the Work of Heron*

Heron, who lived sometime between 100 B.C. and A.D. 100, is of great interest not only from the standpoint of the history of mathematics but also in exhibiting the characteristics of the Alexandrian period. Proclus refers to Heron as *mechanicus*, which might mean a mechanical engineer today, and discusses him in connection with the inventor Ctesibius, his teacher. Heron was also a good surveyor.

The striking fact about Heron's work is his commingling of rigorous mathematics and the approximate procedures and formulas of the Egyptians. On the one hand, he wrote a commentary on Euclid, used the exact results of Archimedes (indeed he refers to him often), and in original works proved a number of new theorems of Euclidean geometry. On the other hand, he was concerned with applied geometry and mechanics and gave all sorts of approximate results without apology. He used Egyptian formulas freely and much of his geometry was also Egyptian in character.

In his *Metrica* and *Geometrica*, the latter known to us only through a book based on his work, Heron gives theorems and rules for plane areas, surface areas, and volumes of a great number of figures. The theorems in these books are not new. For figures with curved boundaries he uses Archimedes' results. In addition he wrote *Geodesy* and *Stereometry* (calculation of volumes of figures), both of which are concerned with the same subjects as the first two books. In all of these works he is primarily interested in numerical results.

In his *Dioptra* (theodolite), a treatise on geodesy, Heron shows how to find the distance between two points of which only one is accessible and between two points that are visible but not accessible. He also shows how to draw a perpendicular from a given point to a line that cannot be reached and how to find the area of a field without entering it. The formula for the area of a triangle, credited to him though due to Archimedes, namely,

$$\sqrt{s(s-a)(s-b)(s-c)}$$

wherein a, b, and c are the sides and s is half the perimeter, illustrates the last-mentioned idea. This formula appears in his *Geodesy*, and the formula and a proof are in both the *Dioptra* and the *Metrica*. In the *Dioptra* he shows how to dig a straight tunnel under a mountain by working simultaneously from both ends.

Though many of the formulas are proven, Heron gives many without proof and also gives many approximate ones. Thus he gives an inexact formula for the area of a triangle along with the above correct one. One reason that Heron gave many Egyptian formulas may be that the exact ones involved square roots or cube roots and the surveyors could not execute these operations. In fact, there was a distinction between pure geometry and geodesy or metrics. The calculation of areas and volumes belonged to geodesy and was not part of a liberal education. It was taught to surveyors, masons, carpenters, and other technicians. There is no doubt that Heron continued and enriched the Egyptian science of field measurements; his writings on geodesy were used for hundreds of years.

Heron applied many of his theorems and rules to the design of theaters, banquet halls, and baths. His applied works include *Mechanics*, *The Construction of Catapults*, *Measurements*, *The Design of Guns*, *Pneumatica* (the theory and use of air pressure), and *On the Art of Construction of Automata*. He gives designs for water clocks, measuring instruments, automatic machines, weight-lifting machines, and war engines.

5. *Some Exceptional Curves*

Though the classical Greeks did introduce and study some unusual curves, such as the quadratrix, the dictate that geometry was to be devoted to figures constructible with line and circle banished those curves to limbo. The Alexandrians, however, felt freer to ignore the restriction; thus Archimedes did not hesitate to introduce the spiral. A number of other curves were introduced in the Alexandrian period.

Nicomedes (*c.* 200 B.C.) is known for his definition of the conchoid. He starts with a point P and a line AB (Fig. 5.9). He then chooses a length a and lays this off on all rays from P that cross AB, the length a starting from the point of intersection of the ray and AB and in the direction away from P. The endpoints so determined are the points of the conchoid. Thus P_1, P_2, and P_3 of the figure are points of the conchoid.

If b is the perpendicular distance from P to AB and if the lengths a are measured along the rays through P and starting from AB but in the direction

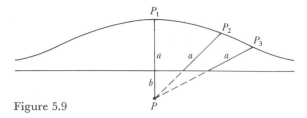

Figure 5.9

of P we get three other curves according as $a > b$, $a = b$, and $a < b$. Hence there are four types of conchoids, all due to Nicomedes. The modern polar equation is $r = a + b \sec \theta$. Nicomedes used the curve to trisect an angle and double the cube.[1]

Nicomedes is supposed to have invented a mechanism to construct the conchoids. The nature of the mechanism is of far less interest than the fact that mathematicians of the period were interested in devising them. The conchoids of Nicomedes, together with the line and circle, are the oldest mechanically constructible curves about which we possess satisfactory information.

Diocles (end of 2nd cent. B.C.), in his book *On Burning-glasses*, solved the problem of doubling the cube by introducing the curve called the cissoid. The curve is defined as follows: AB and CD are perpendicular diameters of a circle (Fig. 5.10) and EB and BZ are equal arcs. Draw ZH perpendicular to CD and then draw ED. The intersection of ZH and ED gives a point P on the cissoid. For Diocles the cissoid is the locus of all points P determined by all positions of E on arc BC and Z on arc BD with arc BE = arc BZ. One

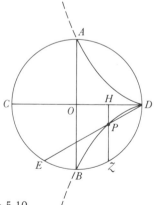

Figure 5.10

1. The method of trisection is given in T. L. Heath: *The Thirteen Books of Euclid's* Elements, Dover (reprint), 1956, Vol. 1, p. 266.

can prove that

$$CH:HZ = HZ:HD = HD:HP.$$

Thus HZ and HD are two mean proportionals between CH and HP. This solves the Delian problem. The equation of the cissoid in rectangular coordinates is $y^2(a + x) = (a - x)^3$, where O is the origin, a the radius of the circle, and OO and OA the coordinate axes. This equation includes the broken-line portions of the curve shown in the figure, which were not considered by Diocles.

6. *The Creation of Trigonometry*

Entirely new in the Alexandrian Greek quantitative geometry was trigonometry, a creation of Hipparchus, Menelaus, and Ptolemy. This work was motivated by the desire to build a quantitative astronomy that could be used to predict the paths and positions of the heavenly bodies and to aid the telling of time, calendar-reckoning, navigation, and geography.

The trigonometry of the Alexandrian Greeks is what we call spherical trigonometry though, as we shall see, the essentials of plane trigonometry were also involved. Spherical trigonometry presupposes spherical geometry, for example the properties of great circles and spherical triangles, much of which was already known; it had been investigated as soon as astronomy became mathematical, during the time of the later Pythagoreans. Euclid's *Phaenomena*, itself based on earlier work, contains some spherical geometry. Many of its theorems were intended to deal with the apparent motion of the stars. Theodosius (*c.* 20 B.C.) collected the then available knowledge of spherical geometry in his *Sphaericae*, but his work was not numerical and so could not be used to handle the fundamental problem of Greek astronomy, namely, to tell time at night by observation of the stars.

The founder of trigonometry is Hipparchus, who lived in Rhodes and Alexandria and died about 125 B.C. We know rather little about him. Most of what we do know comes from Ptolemy, who credits Hipparchus with a number of ideas in trigonometry and astronomy. We owe to him many astronomical observations and discoveries, the most influential astronomical theory of ancient times (Chap. 7, sec. 4), and works on geography. Only one work by Hipparchus, his *Commentary on the* Phaenomena *of Eudoxus and Aratus*, is preserved. Geminus of Rhodes wrote an introduction to astronomy that we do have, and it contains a description of Hipparchus' work on the sun.

Hipparchus' method of approaching trigonometry, as described and used by Ptolemy, is the following. The circumference of a circle is divided into 360°, as was first done by Hypsicles of Alexandria (*c.* 150 B.C.) in his book *On the Risings of the Stars* and by the Babylonians of the last centuries before

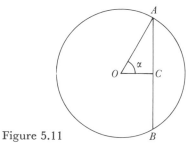

Figure 5.11

Christ, and a diameter is divided into 120 parts. Each part of the circumference and diameter is further divided into 60 parts and each of these into 60 more, and so on according to the Babylonian system of sexagesimal fractions. Then for a given arc AB of some number of degrees, Hipparchus— in a book, now lost, on chords in a circle—gives the number of units in the corresponding chord AB. Just how he calculated these will be described in the discussion of Ptolemy's work, which presents their combined thoughts and results.

The number of units in the chord corresponding to an arc of a given number of degrees is equivalent to the modern sine function. If 2α is the central angle of arc AB (Fig. 5.11), then for us $\sin \alpha = AC/OA$, whereas, instead of $\sin \alpha$, Hipparchus gives the number of units in $2 \cdot AC$ when the radius OA contains 60 units. For example, if the chord of 2α is 40 units, then for us $\sin \alpha = 20/60$, or, more generally,

$$(9) \qquad\qquad \sin \alpha = \frac{1}{60} \cdot \frac{1}{2} \text{ chord } 2\alpha = \frac{1}{120} \text{ chord } 2\alpha.$$

Greek trigonometry reached a high point with Menelaus (c. A.D. 98). His *Sphaerica* is his chief work, but apparently he also wrote *Chords in a Circle* in six books, and a treatise on the setting (or rising) of arcs of the zodiac. The Arabs attribute additional works to him.

The *Sphaerica*, extant in an Arab version, is in three books. In the first book, on spherical geometry, we find the concept of a spherical triangle, that is, the figure formed by three arcs of great circles on a sphere, each arc being less than a semicircle. The object of the book is to prove theorems for spherical triangles analogous to what Euclid proved for plane triangles. Thus the sum of two sides of a spherical triangle is greater than the third side and the sum of the angles of a triangle is greater than two right angles. Equal sides subtend equal angles. Then Menelaus proves the theorem that has no analogue in plane triangles, namely, that if the angles of one spherical triangle equal respectively the angles of another, the triangles are congruent. He also has other congruence theorems and theorems about isosceles triangles.

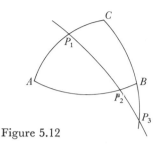

Figure 5.12

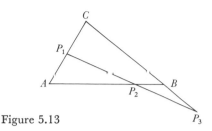

Figure 5.13

The second book of Menelaus' *Sphaerica* is chiefly about astronomy and only indirectly concerned with spherical geometry. The third book contains some spherical trigonometry and bases the development on the first theorem of the book, which supposes that we have a spherical triangle ABC (Fig. 5.12) and any great circle cutting the sides of the triangle (produced where necessary). To state the theorem we shall use our modern sine notion, but for Menelaus the sine of an arc such as AB (or the sine of the corresponding central angle at the center of the sphere) is replaced by the chord of double the arc AB. In terms of our sines, then, Menelaus' theorem says

$$\sin P_1 A \cdot \sin P_2 B \cdot \sin P_3 C = \sin P_1 C \cdot \sin P_2 A \cdot \sin P_3 B.$$

The proof of this theorem rests upon the corresponding theorem for plane triangles, which we still call Menelaus' theorem. For plane triangles the theorem states (Fig. 5.13):

$$P_1 A \cdot P_2 B \cdot P_3 C = P_1 C \cdot P_2 A \cdot P_3 B.$$

Menelaus does not prove the plane theorem. One may conclude that it was already known or perhaps proved by Menelaus in an earlier writing.

The second theorem of Book III, in the notation that arc a lies opposite angle A in triangle ABC, states that if ABC and $A'B'C'$ are two spherical triangles and if $A = A'$ and $C = C'$, or C is supplementary to C', then

$$\frac{\sin c}{\sin a} = \frac{\sin c'}{\sin a'}.$$

Theorem 5 of Book III uses a property of arcs that was presumably known by Menelaus' time, namely (Fig. 5.14), if four great circular arcs emanate from a point O and $ABCD$ and $A'B'C'D'$ are great circles cutting the four, then

$$\frac{\sin AD}{\sin DC} \cdot \frac{\sin BC}{\sin AB} = \frac{\sin A'D'}{\sin D'C'} \cdot \frac{\sin B'C'}{\sin A'B'}.$$

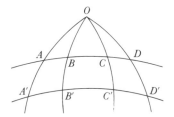

Figure 5.14

We shall find that an expression corresponding to the left or right member reappears under the concept of anharmonic ratio or cross ratio in the work of Pappus and in later work on projective geometry. Many more theorems on spherical trigonometry are due to Menelaus.

The development of Greek trigonometry and its application to astronomy culminated in the work of the Egyptian Claudius Ptolemy (d. A.D. 168), who was a member at least of the royal family of mathematicians though not of the royal house of Egypt. Ptolemy lived in Alexandria and worked at the Museum.

In his *Syntaxis Mathematica* or *Mathematical Collection* (the work was referred to by the Arabs as *Megale Syntaxis, Megiste,* and finally *Almagest*), Ptolemy presents the continuation and completion of the work of Hipparchus and Menelaus in trigonometry and astronomy. Astronomy and trigonometry are commingled in the thirteen books of the *Almagest*, though Book I is largely on spherical trigonometry and the others are devoted largely to astronomy, which will be discussed in Chapter 7.

Ptolemy's *Almagest* is thoroughly mathematical, except where he uses Aristotelian physics to refute the heliocentric hypothesis, which Aristarchus had suggested. He says that since only mathematical knowledge, approached inquiringly, will give its practitioners trustworthy knowledge, he was led to cultivate as far as lay in his power this theoretical discipline. Ptolemy also says he wishes to base his astronomy "on the incontrovertible ways of arithmetic and geometry."

In Chapter IX of Book I Ptolemy begins by calculating the chords of arcs of a circle, thereby extending the work of Hipparchus and Menelaus. As already noted, the circumference is divided into 360 parts or units (he does not use the word "degree") and the diameter into 120 units. Then he proposes, given an arc containing a given number of the 360 units, to find the length of the chord expressed in terms of the number of units which a full diameter contains, that is, 120 units.

He begins with the calculation of the chords of 36° and 72° arcs. In Figure 5.15 *ADC* is a diameter of a circle with center *D* and *BD* is perpendicular to *ADC*. *E* is the midpoint of *DC* and *F* is chosen so that *EF* = *BE*. Ptolemy proves geometrically that *FD* equals a side of the regular inscribed

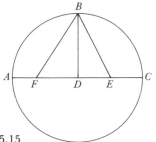

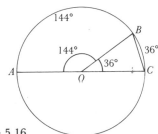

Figure 5.15   Figure 5.16

decagon and *BF*, a side of the regular inscribed pentagon. But *ED* contains 30 units and *BD*, 60 units. Since $EB^2 = ED^2 + BD^2$, $EB^2 = 4500$ and $EB = 67\ 4'55''$ (which means $67 + 4/60 + 55/60^2$ units). Now $EF = EB$ and so he knows *EF*. Then $FD = EF - DE = 67\ 4'55'' - 30 = 37\ 4'55''$. Since *FD* equals the side of the decagon, it is the chord of the 36° arc. Hence he knows the chord of this arc. By using *FD* and the right triangle *FDB*, he can calculate *BF*. It is 70 32'3''. But *BF* is the side of the pentagon. Hence he has the chord of the 72° arc.

Of course for the side of a regular hexagon, since it equals the radius, he has at once that the chord of length 60 belongs to the arc of length 60. Also, since the side of the inscribed square is immediately calculable in terms of the radius, he has the chord of 90°. It is 84 51'10''. Further, since the side of the inscribed equilateral triangle is also immediately calculable in terms of the radius, he gets that the chord of 120° is 103 55'23''.

By using the right triangle *ABC* (Fig. 5.16) on the diameter *AC* one can immediately get the chord of the supplementary arc *AB* if one knows the chord of the arc *BC*. Thus, since Ptolemy knows the chord of 36° he can find the chord of 144°, which turns out to be 114 7'37''.

The relationship established here is equivalent to $\sin^2 A + \cos^2 A = 1$ where *A* is any acute angle. This can be seen as follows. Ptolemy has proved that if *S* is any arc less than 180° then

$$(\text{chord } S)^2 + \left[\text{chord } (180 - S)^2\right] = (120)^2.$$

But by (9) above

$$(\text{chord } S)^2 = (120)^2 \sin^2 \frac{S}{2}.$$

Hence

$$(120)^2 \sin^2 \frac{S}{2} + (120)^2 \sin^2 \left(\frac{180 - S}{2}\right) = 120^2$$

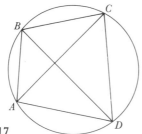

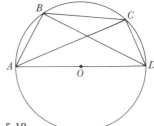

Figure 5.17   Figure 5.18

or

$$\sin^2 \frac{S}{2} + \sin^2 \left(90 - \frac{S}{2}\right) = 1;$$

that is,

$$\sin^2 \frac{S}{2} + \cos^2 \frac{S}{2} = 1.$$

Now Ptolemy proves what he calls a lemma but which is known today as Ptolemy's theorem. Given any quadrilateral inscribed in a circle (Fig. 5.17), he proves that $AC \cdot BD = AB \cdot DC + AD \cdot BC$. The proof is straightforward. He then takes the special quadrilateral $ABCD$ in which AD is a diameter (Fig. 5.18). Suppose we know AB and AC. Ptolemy now shows how to find BC. BD is the chord of the supplement of arc AB and CD is the chord of the supplement of arc AC. If one now applies the lemma, he sees that five of the six lengths involved in it are known and so the sixth length, which in this case is BC, can be calculated. But arc BC = arc AC − arc AB. Hence we can calculate the chord of the difference of two arcs if their chords are known. In modern terms this means that if we know $\sin A$ and $\sin B$ we can calculate $\sin (A - B)$. Ptolemy points out that, since he knows the chords of 72° and 60°, he can calculate the chord of 12°.

He shows next how, given any chord in a circle, he can find the chord of half the arc of the given chord. In modern terms this means finding $\sin A/2$ from $\sin A$. This result is powerful, as Ptolemy points out, because we can start with an arc whose chord is known and find the chords of successive halves of this arc. He also shows that if one knows the chords of arcs AB and BC, then one can find the chord of arc AC. In modern terms, this result is the formula for $\sin (A + B)$. As a special case, he points out that we can calculate, in modern terms, $\sin 2A$ from $\sin A$.

Since Ptolemy can get the chord of 3/4° from the chord of 12° by halving, he can add this arc of 3/4° or subtract it from any arc whose chord is known;

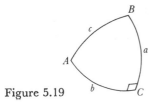

Figure 5.19

and by using the above theorems, he can calculate the chord of the sum or difference of the two arcs. Thus he is in a position to calculate the chords of all arcs in steps of 3/4°. However, he wants to obtain the chords of arcs in steps of 1/2°. Here he is stuck and resorts to clever reasoning with inequalities. The approximate result is that the chord of $1/2° = 0\ 31'25''$.

He is now in a position to make up a table of the chords of arcs for arcs differing by 1/2°, from 0° to 180°. This is the first trigonometric table.

Then Ptolemy proceeds (Chapter XI of Book I) to solve astronomical problems that call for finding arcs of great circles on a sphere. These arcs are sides of spherical triangles, some of whose parts are known through either observation or prior calculation. To determine the unknown arcs Ptolemy proves relationships that are theorems of spherical trigonometry some of which had already been proved in Book III of Menelaus' *Sphaerica*. Ptolemy's basic method is to use Menelaus' theorem for spherical triangles. Thus he proves, in our notation, that in the spherical triangle with right angle at C (Fig. 5.19) and with arc a denoting the side opposite angle A,

$$\sin a = \sin c \sin A$$
$$\tan a = \sin b \tan A$$
$$\cos c = \cos a \cos b$$
$$\tan b = \tan c \cos A.$$

Of course for Ptolemy the various trigonometric functions are chords of arcs. To treat oblique spherical triangles, he breaks them up into right-angled spherical triangles. There is no systematic presentation of spherical trigonometry; he proves just those theorems that he needs to solve specific astronomical problems.

The *Almagest* put trigonometry into the definitive form it retained for over a thousand years. We generally speak of this trigonometry as spherical, but the distinction between plane and spherical trigonometry is of little significance in assessing Ptolemy's material. Ptolemy certainly works with spherical triangles, but in having calculated the chords of arcs, he has really given the theoretical basis for plane trigonometry. For by knowing sin A, and, in effect, cos A for any A from 0° to 90°, one can solve plane triangles.

We should note that trigonometry was created for use in astronomy; and, because spherical trigonometry was for this purpose the more useful tool, it was the first to be developed. The use of plane trigonometry in indirect measurement and in surveying is foreign to Greek mathematics. This may seem strange to us, but historically it is readily understandable, since astronomy was the major concern of the Greek mathematicians. Surveying did come to the fore in the Alexandrian period; but a mathematician such as Heron, who was interested in surveying and would have been capable of developing plane trigonometry, contented himself with applying Euclidean geometry. The uneducated surveyors were not in a position to create the necessary trigonometry.

7. Late Alexandrian Activity in Geometry

Mathematical activity in general, and in geometry in particular, declined in Alexandria from about the beginning of the Christian era. We shall look into possible reasons for the decline in Chapter 8. What we know about the geometrical work of the early Christian era comes from the major commentators Pappus, Theon of Alexandria (end of 4th cent. A.D.), and Proclus.

On the whole, very few original theorems were discovered in this period. The geometers seem to have occupied themselves primarily with the study and elucidation of the works of the great mathematicians who preceded them. They supplied proofs that the original authors had omitted, either because the proofs were regarded as easy enough to be left to the readers or because they were given in treatises that had been lost. These proofs, incidentally, were called lemmas, in an older use of the word.

Both Theon and Pappus report on the work of Zenodorus who lived sometime between 200 B.C. and A.D. 100. Apparently Zenodorus wrote a book on isoperimetric figures, that is, figures with equal perimeters, and in it proved the following theorems:

1. Among n-sided polygons of the same perimeter the regular one has most area.

2. Among regular polygons of equal perimeter, the one having more sides has greater area.

3. The circle has greater area than a regular polygon of the same perimeter.

4. Of all solids with the same surface the sphere has the greatest volume.

The subject of the theorems, which today we would describe as maxima and minima problems, was novel in Greek mathematics.

Late in the Alexandrian period, Pappus' additions to geometry came as a sort of anticlimax. The eight books of his *Mathematical Collection* contain some original material. Pappus' new work was not of the highest order but some of it is worthy of note.

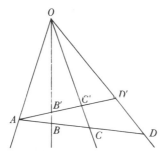

Figure 5.20

Book V gives the proofs, results, and extensions of Zenodorus' work on the areas bounded by curves of equal perimeter. Pappus added the theorem that of all segments of a circle having the same perimeter the semicircle has the greatest area. He also proves that the sphere has more volume than any cone, cylinder, or regular polyhedron with the same surface area.

Proposition 129 of Book VII is a special case of the theorem that the cross ratio (Fig. 5.20)

$$\frac{AB}{AD}\bigg/\frac{BC}{CD}$$

is the same for every transversal cutting the four lines emanating from O. Pappus requires that all the transversals pass through A.

Proposition 130 states, in our language, that if five of the points in which the six sides of a complete quadrilateral (the four sides and the two diagonals) meet a straight line are fixed, then the sixth one is also fixed. Thus if $ABCD$ (Fig. 5.21) is the quadrilateral the six points in which its six sides meet any straight line EK not going through A, B, C or D EK are E, F, G, H, $\mathcal{J}$, and K. If five of these are fixed, the sixth is also. Pappus notes that these six points satisfy the condition

$$\frac{EK}{EH}\bigg/\frac{JK}{JH} = \frac{EK}{EF}\bigg/\frac{GK}{GF}.$$

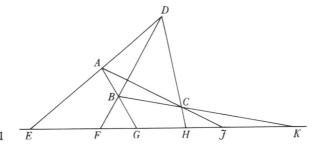

Figure 5.21

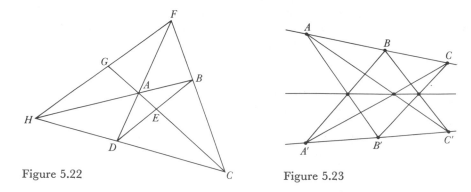

Figure 5.22   Figure 5.23

This condition states that the cross ratio determined by E, K, J, and H equals the cross ratio determined by E, K, G, and F. The condition is equivalent to one we shall find introduced by Desargues, who calls six such points "points of an involution."

Proposition 131 of Book VII amounts to the statement that in each quadrilateral a diagonal is cut harmonically by the other diagonal and by the line joining the points of intersection of the pairs of opposite sides. Thus $ABCD$ is a quadrilateral (Fig. 5.22); CA is a diagonal; CA is cut by the other diagonal BD and by FH which joins the intersection of AD and BC to the intersection of AB and CD. Then the points C, E, A, and G in the figure form a harmonic set; that is, E divides AC internally in the same ratio that G divides AC externally.

Proposition 139 of Book VII gives what is still called Pappus' theorem. If A, B, and C are three points on a line (Fig. 5.23) and A', B', and C' are three points on another, then AB' and $A'B$, BC' and $B'C$, and AC' and $A'C$ meet in three points that lie on one straight line.

One of the last lemmas, Proposition 238, establishes a fundamental property of all conic sections: The locus of all points whose distances from a fixed point (focus) and a fixed line (directrix) are in a constant ratio is a conic section. This basic property of the conics is not in Apollonius' *Conic Sections*, but, as noted in the preceding chapter, Euclid probably knew it.

In the introduction to Book VII Pappus restates Apollonius' assertion that his methods enable one to find the locus of points the product of whose distances from two lines equals a constant times the product of its distances from two other lines. Pappus knows but does not prove that the locus is a conic. He also points out that the problem can be generalized to include five, six, or more lines. We shall say more about this problem in connection with Descartes's work.

Book VIII is of special interest in that it is mainly on mechanics, which, in accordance with the Alexandrian outlook, is regarded as part of mathematics. In fact, Pappus prefaces the book with this contention. He cites Archimedes, Heron, and lesser known figures as the leaders in mathematical mechanics. The center of gravity of a body is defined as the point within the body (it need not be within) such that if the body is suspended from that point it will remain at rest in any position in which it is put. He then explains mathematical methods for determining the point. He also treats the subject of moving a body along an inclined plane and attempts to compare the force required to slide the body along a horizontal plane with that needed to slide it up the inclined plane.

Book VII also contains a famous theorem sometimes called Pappus' theorem and sometimes Guldin's theorem because Paul Guldin (1577–1643) rediscovered it independently. The theorem states that the volume generated by the complete revolution of a plane closed curve that lies wholly on one side of the axis of revolution equals the area bounded by the curve multiplied by the circumference of the circle traversed by the center of gravity. The result is a very general one and Pappus was aware of this. He does not give a proof of the theorem and it is very likely that the theorem and proof were known before his time.

As far as geometry is concerned, the Alexandrian period ends with the works of a number of commentators. Theon of Alexandria wrote a commentary on Ptolemy's *Almagest* and new editions of Euclid's *Elements* and *Optics*. His daughter Hypatia (d. 415), a learned mathematician, wrote commentaries on Diophantus and Apollonius.

Proclus Diadochus, whom we have often mentioned, wrote a commentary on Book I of Euclid's *Elements* (Chap. 3, sec. 2). This commentary is important because Proclus had access to works now lost, including the *History of Geometry* by Eudemus and the book of Geminus that was probably titled the *Doctrine* or the *Theory of Mathematics*.

Proclus received his education in Alexandria and then went to Athens, where he became head of Plato's Academy. He was a foremost neo-Platonist and wrote many books on Plato's work and on philosophy generally; his interests included poetry as well as mathematics. Like Plato he believed that mathematics is a handmaiden to philosophy. It is a propaedeutic because it clears the eye of the soul, removing the impediments that the senses place in the way of knowing universals.

There was another, nonmathematical side to Proclus, who accepted many myths and religious mysteries and was a devout worshiper of Greek and Oriental divinities. Ptolemaic theory he rejected because a Chaldean "whom it is not lawful to disbelieve" thought otherwise. It has been remarked that it was fortunate for Proclus that the Chaldean oracles did not contradict or deny Euclid.

Among many other commentators we shall mention just a few. Simplicius, a commentator on Aristotle, studied at Alexandria and at Plato's Academy, and went to Persia when Justinian closed down the Academy in A.D. 529. He repeated material from Eudemus' *History*, including a long extract on Antiphon's attempt to square the circle and on Hippocrates' quadrature of lunes. Isidorus of Miletus (6th cent. A.D.), who seems to have had a school in Constantinople (which had become the capital of the Eastern Roman Empire and the center of some mathematical activity), wrote commentaries and may have written part of the fifteenth book of Euclid's *Elements*. Eutocius (6th cent. A.D.), probably a pupil of Isidorus, wrote a commentary on Archimedes' work.

Bibliography

Aaboe, Asger: *Episodes from the Early History of Mathematics*, Random House, 1964, Chaps. 3–4.

Ball, W. W. R.: *A Short Account of the History of Mathematics*, Dover (reprint), 1960, Chaps. 4–6.

Cajori, Florian: *A History of Mathematics*, Macmillan, 1919, pp. 29–52.

Dijksterhuis, E. J.: *Archimedes* (English trans.), Ejnar Munksgaard, 1956.

Heath, Thomas L.: *A History of Greek Mathematics*, Oxford University Press, 1921, Vol. 2, Chaps. 13, 15, 17–19, 21.

———: *The Works of Archimedes*, Dover (reprint), 1953.

Pappus d'Alexandrie: *La Collection mathématique*, ed. Paul Ver Eecke, 2 vols., Albert Blanchard, 1933.

Parsons, Edward Alexander: *The Alexandrian Library*, The Elsevier Press, 1952.

Sarton, George: *A History of Science*, Harvard University Press, 1959, Vol. 2, Chaps. 1–3, 5, 18.

Scott, J. F.: *A History of Mathematics*, Taylor and Francis, 1958, Chaps. 3–4.

Smith, David Eugene: *History of Mathematics*, Dover (reprint), 1958, Vol. 1, Chap. 4; Vol. 2, Chaps. 8 and 10.

van der Waerden, B. L.: *Science Awakening*, P. Noordhoff, 1954, Chaps. 7–8.

6

The Alexandrian Period:
The Reemergence of
Arithmetic and Algebra

> Wherever there is number, there is beauty.
>
> PROCLUS

1. The Symbols and Operations of Greek Arithmetic

Let us go back for a moment to pick up the threads of arithmetic in the classical period. The classical Greeks called the art of calculation *logistica*; they reserved the word *arithmetica* for the theory of numbers. The classical mathematicians scorned *logistica* because it was concerned with the practical calculations needed in the trades and commerce. We, however, shall consider both *logistica* and *arithmetica* in order to see what the Alexandrian Greeks had at their disposal.

The classical Greek art of writing and working with numbers did not continue where the Babylonians left off. In *logistica* the Greeks seem to have fashioned their own beginnings. Archaic Greek numerals found in Crete antedate the classical period by about five hundred years. There are no noteworthy features in this scheme, just particular number symbols for 1, 2, 3, 4, 10, 200, 1000, and so forth. Very early in the classical period the Greeks introduced other special symbols for numbers, and used some form of an abacus for calculation. Later, about 500 B.C., they used the Attic system, of which the earliest known occurrence is an inscription of 450 B.C. The Attic system used strokes for the numbers from 1 to 4; Π, the first letter of *penta*, and later Γ were used for 5; Δ from *deka* was 10; H from *hekaton* represented 100; X from *chilioi* stood for 1000; and M from *myrioi* represented 10,000. Combinations of these special symbols produced the in-between numbers. Thus ΓΙ = 6; ΓΔ = 50; ΓH = 500; ΔΓΙΙΙ = 18.

However, no one knows how early classical mathematicians—for example, the Pythagoreans—wrote numbers. They may have used pebbles to calculate, for the word "calculus" means "pebble." The original Greek

meaning of "abacus" was "sand," which suggests that before the introduction of the abacus, and probably afterward, they drew numbers as dots
in sand. In the three hundred years from Thales to Euclid the mathematicians paid no attention to computation, and this art made no progress.
It is significant that the books do not tell us about the practice of arithmetic.

For some unknown reason the classical Greeks changed their way of
writing numbers to the Ionic or Alexandrian system, which uses letters of
the alphabet. This alphabetic system was the most common one in Alexandrian Greek mathematics and is found, in particular, in Ptolemy's
Almagest. It was used also in ancient Syria and Israel.

The details of the Greek system are as follows:

α	β	γ	δ	ϵ	ς	ζ	η	θ
1	2	3	4	5	6	7	8	9

ι	κ	λ	μ	ν	ξ	o	π	ϱ
10	20	30	40	50	60	70	80	90

ρ	σ	τ	υ	ϕ	χ	ψ	ω	T
100	200	300	400	500	600	700	800	900

The intermediate numbers were written by combining the above symbols.
Thus $\iota\alpha = 11$, $\iota\beta = 12$, $\kappa\alpha = 21$, and $\rho\nu\gamma = 153$. The symbols for 6, 90,
and 900 and the symbol M from the Attic system were not in the then-
current Greek alphabet; the first three, now called stigma (or digamma),
koppa, and sampi, belonged to an older alphabet that the Greeks had taken
over from the Phoenicians (who did not, however, use letters for numbers).
From the fact that these older letters were used it is inferred that this system
of writing numbers dates back as far as 800 B.C. and probably came from
Miletus in Asia Minor.

For numbers larger than 1000 the alphabet was repeated, but a stroke
was placed before the letter to avoid confusion. Also, horizontal lines were
drawn over numbers to distinguish them from words. Thus,

$$,\overline{\alpha\tau\epsilon} = 1305.$$

Various Greek writers used minor variations of the above scheme and of
schemes given below.

Greek papyri of the first part of the Alexandrian period (first three
centuries B.C.) contain symbols for zero such as $0 \overline{} \cdot 0$, $\overline{0,}$ $\overline{0}$, and $\ulcorner \overline{0} \urcorner$. The
zero of the Greek Alexandrian period was used, as was the zero of the Seleucid
Babylonian period, to indicate missing numbers. According to Byzantine
manuscripts, which are all we have of Ptolemy's work, he used 0 for zero
both in the middle and at the end of a number.

Archimedes' *Sand-Reckoner* presented a scheme for writing very large
numbers. He sought to show that he could write a number as large as the

number of grains of sand in the universe. He takes the largest number then expressed in Greek numerals, that is 10^8, a myriad myriads, and uses it as the starting point of a new series of numbers that goes up to $10^8 \times 10^8$ or 10^{16}. Then he uses 10^{16} as a new starting point for a series of numbers that goes from 10^{16} to 10^{24}, and so forth. He next estimates the number of grains of sand in the universe and shows it is less than the largest number he can express. What is important in this work of Archimedes is not a practical scheme for actually writing any large number, but the thought that one can construct indefinitely large numbers. Apollonius had a similar scheme.

The arithmetic operations, with the whole numbers written as described above, were like ours. Thus in adding, the Greeks wrote the numbers one below the other to form a unit column, tens column, and so forth, added the numbers in each column, and carried from one column to the other. These methods are a great step forward, compared with the Egyptian ones. The latter, however, were also taught by the Alexandrian Greeks.

As for fractions, there was the special symbol, L'' for $1/2$. Thus (sometimes with one accent), $\alpha L'' = 1\frac{1}{2}$, $\beta L'' = 2\frac{1}{2}$, and $\gamma L'' = 3\frac{1}{2}$. Small fractions were denoted by writing the numerator marked with an accent, then the denominator written once or twice, each time with two accents. Thus $\iota \gamma' \kappa \theta'' \kappa \theta'' = 13/29$. Diophantus often wrote the denominator above the numerator.

The Egyptian scheme of writing fractions whose numerators are larger than 1 as a sum of unit fractions is also found. Thus Heron writes $\frac{163}{224}$ as $\frac{1}{2}\ \frac{1}{7}\ \frac{1}{14}\ \frac{1}{112}\ \frac{1}{224}$, but gives as well $\frac{1}{2}\ \frac{1}{8}\ \frac{1}{16}\ \frac{1}{32}\ \frac{1}{112}$ and other such expressions for the same fraction. He also uses the above Greek form. Ptolemy likewise writes *some* fractions like the Egyptians, for example, $\frac{23}{25}$ as $\frac{1}{2}\ \frac{1}{3}\ \frac{1}{15}\ \frac{1}{50}$. The plus sign was always understood, and of course letters of the Greek alphabet were used where we use numerals.

Common fractions written in the Greek or Egyptian system were too awkward for astronomical calculations. Hence the Alexandrian Greek astronomers adopted the Babylonian sexagesimal fractions. Just when this practice began is not known, but it is used in Ptolemy's *Almagest*. Thus when Ptolemy writes 31 25 he means $\frac{31}{60} + \frac{25}{60^2}$. Ptolemy says that he used sexagesimal fractions to avoid the difficulty of ordinary fractions. He wrote *whole* numbers in the decimal base but not in positional notation. However, large whole numbers occur so rarely in his astronomical calculations that one may say he used sexagesimal positional notation. The use of the sexagesimal system of place value for fractions and of the nonpositional alphabetic numerals for whole numbers seems peculiar and irrational. Yet we write 130°15′17″.5.

As the above discussion implies, the Alexandrians used fractions as numbers in their own right, whereas the mathematicians of the classical period spoke only of a ratio of integers, not of parts of a whole and the ratios were used only in proportions. However, even in the classical period genuine fractions, that is, fractions as entities in their own right, were used in commerce. In the Alexandrian period, Archimedes, Heron, Diophantus, and others used fractions freely and performed operations with them. Though, as far as the records show, they did not discuss the concept of fractions, apparently these were intuitively sufficiently clear to be accepted and used.

The square-root operation, though considered in classical Greece, was in effect bypassed there. There are indications in Plato's writings that the Pythagoreans approximated $\sqrt{2}$ by using 49/25 for 2, thus obtaining 7/5. Likewise, Theodorus probably approximated $\sqrt{3}$ by using 49/16 for 3, obtaining 7/4. The irrational number as such had no mathematical status in classical Greece.

The next information we have on the Greek handling of roots comes from Archimedes. In his *Measurement of a Circle* he seeks primarily to find a good approximation to π, that is, the ratio of the circumference to the diameter of a circle; in the course of the work he operates with large whole numbers and fractions. He also obtains an excellent approximation to $\sqrt{3}$, namely,

$$\frac{1351}{780} > \sqrt{3} > \frac{265}{153},$$

but does not explain how he got this result. Among the many conjectures in the historical literature concerning its derivation the following is very plausible. Given a number A, if one writes it as $a^2 \pm b$ where a^2 is the rational square nearest to A, larger or smaller, and b is the remainder, then

$$a \pm \frac{b}{2a} > \sqrt{a^2 \pm b} > a \pm \frac{b}{2a \pm 1}.$$

Several applications of this procedure do produce Archimedes' result. To obtain the approximation to π, Archimedes first proves, in Proposition 1, that the area of a circle is equal to the area of a right triangle whose base is as long as the circumference of the circle and whose altitude equals the radius. He must now find the circumference. This he approximates more and more closely by using inscribed and circumscribed regular polygons and calculates the perimeters of these polygons. His result for π is

$$3\frac{10}{71} < \pi < 3\frac{1}{7}.$$

Apollonius, too, wrote a book on the quadrature of the circle, called *Okytokion* (The Rapid Delivery), in which he is supposed to have improved on Archimedes' determination of π by using better arithmetic methods. This is the only book in which Apollonius departs from classical Greek mathematics.

Heron approximates square roots frequently by means of

$$\sqrt{A} = \sqrt{a^2 \pm b} \sim a \pm \frac{b}{2a},$$

where a and b have the meaning explained above. He gets this approximation by starting with the approximation $\alpha = (c + A/c)/2$, where c is any guess as to $\sqrt{A}$; if one writes A as $a^2 + b$ and chooses $c = a$, then $\alpha = a + b/2a$. Heron also improves on α by finding $\alpha_1 = (\alpha + A/\alpha)/2$. Clearly the closer α is to $\sqrt{A}$, the better the approximation α_1 will be. Heron's basic expression for α was also used by the Babylonians.

In later Alexandrian times the square root process uses, as ours does, the principle of $(a + b)^2 = a^2 + 2ab + b^2$. The successive approximations are obtained by trial, always keeping the square of the approximation obtained less than the number whose root is desired. Explaining Ptolemy's use of this method, Theon points out that a geometrical figure is used to aid the thinking; this is the figure used in Euclid's *Elements*, Book II, Proposition 4 and is the geometrical way of stating $(a + b)^2$. Thus Ptolemy gives

$$\frac{103}{60} \, (+) \, \frac{55}{60^2} \, (+) \, \frac{23}{60^3}$$

for $\sqrt{3}$, which amounts to 1.7320509 and is correct to six decimal places.

2. *Arithmetic and Algebra as an Independent Development*

We have been reviewing the methods of doing arithmetic employed by the Greeks in both periods but more especially in the Alexandrian period when the geometry and trigonometry became quantitative. But the major development with which this chapter is concerned is the rise of arithmetic and algebra as subjects *independent* of geometry. The arithmetical work of Archimedes, Apollonius, and Ptolemy was a step in this direction, but they used arithmetic to calculate geometric quantities. One might infer that the numbers were meaningful to them because they represented geometric magnitudes and the logic of the operations was guaranteed by the geometrical algebra. But there is no question that Heron, Nichomachus (*c.* A.D. 100), who was probably an Arab from Gerasa in Judea, and Diophantus (*c.* A.D. 250), a Greek of Alexandria, did treat arithmetical and algebraic problems in and for themselves and did not depend upon geometry either for motivation or to bolster the logic.

More significant than Heron's arithmetical work of finding square and cube roots is the fact that he formulated and solved algebraic problems by purely arithmetic procedures. He did not use any special symbols; the account is verbal. For example, he treats the problem: Given a square such that the sum of its area and perimeter is 896 feet, find the side. The problem, in our notation, is to find x satisfying $x^2 + 4x = 896$. Heron completes the square by adding 4 to both sides and takes the square root. He does not prove but merely describes what operations to perform. There are many such problems in his work. Of course this is precisely the old Egyptian and Babylonian style of presentation, and there is no doubt that Heron took much material from the ancient Egyptian and Babylonian texts. There, we may remember, algebra was independent of geometry and, as for Heron, an extension of arithmetic.

In his *Geometrica*, Heron speaks of adding an area, a circumference, and a diameter. In using such words he means, of course, that he wants to add their numerical values. Likewise, when he says that he multiplies a square by a square, he means that he is finding the product of the numerical values. Heron also translated much of the Greek geometrical algebra into arithmetic and algebraic processes.

This work of Heron (as well as his use of approximate Egyptian area and volume formulas) is sometimes evaluated as the beginning of the decline of Greek geometry. It is more fitting to regard it as a Hellenized improvement on Babylonian and Egyptian mathematics. When Heron adds areas and line segments, he is not misapplying classical Greek geometry but merely continuing the practice of the Babylonians, for whom area and length were just words for certain arithmetic unknowns.

More remarkable from the standpoint of the reemergence of an independent arithmetic is the work of Nichomachus, who wrote the *Introductio Arithmetica* in two books. It was the first sizable book in which arithmetic (in the sense of the theory of numbers) was treated entirely independently of geometry. Historically its importance for arithmetic is comparable to Euclid's *Elements* for geometry. Not only was this book itself studied, referred to, and copied by dozens of later writers, but it is known to be typical of many books by other authors of the same period and so reflects the interests of the times. Numbers stood for quantities of objects and were no longer visualized as line segments as in Euclid. Nichomachus uses words throughout, whereas Euclid used a letter, such as A, or two letters such as BC—referring, in the second case, to a line segment—to speak about numbers. Hence Nichomachus' phrasing is clumsier. He treats only whole numbers and ratios of whole numbers

Nichomachus was a Pythagorean; and though the Pythagorean tradition was not dead, he reanimated it. Of the four subjects stressed by Plato—arithmetic, geometry, music, and astronomy—Nichomachus says arithmetic is the mother of the others. This, he maintains is

> not solely because we said that it existed before all the others in the mind
> of the creating God like some universal and exemplary plan, relying upon
> which as a design and archetypal example the creator of the universe sets
> in order his material creations and makes them attain to their proper ends;
> but also because it is naturally prior in birth....

Arithmetic, he continues, is essential to all the other sciences because they could not exist without it. However, if the other sciences were abolished, arithmetic would still exist.

The essence of the *Introductio* is the arithmetical work of the early Pythagoreans. Nichomachus treats even and odd, square, rectangular, and polygonal numbers. He treats also prime and composite numbers and parallelopipedal numbers (of the form $n^2(n + 1)$) and defines many other kinds. He gives the multiplication table for numbers from 1 to 9 precisely as we learn it.

Nichomachus repeats many Pythagorean statements, such as that the sum of two consecutive triangular numbers is a square number, and conversely. He goes beyond the Pythagoreans in seeing, though not proving, general relations. Thus he asserts that the $(n - 1)$st triangular number added to the nth k-gonal number gives the nth $(k + 1)$-gonal number. For example, the $(n - 1)$st triangular number added to the nth square number gives the nth pentagonal number. In our symbols,

$$\frac{n(n - 1)}{2} + n^2 = \frac{n}{2}(3n - 1).$$

Also the nth triangular number, the nth square number, the nth pentagonal number, and so on form an arithmetic progression with the $(n - 1)$st triangular number as the common difference.

He discovered the following proposition: If one writes down the odd numbers

$$1, 3, 5, 7, 9, 11, 13, 15, 17, \ldots$$

then the first is the cube of 1, the sum of the next two is the cube of 2, the sum of the next three is the cube of 3, and so on. There are other propositions on progressions.

Nichomachus gives four perfect numbers, 6, 28, 496, and 8128, and repeats Euclid's formula for perfect numbers. He classifies all sorts of ratios, including $m + 1 : m$, $2m + n : m + n$, and $mn + 1 : n$, and gives them names. These were important for music.

He also studies proportion, which, he says, is very necessary for "natural science, music, spherical trigonometry and planimetry and particularly for the study of the ancient mathematicians." He gives many types of proportion, among them the musical proportion

$$a:\frac{a + b}{2} = \frac{2ab}{a + b}:b.$$

The *Introductio* also gives the sieve of Eratosthenes (about whom we shall say more in Chapter 7); it is a method of obtaining prime numbers quickly. One writes down all the odd numbers from 3 on as far as one wishes, then crosses out all multiples of 3, that is to say every third number after 3. Next one crosses out all multiples of 5, or every fifth number after 5, counting any that may have been crossed out before. Then every seventh number after 7, and so forth. No crossed-out number can be the starting point of a new crossing-out process. The number 2 must now be included with those that are not crossed out. These numbers are the primes.

Nichomachus always uses specific numbers to discuss various categories and proportions. The examples illustrate and explain what he asserts but there is no support for any general assertions beyond the examples. He does not prove deductively.

The *Introductio* had value because it was a systematic, orderly, clear, and comprehensive presentation of the arithmetic of integers and ratios of integers freed of geometry. It was not original as far as ideas were concerned, but was a very useful compilation. It also incorporated speculative, aesthetic, mystical, and moral properties of numbers, but no practical applications. The *Introductio* was the standard text in arithmetic for a thousand years. At Alexandria, from the time of Nichomachus, arithmetic rather than geometry became the favorite study.

At this time, too, algebra came to the fore. Books of problems appeared that were solved by algebraic techniques. Some of these problems were exactly those that appeared in Babylonian texts of 2000 B.C. or in the Rhind papyrus. This Greek algebraic work was written in verbal form; no symbolism was used. Also, no proofs of the procedures were given. From the time of Nichomachus onward, problems leading to equations were a common form of puzzle. Between fifty and sixty of these are preserved in the Palatine Codex of Greek Epigrams (10th cent. A.D.). At least thirty of them are attributed to Metrodorus (*c.* A.D. 500), but are surely older. One is the Archimedes cattle problem, which calls for finding the number of bulls and cows of different colors subject to given information. Another is due to Euclid and involves a mule and a donkey carrying corn. Another calls for the time required for pipes to fill a cistern. There were also age problems such as appear in our algebra texts.

The highest point of Alexandrian Greek algebra is reached with Diophantus. We know almost nothing about his origins or life; he was probably

Greek. One of the algebraic problems found in a Greek collection gives the following facts about his life: His boyhood lasted 1/6 of his life; his beard grew after 1/12 more; he married after 1/7 more; and his son was born 5 years later. The son lived to half of his father's age and the father died 4 years after the son. The problem is to find how long Diophantus lived. The answer is easily found to be 84. His work towers above that of his contemporaries; unfortunately, it came too late to be highly influential in his time because a destructive tide was already engulfing the civilization.

Diophantus wrote several books that are lost in their entirety. Part of a tract *On Polygonal Numbers* is known, in which he states and proves theorems in the deductive manner of Books VII, VIII, and IX of the *Elements*; however, the theorems are not striking. His great work is the *Arithmetica* which, Diophantus says, comprises thirteen books. We have six, which come from a thirteenth-century manuscript that is a Greek copy of an older one and from later versions.

The *Arithmetica*, like the Rhind papyrus, is a collection of separate problems. The dedication says it was written as a series of exercises to help one of his students learn the subject. One of Diophantus' major steps is the introduction of symbolism in algebra. Since we do not have the manuscript written by him but only much later ones, we do not know the precise symbols. It is believed that the symbol he used for the unknown was ς, which served as our x. This ς may have been the same letter as the Greek σ when used at the end of a word, as in $\dot{\alpha}\rho\iota\theta\mu\acute{o}\varsigma$ (*arithmos*), and may have been chosen because it was not a number in the Greek system of using letters for numbers. Diophantus called the unknown "the number of the problem." For our x^2 Diophantus used Δ^Y, the Δ being the first letter of $\delta\acute{v}\nu\alpha\mu\iota\varsigma$ (*dýnamis*, "power"). x^3 is K^Y; the K is from $\kappa\acute{v}\beta o\varsigma$ (*cubos*). x^4 is $\Delta^Y\Delta$; x^5 is ΔK^Y; x^6 is $K^Y K$. In this system K^Y is not clearly the cube of ς as our x^3 is of x. For Diophantus $\varsigma^x = 1/x$. He also uses names for these various powers, e.g. number for x, square for x^2, cube for x^3, square-square (dynamodynamis) for x^4, square-cube for x^5, and cubocube for x^6.[2]

The appearance of such symbolism is of course remarkable but the use of powers higher than three is even more extraordinary. The classical Greeks could not and would not consider a product of more than three factors because such a product had no geometrical significance. On a purely arithmetic basis, however, such products do have a meaning; and this is precisely the basis Diophantus adopts.

Addition is indicated in Diophantus by putting terms alongside one another. Thus

$$\Delta^Y \bar{\gamma} \overset{\circ}{M} \iota \bar{\beta} \quad \text{means} \quad x^2 \cdot 3 + 12.$$

2. Some modern authors use δ, K, and $\bar{v}$ for Δ, K, and Y.

The $\overset{\circ}{M}$ is a symbol for unity and indicates that a pure number not involving the unknown follows. Again

$$\Delta^Y \bar{\alpha} s \beta \overset{\circ}{M} \bar{\gamma} \quad \text{means} \quad x^2 + x \cdot 2 + 3.$$

For subtraction he uses the symbol $\wedge$. Thus for $x^6 - 5x^4 + x^2 - 3x - 2$ he writes

$$K^Y \kappa \bar{\alpha} \Delta^Y \bar{\alpha} \; \wedge \; \Delta^Y \Delta \bar{\epsilon} s \bar{\gamma} \overset{\circ}{M} \beta,$$

putting all the negative terms after the positive ones. There are no symbols for addition, multiplication, or division as operations. The symbol ι^σ is used (at least in the extant versions of the *Arithmetica*) to denote equality. The coefficients of the algebraic expressions are specific numbers; there are no symbols for general coefficients. Because he does use some symbolism, Diophantus' algebra has been called syncopated, whereas that of the Egyptians, the Babylonians, Heron, and Nichomachus is called rhetorical.

Diophantus writes out his solutions in a continuous text, as we write prose. His execution of the operations is entirely arithmetical; that is, there is no appeal to geometry to illustrate or substantiate his assertions. Thus $(x - 1)(x - 2)$ is carried out algebraically as we do it. He also applies algebraic identities such as

$$\left(\frac{p + q}{2}\right)^2 - \left(\frac{p - q}{2}\right)^2 = pq$$

to expressions such as $x + 2$ for p and $x + 3$ for q. That is, he makes steps that use the identities but the identities themselves do not appear.

The first book of the *Arithmetica* consists mainly of problems that lead to determinate equations of the first degree in one or more unknowns. The remaining five books treat mainly indeterminate equations of second degree. But this segregation is not sharply adhered to. In the case of the determinate equations (i.e. equations leading to unique solutions) with more than one unknown involved, he uses given information to eliminate all but one unknown and, at worst, ends with quadratics of the form $ax^2 = b$. Thus Problem 27 of Book I states: Find two numbers such that their sum is 20 and product is 96. Diophantus proceeds thus: Given sum 20, given product 96, $2x$ the difference of the required numbers. Therefore the numbers are $10 + x$, $10 - x$. Hence $100 - x^2 = 96$. Then $x = 2$ and the required numbers are 12, 8.

The most striking feature of Diophantus' algebra is his solution of indeterminate equations. Such equations had been considered before, as for example in the Pythagorean work on solutions of $x^2 + y^2 = z^2$, in the Archimedean cattle problem, which leads to seven equations in eight unknowns (plus two supplementary conditions), and in other odd writings. Diophantus, however, pursues indeterminate equations extensively and is the founder of this branch of algebra now called, in fact, Diophantine analysis

He solves linear equations in two unknowns, e.g.

$$x + y - 5 = 0.$$

In such equations he gives a value to one unknown and then solves for a positive rational value of the other. He recognizes that the value assigned to the first unknown is merely typical. (In modern Diophantine analysis one seeks integral solutions only.) Very little is done with this type of equation, and the work is hardly significant since positive rational solutions are obtainable at once.

He then solves quadratic equations in two unknowns, of which the most general type is (in our notation)

$$(1) \qquad\qquad y^2 = Ax^2 + Bx + C.$$

Diophantus does not write y^2 but says that the quadratic expression must equal a square number (square of a rational number). He considers (1) for special values of A, B, and C and treats these types in separate cases. For example, when C is absent he lets $y = mx/n$, where m and n are specific whole numbers, obtains

$$Ax^2 + Bx = \frac{m^2}{n^2} x^2,$$

and then cancels x and solves. When A and C do not vanish but $A = a^2$, he assumes $y = ax - m$. If $C = c^2$ he assumes $y = (mx - c)$. In all cases m is a specific number.

He also treats the case of simultaneous quadratics, namely,

$$(2) \qquad\qquad y^2 = Ax^2 + Bx + C$$

$$(3) \qquad\qquad z^2 = Dx^2 + Ex + F.$$

Here, too, he undertakes only particular cases, that is, where A, B, $\ldots$, F are special numbers or satisfy special conditions, and his method is to assume expressions for y and z in terms of x and then solve for x.

In effect he is solving determinate equations in one unknown. He realizes, however, that in choosing expressions for y and z in (2) and (3) and for y in (1), he is giving merely typical solutions and that the values assigned to y and z are somewhat arbitrary.

He also has problems in which cubic and higher degree expressions in x must equal a square number, e.g.

$$Ax^3 + Bx^2 + Cx + d^2 = y^2.$$

Here he lets $y = mx + d$ and fixes m so that the coefficient of x vanishes. Since the d^2 terms cancel and he can then divide through by x^2, he obtains a first degree equation in x. There are also special cases of a quadratic in x equaling y^3.

All his quadratics in x reduce to the types

$$ax^2 = bx, \quad ax^2 = b, \quad ax^2 + bx = c, \quad ax^2 + c = bx, \quad ax^2 = bx + c,$$

and he solves each of these types. Only one cubic in x, of no significance, is solved.

The above equations show the *types* of problems Diophantus solves. The actual wording of the problems is illustrated by the following examples:

Book I, Problem 8. To divide a given square number into two squares.

Here he takes 16 as the given square number and obtains 256/25 and 144/25. This is the problem which Fermat generalized and which led to his assertion that $x^m + y^m = z^m$ is not solvable for $m > 2$.

Book II, Problem 9. To divide a given number which is a sum of two squares into two other squares.

He takes 13 or $4 + 9$ as the given number and obtains 324/25 and 1/25.

Book III, Problem 6. To find three numbers such that their sum and the sum of any two is a square.

Diophantus gives 80, 320, and 41 as the three numbers.

Book IV, Problem 1. To divide a given number into two cubes such that the sum of their sides is a given number.

With the given number of 370 and the given sum of the sides as 10, he finds 343 and 27. The sides are the cube roots of the cubes.

Book IV, Problem 29. To express a given number as the sum of four squares plus the sum of their sides.

Given the number 12 he finds 121/100, 49/100, 361/100, 169/100 as the four squares; their sides are the square root of each square.

In Book VI Diophantus solves a number of problems involving the (rational) sides of a right-angled triangle. The use of the geometrical language is incidental even where the term area occurs. He seeks rational numbers a, b, and c such that $a^2 + b^2 = c^2$ and subject to some other condition. Thus the first problem is to find a (rational) right-angled triangle such that the hypotenuse minus each of the sides gives a cube. In this case he happens to obtain the integral answers 40, 96, 104. However, in general he gets rational answers.

Diophantus shows great skill in reducing equations of various types to forms he can handle. We do not know how he arrived at his methods. Since he makes no appeal to geometry, it is not likely that he translated Euclid's methods for solving quadratics. Moreover, indeterminate problems are not in Euclid and as a class are new with Diophantus. Because we lack information on the continuity of thought in the later Alexandrian period, we cannot find traces of Diophantus' work in his predecessors. As far as we can tell, his work in pure algebra is remarkably different from past work.

He accepts only positive rational roots and ignores all others. Even when a quadratic equation has two positive roots he gives only one, the larger one. When an equation, as it is being solved, clearly leads to two negative or imaginary roots he rejects the equation and says it is not solvable. In the case of irrational roots, he retraces his steps and shows how by altering the equation he can get a new one that has rational roots. Here Diophantus differs from Heron and Archimedes. Heron was a surveyor and the geometrical quantities he sought could be irrational. Hence he accepted them, though of course he approximated them to obtain a useful value. Archimedes also sought exact answers, and when they were irrational he obtained inequalities to bound the irrational. Diophantus is a pure algebraist; and since algebra in his time did not recognize irrational, negative, and complex numbers, he rejected equations with such solutions. It is, however, worthy of note that fractions for Diophantus are numbers, rather than just a ratio of two whole numbers.

He has no *general methods*. Each of the 189 problems in the *Arithmetica* is solved by a different method. There are more than 50 different types of problems but no attempt is made to classify them by type. His methods are closer to the Babylonian ones than to those of his Greek predecessors, and there are indications of Babylonian influences. In fact, he does solve some problems just as the Babylonians did. But it has not been established that there was any direct connection between Diophantus' work and Babylonian algebra. His advance in algebra over the Babylonians consists in the use of symbolism and the solution of indeterminate equations. In determinate equations he went no further than they did, but his *Arithmetica* assimilated *logistica*, which Plato, among others, had banned from mathematics.

Diophantus' variety of methods for the separate problems dazzles rather than delights. He was a shrewd and clever virtuoso but apparently not deep enough to see the essence of his methods and thereby attain generality. (It is still true that Diophantine analysis is a maze of separate problems.) Unlike a speculative thinker who seeks general ideas, Diophantus sought only correct answers. There are a few results that might be called general, such as that no prime number of the form $4n + 3$ can be the sum of two squares. Euler did credit Diophantus with illustrating general methods that he could not display as such because he did not have literal coefficients. And there are others who credit Diophantus with recognizing that his material belonged to an abstract and basic science. But this view is not shared by all. His work as a whole, however, is a monument in algebra.

An element of mathematics that is of enormous importance today, which was missing in Greek algebra, is the use of letters to represent a class of numbers, as, for example, coefficients in equations. Aristotle did use letters of the Greek alphabet to indicate an arbitrary time or an arbitrary distance and in discussions of motion employed such phrases as "the half of B." Euclid, too, used letters for classes of numbers in Books VII to IX of the *Elements*, a practice that Pappus followed. However, there was no recognition of the enormous contribution that letters could make in increasing the effectiveness and generality of algebraic methodology.

Another feature of Alexandrian algebra is the absence of any explicit deductive structure. The various types of numbers—whole numbers, fractions, and irrationals—were certainly not defined. Nor was there any axiomatic basis on which a deductive structure could be erected. The work of Heron, Nichomachus, and Diophantus, and of Archimedes as far as his arithmetic is concerned, reads like the procedural texts of the Egyptians and Babylonians, which tell us how to do things. The deductive, orderly proof of Euclid and Apollonius, and of Archimedes' geometry is gone. The problems are inductive in spirit, in that they show methods for concrete problems that presumably apply to general classes whose extent is not specified. In view of the fact that as a consequence of the work of the classical Greeks mathematical results were supposed to be derived deductively from an explicit axiomatic basis, the emergence of an independent arithmetic and algebra with no logical structure of its own raised what became one of the great problems of the history of mathematics. This approach to arithmetic and algebra is the clearest indication of the Egyptian and Babylonian influences in the Alexandrian world. Though the Alexandrian Greek algebraists did not seem to be concerned about this deficiency, we shall find that it did trouble deeply the European mathematicians.

Bibliography

Ball, W. W. R.: *A Short Account of the History of Mathematics*, Dover (reprint), 1960, Chaps. 5 and 7.

Cajori, Florian: *A History of Mathematics*, Macmillan, 1919, pp. 52–62.

Heath, Thomas L.: *Diophantus of Alexandria*, Dover (reprint), 1964.

————: *The Works of Archimedes*, Dover (reprint), 1953, Chaps. 4 and 6 of the Introduction, pp. 91–98, 319–326.

————: *A History of Greek Mathematics*, Oxford University Press, 1921, Vol. 1, Chaps. 1–3; Vol. 2, Chap. 20.

————: *A Manual of Greek Mathematics*, Dover (reprint), 1963, Chaps. 2–3, and 17.

D'Ooge, Martin Luther: *Nichomachus of Gerasa*, University of Michigan Press, 1938.

van der Waerden, B. L.: *Science Awakening*, P. Noordhoff, 1954, pp. 278–86.

7

The Greek Rationalization
of Nature

> Mathematics is the gate and key of the sciences.
>
> ROGER BACON

1. *The Inspiration for Greek Mathematics*

Unfortunately, except for occasional hints, the Greek classics, such as Euclid's *Elements*, Apollonius' *Conic Sections*, and the geometrical works of Archimedes, give no indication of why these authors investigated their subjects. They give only the formal, polished deductive mathematics. In this respect, the Greek texts are no different from modern mathematics textbooks and treatises. Such books seek only to organize and present the mathematical results that have been attained and so omit the motivations for the mathematics, the clues and suggestions for the theorems, and the uses to which the mathematical knowledge is put.

To understand why the Greeks created so much vital mathematics, one must investigate their objectives. It was the urgent and irrepressible desire of the Greeks to understand the physical world that impelled them to create and value mathematics. Mathematics was part and parcel of the investigation of nature and the key to comprehension of the universe, for mathematical laws are the essence of its design.

What evidence do we have that this was the role of mathematics? It is difficult to demonstrate that any one theorem or body of theorems was created for a specific purpose because we do not have enough information about the Greek mathematicians. Ptolemy's direct statement that he created trigonometry for astronomy is an exception. However when one finds that Eudoxus was primarily an astronomer and that Euclid wrote not just the *Elements* but the *Phaenomena* (a work on the geometry of the sphere as applied to the motion of the sphere of stars), the *Optics* and *Catoptrica*, the *Elements of Music*, and small works on mechanics, all of which were mathematical, one cannot escape the conclusion that mathematics was more than an isolated discipline. Knowing how the human mind works and knowing in great detail how men such as Euler and Gauss worked, we may be fairly certain that the investigations in astronomy, optics, and music must have suggested mathematical problems, and it is most likely that the motivation for the mathematics was its application to these other areas. It is also relevant that the geometry of the sphere, known in Greek times as "sphaeric," was studied just as soon as astronomy became mathematical, which happened even before Eudoxus' time. The word "sphaeric" meant "astronomy" to the Pythagoreans.

Fortunately the inferences we may draw from the works of the mathematicians, though reasonable enough, are established beyond doubt by the overwhelming evidence in the writings of the Greek philosophers, many of whom were also prominent mathematicians, and of the Greek scientists. The bounds of mathematics were not mathematics proper. In the classical period mathematics comprised arithmetic, geometry, astronomy, and music; and in the Alexandrian period, as we have already noted in Chapter 5, the divisions of the mathematical sciences were arithmetic (theory of numbers), geometry, mechanics, astronomy, optics, geodesy, canonic (musical harmony), and logistics (applied arithmetic).

2. The Beginnings of a Rational View of Nature

The civilizations that preceded the Greek or were contemporary with it regarded nature as chaotic, mysterious, capricious, and terrifying. The happenings in nature were manipulated by gods. Prayers and magic might induce the gods to be kind and even to perform miracles but the life and fate of man were entirely subject to their will.

From the time our knowledge of Greek civilization and culture begins to be reasonably definite and specific, that is, from about 600 B.C., we find among the intellectuals a totally new attitude toward nature: rational, critical, and secular. Mythology was discarded, as was the belief that the gods manipulate man and the physical world according to their whims. The new doctrine holds that nature is orderly and functions invariably according to a plan. Moreover, the conviction is manifest that the human mind is powerful and even supreme; not only can the ways of nature be learned by man, but he can even predict the occurrences.

It is true that the rational approach was entertained only by the intellectuals, a small group in both the classical and Alexandrian periods. Whereas these men opposed the attribution of events to gods and demons and defied the mysteries and terrors of nature, people in general were deeply religious and believed that the gods controlled all events. They accepted mystical doctrines and superstitions as credulously as did the Egyptians and the Babylonians. In fact Greek mythology was vast and highly developed.

The Ionians began the task of determining the nature of reality. We shall not describe the qualitative theories of Thales, Anaxagoras, and their colleagues, each of whom fixed on a single substance persisting through all apparent change. The underlying identity of this prime substance is conserved but all forms of matter can be explained in terms of it. This natural philosophy of the Ionians was a series of bold speculations, shrewd guesses, and brilliant intuitions rather than the outcome of extensive and careful scientific investigations. They were perhaps a little too eager to see the whole picture and so naïvely jumped to broad conclusions. But they did substitute material and objective explanations of the structure and design of the universe for the older mythical stories. They offered a reasoned approach in place of the fanciful and uncritical accounts of the poets and they defended their contentions by reason. At least these men dared to tackle the universe with their minds and refused to rely upon gods, spirits, ghosts, devils, angels, and other mythical agents.

3. *The Development of the Belief in Mathematical Design*

The decisive step in removing the mystery, mysticism, and arbitrariness from the workings of nature and in reducing the seeming chaos to an understandable ordered pattern was the application of mathematics. The first major group to offer a rational and mathematical philosophy of nature were the Pythagoreans. They did draw some inspiration from the mystical side of Greek religion; their religious doctrines centered about the purification of the soul and its redemption from the taint and prison of the body. The members lived simply and devoted themselves to the study of philosophy, science, and mathematics. New members were pledged to secrecy at least as to religious beliefs and required to join up for life. Membership in the community was open to men and women.

The Pythagoreans' religious thinking was undoubtedly mystical, but their natural philosophy was decidedly rational. They were struck by the fact that phenomena that are most diverse from a qualitative point of view exhibit identical mathematical properties. Hence, mathematical properties must be the essence of these phenomena. More specifically, the Pythagoreans found this essence in number and in numerical relationships. Number was their first principle in the explanation of nature. All objects were made up of points or "units of existence" in combinations corresponding to the various geometrical figures. Since they thought of numbers both as points and as elementary particles of matter, number was the matter and form of the universe and the cause of every phenomenon. Hence the Pythagorean doctrine "All things are numbers." Says Philolaus, a famous fifth-century Pythagorean, "Were it not for number and its nature, nothing that exists would be clear to anybody either in itself or in its relation to other things. . . . You can observe the power of number exercising itself not only in the affairs of demons and gods but in all the acts and the thoughts of men, in all handicrafts and music."

The reduction of music, for example, to simple relationships among numbers became possible for the Pythagoreans when they discovered two facts: first, that the sound caused by a plucked string depends upon the length of the string; and second, that harmonious sounds are given off by equally taut strings whose lengths are to each other as the ratios of whole numbers. For example, a harmonious sound is produced by plucking two equally taut strings, one twice as long as the other. In our language, the interval between the two notes is an octave. Another harmonious combination is formed by two strings whose lengths are in the ratio 3 to 2; in this case the shorter one gives forth a note called the fifth above that given off by the first string. In fact, the relative lengths in every harmonious combination of plucked strings can be expressed as ratios of whole numbers. The Pythagoreans also developed a famous Greek musical scale. Though we shall not devote space to the music of the Greek period, we note that many Greek mathematicians, including Euclid and Ptolemy, wrote on the subject, especially on harmonious combinations of sounds and the construction of scales.

The Pythagoreans reduced the motions of the planets to number relations. They believed that bodies moving in space produce sounds; perhaps this was suggested by the swishing of an object whirled on the end of a string. They believed, further, that a rapidly moving body gives forth a higher note than one that moves slowly. Now according to their astronomy the greater the distance of a planet from the earth the more rapidly it moved. Hence the sounds produced by the planets, which we do not hear because we are accustomed to them from birth, varied with their distances from the earth and all harmonized. But since this "music of the spheres," like all harmony, reduced to no more than number relationships, so did the motions of the planets.

The Pythagoreans and probably Pythagoras himself wanted not just to observe and describe the heavenly motions but to find regularity in them. The idea of uniform circular motion, seemingly obvious in the case of the moon and sun, suggested that all the planetary motions were explainable in terms of uniform circular motions. The later Pythagoreans made a more striking break with tradition; they were the first to believe that the earth was spherical. Moreover, because 10 was their ideal number, they decided that the moving bodies in the heavens must be 10 in number. First, there was a central fire around which the heavenly bodies, *including the earth*, moved. They knew five planets in addition to the earth. These six bodies, the sun, the moon, and the sphere to which the stars were attached made only 9 moving bodies. Hence they asserted the existence of a tenth one, called the counter-earth, which also revolved around the central fire. We cannot see this tenth one because it moves at exactly the same speed as the earth on the opposite side of the central fire and also because the inhabited part of earth faces away from the central fire. Here we have the first theory to put the earth in motion. However, the Pythagoreans did not assert the rotation of the earth; rather, the sphere of fixed stars revolves about the center of the universe.

The belief that the celestial bodies are eternal, divine, perfect, and unchangeable and that the sublunar bodies, that is the earth and (according to the Greeks) the comets, are subject to change, decomposition, decay, and death may also have come from the Pythagoreans. The doctrine of uniform circular motion and the distinction between celestial and sublunar bodies became embedded in Greek thought.

Other features of nature also "reduced" to number. The numbers 1, 2, 3, and 4, the *tetractys*, were especially valued because they added up to 10. In fact the Pythagorean oath is reported to have been: "I swear in the name of the Tetractys which has been bestowed on our soul. The source and roots of the everflowing nature are contained in it." The Pythagoreans asserted that nature was composed of fournesses; for example, point, line, surface, and solid, and the four elements, earth, air, fire, and water. The four elements were also central in Plato's natural philosophy. Because 10 was ideal, 10 represented the universe. The ideality of 10 required that the whole universe be describable in terms of 10 categories of opposites: odd and even, bounded and unbounded, good and evil, right and left, one and many, male and female, straight and curved, square and oblong, light and darkness, and rest and motion.

Clearly, Pythagorean philosophy mingled serious thoughts with what we would consider fanciful, useless, and unscientific doctrines. Their obsession with the importance of numbers resulted in a natural philosophy that certainly had little correspondence with nature. But they did stress the understanding of nature, not, like the Ionians, through a single substance, but through the formal structure of number relationships. Moreover they and the Ionians both saw that underlying mere sense data there must be a harmonious account of nature.

We can now see why the discovery of incommensurable lengths was so disastrous to Pythagorean philosophy: a ratio of incommensurable lengths could not be expressed as a ratio of whole numbers. In addition, they had believed that a line is made up of a finite number of points (which they identified with physical particles); but this could not be the case for a length such as $\sqrt{2}$. Their philosophy, based on the primariness of the whole numbers, would have been shattered if they had accepted irrationals as numbers.

Because the Pythagoreans "reduced" astronomy and music to number, these subjects came to be linked to arithmetic and geometry; these four were regarded as the mathematical subjects. They became and remained part of the school curriculum even into medieval times, when they were called, collectively, "the quadrivium." As we have noted, the Pythagorean interest in arithmetic (i.e. the theory of numbers) was due not to the purely aesthetic value of that subject but to a search for the meaning of natural phenomena in numerical terms; and this value caused the emphasis on special proportions and on triangular, square, pentagonal, and higher forms into which numbers could be arranged. Further, it was the Pythagorean natural philosophy centering about number that gave the subject importance with such men as Nichomachus. In fact, modern science adheres to the Pythagorean emphasis on number—though, as we shall see, in a much more sophisticated form— while the purely aesthetic modern theory of numbers derives from Pythagorean arithmetic per se.

The philosophers who came chronologically between the Pythagoreans and Plato were equally concerned with the nature of reality but did not involve mathematics directly. The arguments and views of men such as Parmenides (5th cent. B.C.), Zeno (5th cent. B.C.), Empedocles (c. 484–c. 424 B.C.), Leucippus (c. 440 B.C.), and Democritus (c. 460–c. 370 B.C.) were, like those of their Ionian predecessors, qualitative. They made broad assertions about reality that were, at best, barely suggested by observation. Nevertheless, each affirmed that nature is intelligible and that reality can be grasped by thought. Each was a link in the chain that led to the mathematical investigation of nature. Leucippus and Democritus are notable because they were the most explicit in affirming the doctrine of atomism. Their common philosophy was that the world is composed of an infinite number of simple, eternal atoms. These differ in shape, size, order, and position, but every object is some combination of these atoms. Though geometrical magnitudes are infinitely divisible, the atoms are ultimate indivisible particles. (The word atom in Greek means indivisible.) Hardness, shape, and size are physically real properties of the atoms. All other properties, such as taste, heat, and color are not in the atoms but in the perceiver; thus sensuous knowledge is unreliable because it varies with the perceiver. Like the Pythagoreans, the atomists asserted that the reality underlying the constantly changing diversity of the physical world was expressible in terms of mathematics and, moreover, that the happenings in this world were strictly determined by mathematical laws.

Plato, the foremost Pythagorean next to Pythagoras, was the most influential propagator of the doctrine that the reality and intelligibility of the physical world can be comprehended only through mathematics. For him there was no question that the world was mathematically designed, for "God eternally geometrizes." The world perceived by the senses is confused and deceptive and in any case imperfect and impermanent. Physical knowledge is unimportant, because material objects change and decay; thus the direct study of nature and purely physical investigations are worthless. The physical world is but an imperfect copy of the ideal world, the one that mathematicians and philosophers should study. Mathematical laws, eternal and unchanging, are the essence of reality.

Plato went further than the Pythagoreans in wishing not merely to understand nature through mathematics but to substitute mathematics for nature itself. He believed that a few penetrating glances at the physical world would supply some basic truths with which reason could then carry on unaided. From that point there would be no nature, just mathematics, which would substitute for physical investigations as it does in geometry.

Plato's attitude toward astronomy illustrates his position on the knowledge to be sought. This science is not concerned with the movements of the visible heavenly bodies. The arrangement of the stars in the heavens and their apparent movements are indeed wonderful and beautiful to behold, but mere observation and explanation of the motions fall far short of true astronomy. Before we can attain to the latter we "must leave the heavens alone," for true astronomy deals with the laws of motion of true stars in a mathematical heaven of which the visible heaven is but an imperfect expression. Plato encourages devotion to a theoretical astronomy, whose problems please the mind, not the eye, and whose objects are apprehended by the mind, not visually. The varied figures that the sky presents to the eye are to be used only as diagrams to assist the search for the higher truths. The uses of astronomy in navigation, calendar-reckoning, and the measurement of time were alien to Plato.

Plato's views on the role of mathematics in astronomy are an integral part of his philosophy, which held that there is an objective, universally valid reality consisting of forms or ideas. These realities were independent of human beings and were immutable, eternal, and timeless. We become aware of these ideas through recollection or anamnesis; although they are present in the soul, it must be stimulated to recall them or fetch them up from its depths. These ideas are the only reality. Included among them but occupying a lesser rank are mathematical ideas, which are regarded as intermediate between the sensible world and such higher ideas as goodness, truth, justice, and beauty. In this comprehensive philosophy, mathematical ideas played a double role; not only were they part of reality themselves but, as we have already pointed out in Chapter 3, they helped train the mind to view eternal ideas. As Plato put it in Book VII of *The Republic*, the study of geometry made easier the vision of the idea of goodness: "Geometry will draw the soul toward truth, and create the spirit of philosophy. . . ."

Aristotle, while deriving many ideas from his teacher Plato, had a quite different concept of the study of the real world and of the relation of mathematics to reality. He criticized Plato's otherworldliness and his reduction of science to mathematics. Aristotle was a physicist; he believed in material things as the primary substance and source of reality. Physics and science generally must study the physical world to obtain truths; genuine knowledge is obtained from sense experience by intuition and abstraction. Then reason can be applied to the knowledge so obtained.

Matter alone is not significant. As such it is indeterminate, simply the potentiality of form; matter becomes significant when it is organized into various forms. Form and the changes in matter that give rise to new forms are the interesting features of reality and the real concern of science.

According to Aristotle matter is not, as some earlier Greeks believed, composed of one primitive substance. The matter we see and touch is composed of four basic elements: earth, water, fire, and air. Also, each element has its own characteristic qualities. Earth is cold and dry; water is cold and moist; air is hot and moist; and fire is hot and dry. Hence the qualities of any given object depend upon the proportions of the elements that enter into it; and thereby solidity, hardness, coarseness, and other qualities are determined.

The four elements have other qualities. Earth and water have gravity; air and fire have levity. Gravity causes an element to seek to be at rest at the center of the earth; levity causes it to seek the heavens. Thus by knowing the proportions of the elements that enter into a given object, one can also determine its motion.

Aristotle regarded solids, fluids, and gases as three different types of matter, distinguished by the possession of different substantial qualities. The transition from solid to fluid, for example, meant the loss of one quality and the substitution of another. Thus changing mercury into rigid gold involved taking from mercury the substance that possessed fluidity and substituting some other substance.

Science also had to consider the causes of change. For Aristotle there were four types of causes. The first was the material or immanent cause; for a statue made of bronze, bronze is the immanent cause. The second was the formal cause; for the statue it is the design or shape. The formal cause of harmony is the pattern of 2 to 1 in the octave. The third cause was the effective cause, the agent or doer; the artist and his chisel are the effective cause of the statue. The fourth was the final cause, or the purpose that the phenomenon served; the statue serves to please people, to offer beauty. Final cause was the most important of the four because it gave the ultimate reasons for events or phenomena. Everything had a final cause.

Where was mathematics in this scheme of things? The physical sciences were fundamental to the study of nature, and mathematics helped by describing formal properties such as shape and quantity. It also provided explanations of facts observed in material phenomena. Thus geometry provided the reasons for facts provided by optics and astronomy, and arithmetical proportions could give the reasons for harmony. But mathematics was definitely an abstraction from the real world, since mathematical objects are not independent of or prior to experience. They exist in human minds as a class of ideas mediating between the sensible objects themselves and the essence of objects. Because they are abstracted from the physical world, they are applicable to it; but they have no reality apart from visible and tangible things. Mathematics alone can never provide an adequate definition of substance. Qualitative differences, as among colors, cannot be reduced to differences in geometry. Hence in the study of causes, mathematics can provide at best some knowledge of the formal cause—that is, a description. It can describe what happens in the physical world, can correlate concomitant variations, but can say nothing about the efficient and final causes of movement or change. Thus Aristotle distinguished sharply between mathematics and physics and assigned a minor role to mathematics. He was not interested in prediction.

From this survey we may see that all the philosophers who forged and molded the Greek intellectual world stressed the study of nature for comprehension and appreciation of its underlying reality. From the time of the Pythagoreans, practically all asserted that nature was designed mathematically. During the classical period, the doctrine of the mathematical design of nature was established and the search for the mathematical laws instituted. Though this doctrine did not motivate all of the mathematics subsequently created, once established it was accepted and consciously pursued by most of the great mathematicians. During the time that this doctrine held sway, which was until the latter part of the nineteenth century, the search for the mathematical design was identified with the search for truth. Though a few Greeks—for example, Ptolemy—realized that mathematical theories were merely human attempts to provide a coherent account, the belief that mathematical laws were the truth about nature attracted some of the deepest and noblest thinkers to mathematics.

We should also note, in order to appreciate more readily what happened in the seventeenth century, the Greek emphasis on the power of the mind. Because the Greek philosophers believed that the mind was the most powerful agent in comprehending nature, they adopted first principles that appealed to the mind. Thus the belief that circular motion was the basic type, defended by Aristotle on the ground that the circle is complete whereas a rectilinear figure, because it is bounded by many curves (line segments), is incomplete and therefore secondary in importance, appealed to the mind on aesthetic grounds. That the heavenly bodies should move with only constant or uniform velocity, was a conception which appealed to the mind perhaps because it was simpler than nonuniform motion. The combination of uniform and circular motion seemed to befit heavenly bodies. That the sublunar bodies should be different from the planets, sun, and stars seemed reasonable also, because the heavenly bodies preserved a constant appearance whereas change on earth was evident. Even Aristotle, who stressed abstractions only insofar as they helped to understand the observable world, said that we must start from principles that are known and manifest to the mind and then proceed to analyze things found in nature. We proceed, he said, from universals to particulars, from man to men, just as children call all men father and then learn to distinguish. Thus even the abstractions made from concrete objects presuppose some general principles emanating from the mind. This doctrine, the power of the mind to yield first principles, was overthrown in the seventeenth century.

4. Greek Mathematical Astronomy

Let us examine now what the Greeks produced in the mathematical description of natural phenomena. It is from the time of Plato that the several sciences created by the Greeks take on significant content and direction. Though we intend to review Greek astronomy, let us note in passing one aspect of Euclidean geometry. We have already observed that spherical geometry was developed for astronomy. Geometry was in fact part of the larger study of cosmology. Geometric principles were, to the Greeks, embodied in the entire structure of the universe, of which space was the primary component. Hence the study of space itself and of figures in space was of importance to the larger goal. Geometry, in other words, was in itself a science, the science of physical space.

It was Plato who, though fully aware of the impressive number of astronomical observations made by the Babylonians and Egyptians, emphasized the lack of an underlying or unifying theory or explanation of the seemingly irregular motions of the planets. Eudoxus, who was for a while a student at the Academy, took up Plato's problem of "saving the appearances." His answer is the first reasonably complete astronomical theory. He wrote four books on astronomy, *Mirror, Phenomena, Eight-Year Period*, and *On Speeds*, only fragments of which are known. From these fragments and accounts by other writers, we know the essence of Eudoxus' theory.

The motions of the sun and moon, as viewed from the earth, can be crudely described as circular with constant speed. However, their deviations from circular orbits are great enough to have been observed and to require explanation. The motions of the planets as seen from the earth are even more complex, for during any one revolution they reverse their course, go backward for a while, and then go forward again. Moreover, their speeds on these paths are variable.

To show that the actual, rather complicated, and apparently lawless motions could be understood in terms of simple circular geometrical motions, Eudoxus proposed the following scheme: For any one heavenly body there was a set of three or four spheres, all concentric with the earth as the center, each rotating about an axis. The innermost sphere carried the body, which moved along what can be called the equator of that sphere; that is, the axis of rotation was perpendicular to the circular path of the body. However, while rotating on its axis, this sphere was being carried along by the rotation of the next of the concentric spheres by the following device. Imagine that the axis of rotation of the first sphere is extended at each end to reach the second sphere and that its endpoints are fixed on that second sphere. If, now, the second sphere rotates about an axis of its own, then this sphere will carry the axis of the first sphere around while the latter rotates about that axis. The axis of the second sphere is in turn carried around by the rotation of a third sphere on its own axis. Eudoxus found that, for the sun and moon, a combination of three spheres sufficed to reproduce the actual motions as viewed from the earth. For each planet a fourth sphere was required, to which the third was related in the manner just described. The outermost sphere of each combination rotated on an axis through the celestial poles once in each 24 hours. In all, Eudoxus used 27 spheres. Their axes of rotation, speeds of rotation, and radii were chosen to make the theory fit as well as possible the observations available in his time.

Eudoxus' scheme was mathematically elegant and remarkable in many ways. The very idea of using combinations of spheres was ingenious; and the task of choosing the axes, speeds, and radii to make the resultant motion of the heavenly body fit the actual observations called for tremendous mathematical skill in working with surfaces and curves (i.e. the paths of the planets) in space.

It is especially worthy of note that Eudoxus' theory is purely mathematical. His spheres, except for the "sphere" of fixed stars, were not material bodies but mathematical constructions. Nor did he try to account for the forces that would make the spheres rotate as he said they did. His theory is thoroughly modern in spirit, for today mathematical description and not physical explanation is the goal in science.

The Eudoxian system had serious shortcomings. It did not include the variable velocity of the sun and was slightly in error about its actual path. His theory did not fit at all well the actual motion of Mars and was unsatisfactory for Venus. That Eudoxus tolerated such inadequacies may be accounted for by the fact that he might not have had at his command a sufficient number of observations. He had probably learned in Egypt only the main facts about the stationary points, retrogressions, and periods of revolution of the outer planets (Mars, Jupiter, and Saturn). Perhaps also for this reason his values for the sizes and distances of the celestial bodies were very crude. Aristarchus says that Eudoxus believed the diameter of the sun to be nine times that of the moon.

Aristotle did not appreciate a purely mathematical scheme and hence was not satisfied with Eudoxus' solution. To devise a real mechanism that made one sphere force another to rotate, he added 29 spheres. These were

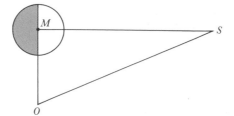

Figure 7.1

inserted among Eudoxus' so as to make the motion of one sphere drive another through actual contact and so that all derived their motive power from the outermost one. In some Aristotelian writings the sphere of stars, itself moving, is the prime mover of the other spheres. In others there is an unmoved mover behind the sphere of stars. His 56 spheres so complicated the system that it was discredited by scientists, though it was popular among educated laymen in medieval Europe. Aristotle, too, believed that the earth is spherical, for reasons of symmetry and equilibrium and because the shadow of the earth on the moon, seen in lunar eclipses, is circular.

Writings on astronomy continue after Aristotle in an almost unbroken sequence. After the works of Autolycus (Chap. 3, sec. 10) and the *Phaenomena* of Euclid (Chap. 4, sec. 11) the next major astronomical works are Alexandrian. Chaldean observations, observations made by the Babylonians of the Seleucid period, and measurements made by the Alexandrians themselves increased enormously the number and accuracy of the available data.

The first great Alexandrian astronomer is Aristarchus (*c*. 310–230 B.C.), who was learned in geometry, astronomy, music, and other branches of science. His book *On the Sizes and Distances of the Sun and Moon*, manuscripts of which are extant in Greek and Arabic, is the first major attempt to measure the distances of the sun and moon from the earth and the relative sizes of these bodies. These calculations are, incidentally, another example of the quantitative interests of the Alexandrians. Aristarchus did not have trigonometry at his disposal, nor did he have a good value for π (Archimedes' work on this came later), but he used Euclidean geometry very capably.

He knew that the moon's light is reflected light. When exactly half of the moon is illuminated, the angle at M (Fig. 7.1) is a right angle. The observer at O can measure the angle there and then the *relative* distances OM and OS can at least be estimated. Aristarchus' measurement of the angle was $87°$; it is, to the nearest minute, $89°52'$. Hence he estimated that the sun is more than 18 times and less than 20 times more distant than the moon. The correct ratio is 346 times more distant.

Knowing the relative distances, Aristarchus calculated the relative sizes by measuring the sizes of the discs the sun and moon present to the earth. He concluded that the volume of the sun was 7000 times larger than that of the moon. He was far from the truth here; the correct number is 64,000,000. He also found the ratio of the diameter of the sun to the diameter of the earth to be between 19/3 and 43/6; but the correct ratio is about 109.

Aristarchus is equally famous for having been the first to propose the heliocentric hypothesis—that the earth and the planets all revolve in circles around the fixed sun. The stars too are fixed and their apparent motion is due to the rotation of the earth on its axis. The moon revolves about the earth. Though, as we know today, Aristarchus had the right idea, it was not accepted for many reasons. For one thing, Greek mechanics (see below), already well developed by Aristotle, could not account for objects staying on a moving earth. According to Aristotle, heavy objects seek the center of the universe. This principle accounted for the fall of objects to earth as long as it was the center of the universe; but if it moved, the objects would stay behind. That argument was used by Ptolemy against Aristarchus and, in fact, was used later against Copernicus because the prevailing system of mechanics was still Aristotle's. Ptolemy also said that the clouds would lag behind a moving earth. Further, Aristotle's mechanics required a force to keep earthly objects in motion and there was no apparent force. We do not know how Aristarchus answered these arguments.

Another argument advanced against Aristarchus was that if the earth were in motion its distance from the fixed stars would vary, but it apparently does not. To this Aristarchus gave the correct rebuttal; he said that the radius of the sphere of fixed stars is so large that the earth's orbit is too small to matter. Aristarchus' heliocentric idea was rejected by many because it was impious to identify the corruptible matter of the earth with the incorruptible matter of the heavenly bodies. The hypothesis that the planets move about the sun in circles is, of course, unsatisfactory, because the motion is actually more complicated. But the heliocentric idea could have been refined, as Copernicus did later. It was, however, too radical for Greek thought.

The founder of quantitative mathematical astronomy is Apollonius. He was called Epsilon because the symbol ϵ was used to denote the moon, and much of his astronomy was devoted to the motion of that body. Before considering his work and that of Hipparchus and Ptolemy, to which it is closely related, we shall examine the basic scheme that had entered Greek astronomy between the times of Eudoxus and Apollonius, the scheme of epicycle and deferent. In this scheme a planet P moves at a constant speed on a circle (Fig. 7.2) with center S, while S itself moves with constant speed on a circle with center at the earth E. The circle on which S moves is called the deferent; the circle on which P moves is called the epicycle. The point S for some planets is the sun but in other cases it is just a mathematical point. The direction of the motion of P may agree with or be opposite to the direction of motion of S. The latter is the case for the sun and moon.

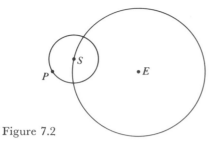

Figure 7.2

Apollonius is supposed to have known thoroughly the scheme of epi-cyclic motion and the details which made possible the representation of the motions of the planets, moon, and sun. Ptolemy credits Apollonius specifically with determining the points at which a planet appears to be stationary and reverses its direction of motion.

The climax of Greek astronomy is the work of Hipparchus and Ptolemy. The scheme of deferent and epicycle was taken over by Hipparchus (died *c.* 125 B.C.) and applied to the motion of the five planets then known, the moon, the sun, and the stars. We know Hipparchus' work through Ptolemy's *Almagest,* though it is difficult to distinguish what is due to Hipparchus and what to Ptolemy. After making observations at Rhodes for thirty-five years and utilizing Babylonian observations, Hipparchus worked out the details for the epicyclic theory of motion. By properly selecting the radii of the epicycle and deferent and the speeds of a body on its epicycle and the center of the epicycle on the deferent, he was able to get an improved description of the motions. He was quite successful for the sun and moon but only partially so for the planets. From the time of Hipparchus, an eclipse of the moon could be predicted to within an hour or two, though eclipses of the sun were predicted less accurately. This theory also accounted for the seasons.

Hipparchus' most original contribution is his discovery of the precession of the equinoxes. To understand this phenomenon we suppose that the earth's axis of rotation extends up to the stars. The point in which it hits the sphere of stars moves in a circle and takes 26,000 years to traverse the circle. In other words the earth's axis continually changes its direction relative to the stars and the motion is periodic. The star to which it points at any one time is called the polar star. The angle subtended by a diameter of the above-mentioned circle at the earth is 45°.

Hipparchus made many other contributions to astronomy, such as the construction of instruments for observation, the determination of the angle of the ecliptic, measurements of irregularities in the motion of the moon, improvement of the determination of the length of the solar year (which he estimated to be 365 days, 5 hours, 55 minutes, and 12 seconds—or about

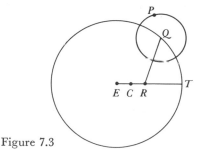

Figure 7.3

6 1/2 minutes too long), and a catalogue of about a thousand stars. He found that the ratio of the distance to the moon to the radius of the earth was 67.74; the modern figure is 60.3. He calculated the moon's radius to be 1/3 the earth's radius; the present figure is 27/100.

Ptolemy extended the work of Hipparchus by further improving the mathematical descriptions of the motions of all the heavenly bodies. The extended theory, presented in the *Almagest*, offers a complete exposition of the geocentric theory of deferent and epicycle, which has come to be known as the Ptolemaic theory.

To make the geometrical account fit observations, Ptolemy also introduced a variation on epicyclic motion known as uniform equant motion. In this scheme (Fig. 7.3) the planet moves on an epicycle with center Q while Q moves on a circle with center C which is not the earth but somewhat offset. He fixed the speed of Q by introducing the point R such that $EC = CR$ and required that $\measuredangle QRT$ increase uniformly. Thus Q moves with uniform angular velocity, but not with uniform linear velocity.

The approach and understanding that the Greek astronomers attained are thoroughly modern. The Alexandrian Greek astronomers, notably Hipparchus and Ptolemy, made observations of their own; in fact, Hipparchus did not trust many of the older Egyptian and Chaldean observations and repeated them. The classical and Alexandrian astronomers not only constructed theories but fully realized that these theories were not the true design but just descriptions that fit the observations. Ptolemy says in the *Almagest*[1] that in astronomy one ought to seek as simple as possible a mathematical model. These men, unlike other Greeks, did not look for some physical explanation of the motions. On this point Ptolemy says,[2] "After all, generally speaking, the cause of the first principles is either nothing or hard to interpret in its nature." But his own mathematical model was later taken as the literal truth by the Christian world.

1. Book XIII, Chap. 2, last paragraph.
2. *Almagest*, Book IX.

Ptolemaic theory offered the first reasonably complete evidence of the uniformity and invariability of nature and was the final Greek answer to Plato's problem of rationalizing the apparent motions of the heavenly bodies. No other product of the entire Greek era rivaled the *Almagest* for its profound influence on conceptions of the universe and none, except Euclid's *Elements*, achieved such unquestioned authority.

This brief account of Greek astronomy has not revealed the full depth and extent of the work done even by the men discussed here, and omits many other contributions. Almost every Greek mathematician, including Archimedes, devoted himself to the subject. Greek astronomy was masterful and comprehensive and it employed a vast amount of mathematics.

5. *Geography*

Another science that received its foundation in Greek times is geography. Though a few classical Greeks such as Anaximander and Hecataeus of Miletus (died *c*. 475 B.C.) made maps of the earth as it was known at that time, it was the Alexandrians who made the great strides in geography. They measured or calculated distances along the earth, the heights of mountains, the depths of valleys, and the extent of the seas. The Alexandrians were especially stimulated to study geography because the Greek world had widened.

The first great Alexandrian geographer was Eratosthenes of Cyrene (*c*. 284–*c*. 192 B.C.), director of the library at Alexandria, a mathematician, poet, philosopher, historian, philologist, chronologist, and by reputation one of the most learned men of antiquity. He studied in Plato's school in Athens and was invited to Alexandria by Ptolemy Euergetes. Eratosthenes worked at Alexandria until blindness overtook him in his old age; because of this affliction he starved himself to death.

Eratosthenes collected the available geographical knowledge and made numerous calculations of distances on the earth between significant places (such as cities). His most famous calculation is the length of the circumference of the earth. At noon on the summer solstice, the sun was observed to be practically overhead at Syene, the city that today is called Aswan (Fig. 7.4). (This was confirmed by observing that the sun shone directly down a well there.) At the same time, in Alexandria, which is (within 3°) on the same meridian as Syene but north of it, the angle between the overhead direction for that location (*OB* in the figure) and the direction of the sun (*AD* in the figure) was observed to be 1/50 of 360°. The sun is so far from the earth that *SE* and *AD* may be considered parallel. Hence $\sphericalangle SOA$ is 1/50 × 360°. This means that arc *SA* is 1/50 of the circumference of the earth. Eratosthenes estimated the distance from Alexandria to Syene by using the fact that

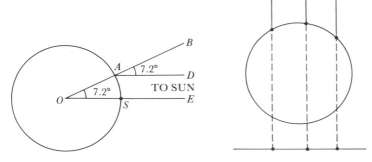

Figure 7.4   Figure 7.5

camel trains, which usually traveled 100 stadia a day, took 50 days to reach Syene. Hence this distance is 5000 stadia and the circumference of the earth is 250,000 stadia. It is believed that a stadium was 157 meters so that Eratosthenes' result is 24,662 miles. This result was far more accurate than all previous estimates.

Eratosthenes wrote the *Geography*, in which he incorporated the methods and results of his measurements and calculations. It includes explanations of the nature and causes of the changes that had taken place on the earth's surface. He also made a map of the world.

Scientific map-making became part of the work of geography. Hipparchus is generally credited with introducing latitude and longitude, though the scheme was known earlier. The use of latitude and longitude, of course, permitted accurate description of locations on the earth. Hipparchus did invent orthographic projection, in which "light rays" from infinity project the earth on a plane (Fig. 7.5). Our view of the moon, for example, is practically orthographic. This method enabled him to map a portion of the earth onto a flat surface.

Ptolemy in his *Planisphaerium*, and probably Hipparchus before him, used the method of stereographic projection. A line from O (Fig. 7.6) through P on the earth's surface is continued until it hits the equatorial plane or a tangent plane at the opposite pole. Hipparchus supposedy used

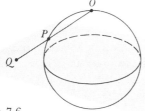

Figure 7.6

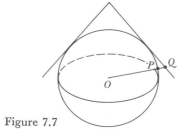

Figure 7.7

the latter and Ptolemy the equatorial plane. Thus points on the sphere are transferred to a plane. In this scheme all points on the map have true direction from the center of the map. Also angles are preserved locally (conformal mapping), though Ptolemy does not mention this. The meridians and the parallels of latitude are therefore at right angles. Circles on the sphere become circles on the plane, but area is not preserved. Ptolemy himself invented conical projection, that is, the projection of a region on the earth from the center of the sphere onto a tangent cone (Fig. 7.7).

In his *Geographia*, a work in eight books, Ptolemy teaches methods of map-making. Chapter 24 of Book I is the oldest work extant that is devoted by title and contents to mapping a sphere onto a plane. The entire *Geographia* is the first atlas and gazeteer. It gives the latitude and longitude of 8000 places on the earth and was the standard reference for hundreds of years.

6. *Mechanics*

The Greeks initiated the science of mechanics. In his *Physics*, Aristotle put together a theory of motion that is the high point of Greek mechanics. Like all of his physics, his mechanics is based on rational, seemingly self-evident principles only slightly checked by or drawn from observation and experiment.

There are, according to Aristotle, two kinds of motion, natural and violent or man-made. The heavenly spheres possess only natural motion, which is circular. As for earthly objects, he taught that a natural motion (as opposed to violent motions caused by throwing or pulling a body from one place to another) arises from the fact that each body has a natural place in the universe where it is in equilibrium or at rest. Heavy bodies have their natural place at the center of the universe, which is the center of the earth. Light bodies, such as gases, have their natural place in the sky. Natural motion results when a body seeks its natural place. In natural motions earthly objects move up or down in straight lines. If an earthly object were not in its natural place it would seek that place as soon as possible. Violent motions, that is, man-made ones, are composed of circular and rectilinear parts. Thus a stone thrown out and up follows a straight line path up and a straight line path down.

Any body in motion is subject to a force and a resistance. In the case of natural motion, the force is the weight of the body, and the resistance comes from the medium in which the body moves. In violent motion the force is applied by the hand or by some mechanism and the resistance comes from the body's weight. Without force there could be no motion; without resistance the motion would be accomplished in an instant. The velocity of any motion, then, depends upon the force and the resistance. These principles can be summed up in modern form by the formula $V \propto F/R$; that is, the velocity depends directly upon the force and inversely upon the resistance.

Since in violent motion the resistance is furnished by the weight, for lighter bodies the resistance, R, is smaller. By the above formula, the velocity V must then be larger; that is, lighter bodies move faster under the same force. In the case of natural motion, the force is the weight, and so heavier bodies fall faster. Since in natural motion the resistance is furnished by the medium, in a vacuum the velocity would be infinite. Hence a vacuum is impossible.

Aristotle had difficulty accounting for some phenomena. To explain the increasing velocities of falling bodies he said that the speed increases as the body comes closer to its natural place because the body moves more jubilantly; but this is not consistent with the speed being dependent on the fixed weight. In the case of an arrow shot from a bow, Aristotle said the arrow continued to move, although no longer in contact with the bow, because the hand or bowstring communicated a power of movement to the air nearby and this air to the next layer of air, and so on. Alternatively the air in front of the arrow is compressed and rushes around to the rear of the arrow to prevent a void and so the arrow is pushed forward. He did not explain the decay of the motive power.

The greatest mathematical physicist of Greek times is Archimedes. He, more than any other Greek writer, tied geometry to mechanics and used geometrical arguments ingeniously to make his proofs. In mechanics he wrote *On the Equilibrium of Planes or The Centers of Gravity of Planes*, a work in two books. By the center of gravity of a body or of a collection of bodies rigidly attached to each other he means, as we do, the point at which the body or collection of bodies can be supported so as to be in equilibrium under the pull of gravity. He starts with postulates about the lever and center of gravity. For example (the numbering follows Archimedes):

1. Equal weights at equal distances are in equilibrium and equal weights at unequal distances are not in equilibrium but incline towards the weight which is at greater distance.

2. If, when weights at certain distances are in equilibrium, something be added to one of the weights, they are not in equilibrium but incline towards the weight to which the addition was made.

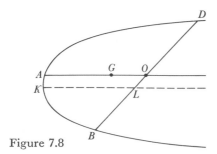

Figure 7.8

5. In figures which are unequal but similar the centers of gravity will be similarly situated....

7. In any figure whose perimeter is concave in the same direction the center of gravity must be within the figure.

He follows these postulates with a number of propositions, some of whose proofs depend upon results in a lost treatise, *On Levers*:

Proposition 4. If two equal weights have not the same center of gravity, the center of gravity of both taken together is at the middle point of the lines joining their centers of gravity.

Propositions 6 and 7. Two magnitudes whether commensurable or incommensurable balance at distances reciprocally proportional to the magnitudes.

Proposition 10. The center of gravity of any parallelogram is the point of intersection of its diagonals.

Proposition 14. The center of gravity of any triangle is at the intersection of the lines drawn from any two vertices to the middle points of the opposite sides respectively.

Book II deals with the center of gravity of a parabolic segment. Among the principal theorems are:

Proposition 4. The center of gravity of any parabolic segment cut off by a straight line lies on the diameter of the segment.

The diameter is AO (Fig. 7.8), where O is the midpoint of BD and AO is parallel to the axis of the parabola. The proof uses results obtained in his *Quadrature of the Parabola*.

Proposition 8. If AO be the diameter of a parabolic segment and G its center of gravity, then $AG = (3/2)GO$.

Work on centers of gravity is found in many books of the Alexandrian Greek period. Examples are Heron's *Mechanica* and Book VIII of Pappus' *Mathematical Collection* (Chap. 5, sec. 7).

The science of hydrostatics—the study of fluid pressure when the fluid is at rest—was founded by Archimedes. In his book *On Floating Bodies* he is interested in the pressure exerted by water on objects placed in it. He gives two postulates. The first is to the effect that the pressure exerted by any part of a fluid on the fluid is downward. The second postulate states that the pressure of the fluid on a body placed in it is exerted upward along the perpendicular through the center of gravity of the body. Some of the theorems he proves in Book I are:

Proposition 2. The surface of any fluid at rest is the surface of a sphere whose center is the same as that of the earth.

Proposition 3. Of solids those which, size for size, are of equal weight with a fluid will, if let down into the fluid, be immersed so that they do not project above the surface but do not sink lower.

Proposition 5. Any solid lighter than a fluid will, if placed in the fluid, be so far immersed that the weight of the solid [in air] will be equal to the weight of the fluid displaced.

Proposition 7. A solid heavier than a fluid will, if placed in it, descend to the bottom of the fluid, and the solid will, when weighed in the fluid, be lighter than its true weight by the weight of the fluid displaced.

This last proposition is believed to be the one Archimedes used to determine the contents of the famous crown (Chap. 5, sec. 3). He must have argued as follows: Let W be the weight of the crown. Take a crown of pure gold and of weight W and weigh it in the fluid. It will weigh less by some amount, F_1, which is the weight of the displaced water. Likewise a weight W of pure silver will displace water of weight F_2 which can be measured by weighing the silver in the water. Then if the original crown contains weight w_1 of gold and weight w_2 of silver, the original crown will displace a weight of water equal to

$$\frac{w_1}{W} F_1 + \frac{w_2}{W} F_2.$$

Let F be the actual weight of the water displaced by the crown. Then

$$\frac{w_1}{W} F_1 + \frac{w_2}{W} F_2 = F$$

or

$$w_1 F_1 + w_2 F_2 = (w_1 + w_2)F$$

or

$$\frac{w_1}{w_2} = \frac{F_2 - F}{F - F_1}.$$

Thus Archimedes was able to determine the ratio of the gold to the silver in the crown without destroying the crown. Vitruvius' account of this story is that Archimedes used *volumes* of displaced water instead of weights. In this case, the F, F_1, and F_2 above are, respectively, the volumes of water displaced by the crown, a weight W of pure gold, and a weight W of pure silver. The same algebra follows but Proposition 7 is not involved. Archimedes did find that the gold had been debased with silver.

To gain some appreciation of the mathematical and physical complexity of the problems treated by Archimedes in this work, we shall cite one of the simple propositions of Book II.

Proposition 2. If a right segment of a paraboloid of revolution whose axis is not greater than $3p/4$ [p is the principal parameter or latus rectum of the generating parabola], and whose specific gravity is less than that of a fluid, be placed in the fluid with its axis inclined to the vertical at any angle, but so that the base of the segment does not touch the surface of the fluid (Fig. 7.9), the segment of the paraboloid will not remain in that position but will return to the position in which its axis is vertical.

The subject Archimedes is treating is the stability of bodies placed in water. He shows under what conditions a body will, when placed in water, either turn to or remain in a position of equilibrium. The problems are clearly idealizations of how ships would behave when obliged to assume different tilts in water.

7. *Optics*

Next to astronomy, optics has been the most constantly pursued and most successful of the mathematical sciences. It was founded by the Greeks. Almost all of the Greek philosophers, beginning with the Pythagoreans, speculated on the nature of light, vision, and color. Our concern, however, is with the mathematical accomplishments. The first of these is the assertion on a priori grounds by Empedocles of Agrigentum in Sicily (c. 490 B.C.) that light travels with finite velocity.

The first systematic treatments that we have are Euclid's *Optics* and *Catoptrica*. The *Optics* is concerned with the problem of vision and with the use of vision to determine sizes of objects. Euclid begins with definitions (which are really postulates), the first of which states (as had Plato) that vision is possible because rays of light emitted by the eye travel along straight lines and impinge on the object seen. Definition 2 states that the figure formed by

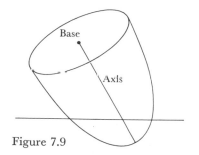

Figure 7.9

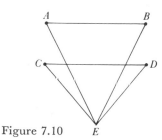

Figure 7.10

the visual rays is a cone that has its vertex in the eye and its base at the extremities of the object seen. Definition 4 says that of two objects the one that determines a greater vertex angle of the cone appears greater. Then in Proposition 8 Euclid proves that the apparent sizes of two equal and parallel objects (*AB* and *CD* in Fig. 7.10) are not proportional to their distances from the eye. Propositions 23 to 27 prove that the eye looking at a sphere really sees less than half of it, and that the contour of what is seen is a circle. Propositions 32 to 37 point out that the eye looking at a circle will see a circle only if the eye is located on the perpendicular to the plane of the circle and the perpendicular strikes the center. Euclid also shows how to calculate the sizes of objects that are seen as images in a plane mirror. There are 58 propositions in the book.

The *Catoptrica* (theory of mirrors) describes the behavior of light rays reflected from plane, concave, and convex mirrors and the effect of this behavior on what we see. Like the *Optics*, it starts with definitions that are really postulates. Theorem 1, the law of reflection, is now fundamental in what is called geometrical optics. It says that the angle *A*, which the incident ray makes with the mirror (Fig. 7.11), equals the angle *B*, which the reflected ray makes with the mirror. It is more customary today to say that $\angle C = \angle D$ and to speak of $\angle C$ as the angle of incidence and $\angle D$ as the angle of reflection. Euclid also proves the law for a ray striking a convex or a concave mirror by substituting a tangent for the mirror at the point where the ray strikes.

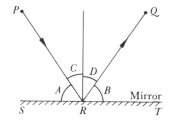

Figure 7.11

From the law of reflection Heron drew an important consequence. If P and Q (Fig. 7.11) are any two points on one side of the line ST, then of all the paths one could follow in going from point P to the line and then to point Q the shortest path is by way of the point R such that the two line segments PR and QR make equal angles with the line—which is exactly the path a light ray takes. Hence, the light ray takes the shortest path in going from P to the mirror to Q. Apparently nature is well acquainted with geometry and employs it to full advantage. This proposition appears in Heron's *Catoptrica*, which also treats concave and convex mirrors and combinations of mirrors.

Any number of works were written on the reflection of light by mirrors of various shapes. Among these are the now lost *Catoptrica* of Archimedes and the two works, both called *On Burning-Mirrors*, by Diocles and Apollonius. Burning-mirrors were undoubtedly concave mirrors shaped like spheres, paraboloids of revolution, and ellipsoids, the last formed by revolving an ellipse about its major axis. Apollonius undoubtedly knew that a paraboloidal mirror will reflect light emanating from the focus into a beam parallel to the axis of the mirror. Conversely, rays coming in parallel to the axis will, after reflection, be concentrated at the focus. The sun's rays thus concentrated produce great heat at the focus; hence the term burning-mirror. This is the property of the paraboloidal mirror that Archimedes is supposed to have used to concentrate the sun's rays on the Roman ships and set them afire. Apollonius also knew the reflection properties of the other conic sections—for example, that all rays emanating from one focus of the ellipsoidal mirror will be reflected to the other focus. He gives the relevant geometrical properties of the ellipse and hyperbola in Book III of his *Conic Sections* (Chap. 4, sec. 12). Later Greeks, Pappus in particular, certainly knew the focusing property of the paraboloid.

The phenomenon of refraction of light—that is, the bending of light rays as they pass through a medium whose properties change from point to point, or the sudden change in the direction of a light ray as it passes from one medium into another, say from air to water, was studied by the Alexandrian Greeks. Ptolemy noted the effect of refraction by the atmosphere on rays coming from the sun and stars and attempted, unsuccessfully, to find the correct law of refraction when light passes from air to water or air to glass. His *Optics*, which deals with both mirrors and refraction, is extant.

8. *Astrology*

Though astrology is not accepted as a science today, in earlier civilizations it did have that standing. The astrology developed by the Alexandrian Greeks of about the second century B.C. was different from the Babylonian astrology of the Assyrian period. The latter merely drew from observations of the positions of the planets some conclusions about the king and affairs of state. There were no computations, and the appearance of the sky at the moment of birth played no role. However, Hellenistic or Alexandrian astrology was personal; it predicted the future and fate of specific individuals on the basis of the computed positions of the sun, moon, and the five planets in the zodiac at the moment of birth. To evaluate these data an enormous body of doctrines was built up.

Certainly the science was taken seriously by the Alexandrian Greeks. Ptolemy wrote a well-known work on the subject, the *Quadripartite* or *Tetrabiblos*, or *Four Books Concerning the Influence of the Stars*, in which he gave rules for astrological predictions that were used for a thousand years.

The importance of astrology in the history of science lies in the motivation it supplied for the study of astronomy, not only in Greece but in India, Arabia, and medieval Europe. Astrology nourished astronomy much as alchemy nourished chemistry. Curiously, errors in astrological predictions were ascribed to errors in astronomy, not to the unreliability of the astrological doctrines.

Alexandrian Greece witnessed the beginnings of the application of mathematics to medicine, peculiarly enough through the medium of astrology. Doctors, called iatromathematicians, employed astrological signs to decide courses of treatment. Galen, the great physician of Greek times, was a firm believer in astrology, perhaps excusably so since Ptolemy, the most renowned astronomer, was also. This connection between mathematics and medicine became stronger in the Middle Ages.

Our account of Greek science, in view of our concern with mathematics, has dealt with the mathematical sciences. The Greeks carried on other investigations, in areas where mathematics, at least at that time, played no role. Moreover, they performed experiments and made observations, the latter particularly in astronomy. Nevertheless their vital achievement is that they established the value of mathematics in the scientific enterprise. The Platonic dialogue *Philebus* first gave expression to the thought that each science is a science only so far as it contains mathematics; this doctrine gained immense support from the Greek accomplishments. Moreover, the Greeks produced ample evidence that nature is mathematically designed. It was their vision of nature and their initiation of the mathematical investigation of nature that inspired the creation of mathematics, in Greek times and in all succeeding centuries.

Bibliography

Apostle, H. G.: *Aristotle's Philosophy of Mathematics*, University of Chicago Press, 1952.

Berry, Arthur: *A Short History of Astronomy*, Dover (reprint), 1961, Chaps. 1–2.

Clagett, Marshall: *Greek Science in Antiquity*, Abelard-Schuman, 1955.

Dreyer, J. L. E.: *A History of Astronomy from Thales to Kepler*, Dover (reprint), 1953, Chaps. 1–9.

Farrington, Benjamin: *Greek Science*, 2 vols., Penguin Books, 1944 and 1949.

Gomperz, Theodor: *Greek Thinkers*, 4 vols., John Murray, 1920.

Heath, Thomas L.: *Greek Astronomy*, J. M. Dent and Sons, 1932.

————: *Aristarchus of Samos*, Oxford University Press, 1913.

————: *The Works of Archimedes*, Dover (reprint), 1953, pp. 189–220 and 253–300.

Jaeger, Werner: *Paideia*, 3 vols., Oxford University Press, 1939–44.

Pannekoek, A.: *A History of Astronomy*, John Wiley and Sons, 1961.

Sambursky, S.: *The Physical World of the Greeks*, Routledge and Kegan Paul, 1956.

Santillana, G. de: *The Origins of Scientific Thought from Anaximander to Proclus, 600 B.C. to 300 A.D.*, University of Chicago Press, 1961.

Sarton, George: *A History of Science*, Harvard University Press, 1952 and 1959, Vols. 1 and 2.

Ver Eecke, Paul: *Euclide, L'Optique et la catoptrique*, Albert Blanchard, 1959.

Wedberg, Anders: *Plato's Philosophy of Mathematics*, Almqvist and Wiksell, 1955.

8

The Demise
of the Greek World

He who understands Archimedes and Apollonius will admire
less the achievements of the foremost men of later times.

G. W. LEIBNIZ

1. *A Review of the Greek Achievements*

Though the Alexandrian Greek civilization lasted until A.D. 640, when it was
finally destroyed by the Mohammedans, it is apparent, from its decreasing
productiveness, that the civilization was already declining in the early
centuries of the Christian era. Before investigating the reasons for the decline,
we shall summarize the achievements and deficiencies of Greek mathematics
and note the problems it left for subsequent generations. The Greeks accomplished so much, and the pursuit of mathematics when taken over by the
Europeans, after minor interpolations by the Hindus and Arabs, was so
completely determined by what the Greeks bequeathed, that it is important
to be clear as to where mathematics stood.

The Greeks are to be credited with making mathematics abstract. This
major contribution is of immeasurable significance and value, for the fact
that the same abstract triangle or algebraic equation may apply to hundreds
of different physical situations has proved to be the secret of the power of
mathematics.

The Greeks insisted on deductive proof. This was indeed an extraordinary step. Of the hundreds of civilizations that have existed a number
did develop some crude arithmetic and geometry. However, no civilization
but the Greeks conceived the idea of establishing conclusions exclusively by
deductive reasoning. The decision to require deductive proof is entirely at
odds with the methods mankind has utilized in all other fields; it is, in fact,
almost irrational, since so much highly reliable knowledge is acquired by
experience, induction, reasoning by analogy, and experimentation. But the
Greeks wanted truths, and saw that they could obtain them only by the
unquestionable methods of deductive reasoning. They also realized that to

secure truths they had to start from truths, and be sure not to assume any unwarranted facts. Hence they stated all their axioms explicitly and in addition adopted the practice of stating them at the very outset of their work so that they could be examined critically at once.

Beyond conceiving this highly remarkable plan to establish secure knowledge, the Greeks showed a sophistication one could hardly expect from innovators. Their recognition that the concepts must be free of contradiction and that one cannot build a consistent structure by working with nonexistent figures (such as a ten-faced regular polyhedron) shows an almost superhuman and certainly unprecedented keenness of thought. As we now know, their method of establishing the existence of concepts, so they could deal with them, was to demonstrate their constructibility with straightedge and compass.

The power of the Greeks to divine theorems and proofs is attested to by the fact that Euclid's *Elements* contains 467 propositions and Apollonius' *Conic Sections* contains 487, all derived from the 10 axioms in the *Elements*. No doubt secondary in importance, and perhaps secondary in intent, is the coherence that the deductive structures provided. Were the same results obtainable from numerous different—though equally reliable—sets of axioms, they might have presented far less manageable and assimilable knowledge.

The Greek contribution to the content of mathematics—plane and solid geometry, plane and spherical trigonometry, the beginnings of the theory of numbers, the extension of Babylonian and Egyptian arithmetic and algebra —is enormous, especially in view of the small number of people involved and the few centuries of extensive activity. To these contributions we must add the geometrical algebra, which awaited only the recognition of irrational numbers and translation into symbolic language to become the basis of considerable elementary algebra. The treatment of curvilinear figures by the method of exhaustion, though part of their geometry, also warrants special mention because it is the beginning of the calculus.

An equally vital contribution and inspiration to later generations was the Greek conception of nature. The Greeks identified mathematics with the reality of the physical world and saw in mathematics the ultimate truth about the structure and design of the universe. They founded the alliance between mathematics and the disinterested study of nature, which has since become the very basis of modern science. Moreover, they went far enough in rationalizing nature to establish firmly the conviction that the universe is indeed mathematically designed, that it is controlled, lawful, and intelligible to man.

The aesthetic appeal of mathematics was by no means overlooked. In Greek times the subject was also valued as an art; beauty, harmony, simplicity, clarity, and order were recognized in it. Arithmetic, geometry, and astronomy were considered the art of the mind and music for the soul. Plato delighted in geometry; Aristotle would not divorce mathematics from aesthetics, for order and symmetry were to him important elements of beauty, and these he found in mathematics. Indeed, rational and aesthetic, as well as moral, interests can hardly be separated in Greek thought. Repeatedly one reads that the sphere has the most beautiful shape of all bodies and is therefore divine and good. The circle shared aesthetic appeal with the sphere; it therefore seemed obvious that the circle should be the path of those bodies that represented the changeless, eternal order of the heavens, whereas straight-line motion prevailed on the imperfect earth. There is no doubt that it was the aesthetic appeal of the subject that caused Greek mathematicians to carry the exploration of particular topics beyond their use in understanding the physical world.

2. The Limitations of Greek Mathematics

Despite its marvelous achievements, Greek mathematics was flawed. Its limitations indicate the avenues of progress that were yet to be opened up.

The first limitation was the inability to grasp the concept of the irrational number. This meant not only a restriction of arithmetic and algebra but a turn to and an emphasis on geometry, because geometrical thinking avoided explicit confrontation of the irrational as a number. Had the Greeks faced the irrational number, they might have furthered the development of arithmetic and algebra; and even if they themselves had not done so, they would not have hindered later generations, which were induced to think that only geometry offered a secure foundation for the treatment of any magnitude whose values might include irrationals. Archimedes, Heron, and Ptolemy started to work with irrationals as numbers, but did not alter the tenor of Greek mathematics or the subsequent impress of Greek thought. The Greek concentration on geometry blurred the vision of later generations by masking the intimate correspondence between geometric and arithmetic concepts and operations. The failure to define, accept, and conceptualize the irrational as a number forced a distinction between number and magnitude. Consequently algebra and geometry were regarded as unrelated disciplines.

Had the Greeks been less concerned to be logical and precise they might have casually accepted and operated with irrational numbers as had the Babylonians, and as civilizations succeeding the Greeks did. But the intuitive basis of the idealization was not clear, and the logical construction was not quite within their power. The Greek virtue of insisting on exact concepts and proofs was a defect so far as creative mathematics was concerned.

The restriction of rigorous mathematics (apart from the theory of numbers) to geometry imposed another serious disadvantage. The use of geometric methods led to more and more complicated proofs as mathematics was extended, particularly in the area of solid geometry. Moreover, even in the simpler proofs there is a lack of general methods, which is clear to us now that we have analytic geometry and the calculus. When one considers how hard Archimedes worked to find the area of a parabolic segment or the area under an arc of his spiral and compares this with the modern calculus treatment, one appreciates the effectiveness of the calculus.

Not only did the Greeks restrict mathematics largely to geometry; they even limited that subject to figures that could be obtained from the line and circle. Accordingly, the only surfaces admitted were those obtained by revolving lines and circles about an axis, for example the cylinder, cone, and sphere, formed by the revolution of a rectangle, triangle, and circle, respectively, about a line. A few exceptions were made: the plane, which is the analogue of a line; the prism, which is a special cylinder; and the pyramid, which results from decomposition of a prism. The conic sections were introduced by cutting cones by a plane. Curves such as the quadratrix of Hippias, the conchoid of Nicomedes, and the cissoid of Diocles were kept on the fringe of geometry; they were called mechanical, rather than geometrical.

Pappus' classification of curves shows an attempt to keep within fixed bounds. The Greeks, according to Pappus, distinguished curves as follows: Plane loci or plane curves were those constructed from straight lines and circles; the conics were called solid loci because they originated from the cone; the linear curves, such as quadratrices, conchoids, cissoids, and spirals constituted the third class. Similarly, they distinguished among plane, solid, and linear problems. Plane problems were solved by means of straight lines and circles; solid problems were solved by one or more of the conic sections. Problems that could not be solved by means of straight lines, circles, or conics were called linear, because they used lines (curves) having a more complicated or less natural origin than the others. Pappus stressed the importance of solving problems by means of plane or solid loci because then the criterion for the possibility of a real solution can be given.

Why did the Greeks limit their geometry to line and circle and figures readily derived from them? One reason was that this solved the problem of establishing the existence of geometrical figures. As we saw, Aristotle, in particular, pointed out that we must be sure that the concepts introduced are not self-contradictory; that is, they must be shown to exist. To settle this point, the Greeks, in principle at least, admitted only those concepts that were constructible. Line and circle were accepted as constructible in the postulates but all other figures had to be constructed with line and circle.

However, the use of constructions to establish existence was not carried over to figures in three dimensions. Here the Greeks apparently accepted what was intuitively clear, for example the existence of figures of revolution such as sphere, cylinder, and cone. Sections of these figures by planes produced curves such as the conic sections; thus even figures in the plane whose existence was not established were accepted—but reluctantly. Descartes notes this point near the beginning of Book II of *La Géométrie*:

"It is true that the conic sections were never freely received into ancient geometry. . . ."

Another reason for the limitation to line, circle, and figures derived from them stems from Plato, according to whom ideas had to be clear to be acceptable. While the whole number seemed to be acceptable as a clear idea in itself, even though never explicitly defined by the Greeks, the geometric figures had to be made precise. Lines and circles and figures derived from them were clear, whereas curves introduced by mechanical instruments (other than straightedge and compass) were not, and so were not admissible. The restriction to clear figures produced a simple, well-arranged, harmonious, and beautiful geometry.

By insisting on a unity, a completeness, and a simplicity for their geometry, and by separating speculative thought from utility, classical Greek geometry became a limited accomplishment. It narrowed people's vision and closed their minds to new thoughts and methods. It carried within itself the seeds of its own death. The narrowness of its field of action, the exclusiveness of its point of view, and the aesthetic demands on it might have arrested its development, had not the influences of the Alexandrian civilization broadened the outlook of Greek mathematicians.

The philosophical doctrines of the Greeks limited mathematics in another way. Throughout the classical period they believed that man does not create the mathematical facts: they preexist. He is limited to ascertaining and recording them. In the *Theaetetus* Plato compares the search for knowledge to a bird-hunt pursued in an aviary. The birds, already captive, need only to be seized. This belief about the nature of mathematics did not prevail.

The Greeks failed to comprehend the infinitely large, the infinitely small, and infinite processes. They "shrank before the silence of infinite spaces." The Pythagoreans associated good and evil with the limited and unlimited, respectively. Aristotle says the infinite is imperfect, unfinished, and therefore unthinkable; it is formless and confused. Only as objects are delimited and distinct do they have a nature.

To avoid any assertion about the infinitude of the straight line, Euclid says a line segment (he uses the word "line" in this sense) can be extended as far as necessary. Unwillingness to involve the infinitely large is seen also in Euclid's statement of the parallel axiom. Instead of considering two lines that extend to infinity and giving a direct condition or assumption under which parallel lines might exist, his parallel axiom gives a condition under which two lines will meet at some finite point.

The concept of the infinitely small is involved in the relation of points to a line or the relation of the discrete to the continuous, and Zeno's paradoxes may have caused the Greeks to shy away from this subject. The relationship of point to line bothered the Greeks and led Aristotle to separate the two. Though he admits points are on lines, he says that a line is not made up of points and that the continuous cannot be made up of the discrete (Chap. 3, sec. 10). This distinction contributed also to the presumed need for separating number from geometry, since to the Greeks numbers were discrete and geometry dealt with continuous magnitudes.

Because they feared infinite processes they missed the limit process. In approximating a circle by a polygon they were content to make the difference smaller than any given quantity, but something positive was always left over. Thus the process remained clear to the intuition; the limit process, on the other hand, would have involved the infinitely small.

3. The Problems Bequeathed by the Greeks

The limitations of Greek mathematical thought almost automatically imply the problems the Greeks left to later generations. The failure to accept the irrational as a number certainly left open the question of whether number could be assigned to incommensurable ratios so that these could be treated arithmetically. With the irrational number, algebra could also be extended. Instead of turning to geometry to solve quadratic and other equations that might have irrational roots, these problems could be treated in terms of number, and algebra could develop from the stage where the Egyptians and Babylonians or where Diophantus, who refused to consider irrationals, left it.

Even for whole numbers and ratios of whole numbers, the Greeks gave no logical foundation; they supplied only some rather vague definitions, which Euclid states in Books VII to IX of the *Elements*. The need for a logical foundation of the number system was aggravated by the Alexandrians' free use of numbers, including irrationals; in this respect they merely continued the empirical traditions of the Egyptians and Babylonians. Thus the Greeks bequeathed two sharply different, unequally developed branches of mathematics. On the one hand, there was the rigorous, deductive, systematic geometry and on the other, the heuristic, empirical arithmetic and its extension to algebra.

The failure to build a deductive algebra meant that rigorous mathematics was confined to geometry; indeed, this continued to be the case as late as the seventeenth and eighteenth centuries, when algebra and the calculus had already become extensive. Even then rigorous mathematics still meant geometry.

The limitation of Euclidean geometry to concepts constructible with straightedge and compass left mathematics with two tasks. The first was specific, to show that one could square the circle, trisect any angle, and double the cube with straightedge and compass. These three problems exerted an enormous fascination and even today intrigue people although, as we shall see, they were disposed of in the nineteenth century.

The second task was to broaden the criteria for existence. Constructibility as a means of proving existence turned out to be too restrictive of the concepts with which mathematics should (and later did) deal. Moreover, since some lengths are not constructible, the Euclidean line is incomplete; that is, strictly it does not contain those lengths that are not constructible. To be internally complete and more useful in the study of the physical world, mathematics had to free itself of a narrow technique for establishing existence.

As we saw, the attempt to avoid a direct affirmation about infinite parallel straight lines caused Euclid to phrase the parallel axiom in a rather complicated way. He realized that, so worded, this axiom lacked the self-evidency of the other nine axioms, and there is good reason to believe that he avoided using it until he had to. Many Greeks tried to find substitute axioms for the parallel axiom or to prove it on the basis of the other nine. Ptolemy wrote a tract on this subject; Proclus, in his commentary on Euclid, gives Ptolemy's attempt to prove the parallel postulate and also tries to prove it himself. Simplicius cites others who worked on the problem and says further that people "in ancient times" objected to the use of the parallel postulate.

Closely related to the problem of the parallel postulate is the problem of whether physical space is infinite. Euclid assumes in Postulate 2 that a straight-line segment can be extended as far as necessary; he uses this fact, but only to obtain a larger finite length—for example in Book I, Propositions 11, 16, and 20. For these theorems Heron gave new proofs that avoid extending the lines, in order to meet the objection of anyone who would deny that the space was available for the extension. Aristotle had considered the question of whether space is infinite and gave six nonmathematical arguments to prove that it is finite; he foresaw that this question would be troublesome.

Another major problem left to posterity was the calculation of areas bounded by curves and volumes bounded by surfaces. The Greeks, notably Eudoxus and Archimedes, had not only tackled such problems but had, as we have seen, made significant progress by using the method of exhaustion. But the method was deficient in at least two respects. First, each problem called for some ingenious scheme of approximating the area or volume in question; but human inventiveness simply could not have handled in this manner the areas and volumes that had to be calculated later. Second, the Greek result was usually the equivalence of the desired area or volume to the area or volume of some simpler figure whose size still was not known quantitatively. But it is quantitative knowledge that applications require.

4. The Demise of the Greek Civilization

Beginning approximately with the start of the Christian era, the vitality of Greek mathematical activity declined steadily. The only important contributions in the new era were those of Ptolemy and Diophantus. The great commentators Pappus and Proclus warrant attention, but they merely close the record. The decline of this civilization, which for five or six centuries made contributions far surpassing in extent and brilliance those of any other, calls for explanation.

Unfortunately, mathematicians are as subject to the forces of history as the lowliest peasant. One has only to familiarize himself with the most superficial facts about the political history of the Alexandrian Greeks of the Christian era to see that not only mathematics but every cultural activity was destined to suffer. As long as the Alexandrian Greek civilization was ruled by the dynasty of the Ptolemies, it flourished. The first disaster was the advent of the Romans, whose entire role in the history of mathematics was that of an agent of destruction.

Before discussing their impact on the Alexandrian Greek civilization, let us note a few facts about Roman mathematics and the nature of the Roman civilization. Roman mathematics hardly warrants mention. The period during which the Romans figured in history extends from about 750 B.C. to A.D. 476, roughly the same period during which the Greek civilization flourished. Moreover, as we shall see, from at least 200 B.C. onward, the Romans were in close contact with the Greeks. Yet in all of the eleven hundred years there was not one Roman mathematician; apart from a few details this fact in itself tells virtually the whole story of Roman mathematics.

The Romans did have a crude arithmetic and some approximate geometric formulas that were later supplemented by borrowings from the Alexandrian Greeks. Their symbols for the whole numbers are familiar to us. To calculate with whole numbers, they used various forms of the abacus. Calculations were also done with the fingers and with the aid of specially prepared tables.

Roman fractions were in base 12. Special symbols and words were used for 1/12, 2/12, ..., 11/12, 1/24, 1/36, 1/48, 1/96, ... The origin of the base 12 may be the relation of the lunar month to the year. The unit of weight, incidentally, was the *as*; one twelfth of this was the *uncia*, from which we get our ounce and inch.

The principal use of arithmetic and geometry by the Romans was in surveying, to fix city boundaries and plots of land for homes and temples. Surveyors could calculate most of the quantities they sought using only simple instruments and congruent triangles.

We do owe to the Romans an improvement in the calender. Up to the time of Julius Caesar (100–44 B.C.), the basic Roman year had 12 months, totalling 355 days. An intercalary month of 22 or 23 days was used every other year so that the average year consisted of 366 1/4 days. To improve this calendar, Caesar called in Sosigenes, an Alexandrian, who advised a year of 365 days and a leap year every 4 years. The Julian calendar was adopted in 45 b.c.

From about 50 b.c. on, the Romans wrote their own technical books; all of the basic material, however, was taken from Greek sources. The most famous of these technical works is Vitruvius' ten books on architecture, which date from 14 b.c. Here, too, the material is Greek Peculiarly, Vitruvius says that the three greatest mathematical discoveries are the 3, 4, 5 right triangle, the irrationality of the diagonal of the unit square, and the Archimedean solution of the crown problem. He does give other facts involving mathematics, such as the proportions of the parts of the ideal human body, some harmonious arithmetical relationships, and arithmetic facts about the capacities of catapults.

Among the Romans the term "mathematics" came into disrepute because the astrologers were called *mathematicii*, and astrology was condemned by the Roman emperors. The emperor Diocletian (a.d. 245–316) distinguished between geometry and mathematics. The former was to be learned and applied in the public service; but the "art of mathematics"—that is, astrology—was damnable and forbidden in its entirety. The "code of mathematics and evil deeds," the Roman law forbidding astrology, was also applied in Europe during the Middle Ages. The Roman and Christian emperors nonetheless employed astrologers at their courts on the chance that there might be some truth in their prophecies. The distinction between the terms "mathematician" and "geometer" lasted until well past the Renaissance. Even in the seventeenth and eighteenth centuries, "geometer" meant what we mean by "mathematician."

The Romans were practical people and they boasted of their practicality. They undertook and completed vast engineering projects—viaducts, magnificent roads that survive even today, bridges, public buildings, and land surveys—but they refused to consider any ideas beyond the particular concrete applications they were making at the moment. The Roman attitude toward mathematics is stated by Cicero: "The Greeks held the geometer in the highest honor; accordingly, nothing made more brilliant progress among them than mathematics. But we have established as the limit of this art its usefulness in measuring and counting."

The Roman emperors did not support mathematics as did the Ptolemys of Egypt. Nor did the Romans understand pure science. Their failure to develop mathematics is striking, because they ruled a worldwide empire and because they did seek to solve practical problems. The lesson one can learn from the history of the Romans is that people who scorn the highly theoretical work of mathematicians and scientists and decry its usefulness are ignorant of the manner in which important practical developments have arisen.

Let us turn to the role the Romans played in the political and military history of Greece. After having secured control of central and northern Italy, they conquered the Greek cities in southern Italy and Sicily. (Archimedes, we recall, contributed to the defense of Syracuse when the Romans attacked the city and was killed by a Roman soldier.) The Romans conquered Greece proper in 146 B.C., and Mesopotamia in 64 B.C. By intervening in the internal strife in Egypt between Cleopatra, the last of the Ptolemy dynasty, and her brother, Caesar managed to secure a hold on the country. In 47 B.C., Caesar set fire to the Egyptian fleet riding at anchor in the harbor of Alexandria; the fire spread to the city and burned the library. Two and a half centuries of book-collecting and half a million manuscripts, which represented the flower of ancient culture, were wiped out. Fortunately, an overflow of books that could no longer be accommodated in the overcrowded library was by this time stored in the temple of Serapis and these were not burned. Also, Attalus III of Pergamum, who died in 133 B.C., had left his great collection of books to Rome. Mark Anthony gave this collection to Cleopatra and it was added to the books in the temple. The total collection was once more enormous.

The Romans returned at the death of Cleopatra in 31 B.C., and from that time on controlled Egypt. Their interest in extending their political power did not include spreading their culture. The subjugated areas became colonies, from which great wealth was extracted by expropriation and taxation. Since most of the Roman emperors were self-seekers, they ruined every country they controlled. When uprisings occurred, as they did, for example, in Alexandria, the Romans did not hesitate to starve and, when finally victorious, to kill off thousands of inhabitants.

The late history of the Roman Empire is also relevant. The Emperor Theodosius (ruled 379–95) divided the extensive empire between his two sons, Honorius, who was to rule Italy and western Europe, and Arcadius, who was to rule Greece, Egypt, and the Near East. The western part was conquered by the Goths in the fifth century A.D. and its subsequent history belongs to that of medieval Europe. The eastern part, which included Egypt (for a while), Greece, and what is now Turkey, preserved its independence until it was conquered by the Turks in 1453. Since the Eastern Roman Empire, known also as the Byzantine Empire, included Greece proper, Greek culture and works were to some extent preserved.

From the standpoint of the history of mathematics, the rise of Christianity had unfortunate consequences. Though the Christian leaders adopted many Greek and Oriental myths and customs with the intent of making Christianity more acceptable to converts, they opposed pagan learning and ridiculed mathematics, astronomy, and physical science; Christians were forbidden to contaminate themselves with Greek learning. Despite cruel persecution by the Romans, Christianity spread and became so powerful that the emperor Constantine (272–337) was obliged to consign it a privileged position in the Roman Empire. The Christians were now able to effect even greater destruction of Greek culture. The emperor Theodosius proscribed the pagan religions and, in 392, ordered that the Greek temples be destroyed. Many of these were converted to churches, though often still adorned with Greek sculpture. Pagans were attacked and murdered throughout the empire. The fate of Hypatia, an Alexandrian mathematician of note and the daughter of Theon of Alexandria, symbolizes the end of the era. Because she refused to abandon the Greek religion, Christian fanatics seized her in the streets of Alexandria and tore her to pieces.

Greek books were burned by the thousands. In the year that Theodosius banned the pagan religions, the Christians destroyed the temple of Serapis, which still housed the only extensive collection of Greek works. It is estimated that 300,000 manuscripts were destroyed. Many other works written on parchment were expunged by the Christians so that they could use the parchment for their own writings. In 529 the Eastern Roman emperor Justinian closed all the Greek schools of philosophy, including Plato's Academy. Many Greek scholars left the country and some—for example, Simplicius—settled in Persia.

The final blow to Alexandria was the conquest of Egypt by the upsurging Moslems in A.D. 640. The remaining books were destroyed on the ground given by Omar, the Arab conqueror: "Either the books contain what is in the Koran, in which case we do not have to read them, or they contain the opposite of what is in the Koran, in which case we must not read them." And so for six months the baths of Alexandria were heated by burning rolls of parchment.

After the capture of Alexandria by the Mohammedans, the majority of the scholars migrated to Constantinople, which had become the capital of the Eastern Roman Empire. Though no activity along the lines of Greek thought could flourish in the unfriendly Christian atmosphere of Byzantium, this flux of scholars and their works to comparative safety increased the treasury of knowledge that was to reach Europe eight hundred years later.

It is perhaps pointless to contemplate what might have been. But one cannot help observe that the Alexandrian Greek civilization ended its active scientific life on the threshold of the modern age. It had the unusual combination of theoretical and practical interests that proved so fertile a thousand years later. Until the last few centuries of its existence, it enjoyed freedom of thought, which is also essential to a flourishing culture. And it tackled and made major advances in several fields that were to become all-important in the Renaissance: quantitative plane and solid geometry; trigonometry; algebra; calculus; and astronomy.

It has often been said that man proposes and God disposes. It is more accurate to say of the Greeks that God proposed them and man disposed of them. The Greek mathematicians were wiped out. But the fruits of their work did reach Europe in a way we have yet to relate.

Bibliography

Cajori, Florian: *A History of Mathematics*, Macmillan, 1919, pp. 63–68.
Gibbon, Edward: *The Decline and Fall of the Roman Empire* (many editions), Chaps. 20, 21, 28, 29, 32, 34.
Parsons, Edward Alexander: *The Alexandrian Library*, The Elsevier Press, 1952.

9
The Mathematics
of the Hindus and Arabs

> As the sun eclipses the stars by its brilliancy, so the man of
> knowledge will eclipse the fame of others in assemblies of the
> people if he proposes algebraic problems, and still more if he
> solves them.   BRAHMAGUPTA

1. *Early Hindu Mathematics*

The successors of the Greeks in the history of mathematics were the Hindus
of India. Though Hindu mathematics became significant only after it was
influenced by Greek achievements, there were earlier, indigenous developments that are worth noting.

The Hindu civilization dates back to at least 2000 B.C. but as far as we
know there was no mathematics prior to 800 B.C. During the Śulvasūtra
period, from 800 B.C. to A.D. 200, the Hindus did produce some primitive
mathematics. There were no separate mathematical documents but a few
facts can be gleaned from other writings and from coins and inscriptions.

From about the third century B.C., number symbols appear, which
varied considerably from one century to another. Typical are the Brahmi
symbols:

$$- \quad = \quad \equiv \quad \text{Y} \quad \text{Γ} \quad \text{6} \quad \text{7} \quad \text{5} \quad \text{?} \quad \alpha \quad \text{o} \quad \text{J} \quad \text{Ѫ} \quad \text{J} \quad \text{ᒋ}$$
$$1 \quad 2 \quad 3 \quad 4 \quad 5 \quad 6 \quad 7 \quad 8 \quad 9 \quad 10 \quad 20 \quad 30 \quad 40 \quad 50 \quad 60.$$

What is significant in this set is a separate, single symbol for each number
from 1 to 9. There was no zero and no positional notation as yet. The wisdom
of using separate symbols was undoubtedly not foreseen by this mathematically illiterate people; the practice may have arisen from using the first
letters of the words for these numbers.

Among the religious writings was a class called Śulvasūtras (rules of the
cord), containing instructions for the construction of altars. In one of the
Śulvasūtras of the fourth or fifth century B.C. an approximation to $\sqrt{2}$ is

given, but there is no indication that it is just an approximation. Almost nothing else is known about the arithmetic of the period.

The geometry of this ancient Hindu period is somewhat better known. The rules contained in the Śulvasūtras prescribed conditions for the shapes and sizes of altars. The three shapes most commonly used were square, circle, and semicircle; and no matter which of these shapes was used, the areas had to be the same. Hence the Hindus had to construct circles equal in area to the squares, or twice as large so that the semicircle could be used. Another shape used was an isosceles trapezoid; here it was permissible to use a similar shape. Hence additional geometrical problems were involved in constructing the similar figure.

In designing the permissible altars the Hindus got to know a few basic geometrical facts, such as the Pythagorean theorem, in the form: "The diagonal of an oblong [rectangle] produces by itself both the areas which the two sides of the oblong produce separately." In general, the geometry of this period consisted of a disconnected set of approximate, verbal rules for areas and volumes. Āpastamba (4th or 5th cent. B.C.) gave a construction for a circle equal in area to a square which, in effect, used the value of 3.09 for π; but he thought the construction was exact. In all of the geometry of this early period there were no proofs; the rules were empirical.

2. *Hindu Arithmetic and Algebra of the Period* A.D. 200–1200

The second period of Hindu mathematics, the high period, may be roughly dated from about A.D. 200 to 1200. During the first part of this period, the civilization at Alexandria definitely influenced the Hindus. Varāhamihira (*c.* 500), an astronomer, says, "The Greeks, though impure [anyone having a different faith is impure], must be honored, since they were trained in the sciences and therein excelled others. What, then, are we to say of a Brahman if he combines with his purity the height of science?" The geometry of the Hindus was certainly Greek, but they did have a special gift for arithmetic. As to algebra, they may have borrowed from Alexandria and possibly directly from Babylonia; but here, too, they went far on their own. India was also somewhat indebted to China.

The most important mathematicians of the second period are Āryabhata (b. 476), Brahmagupta (b. 598), Mahāvīra (9th cent.), and Bhāskara (b. 1114). Most of their work and that of Hindu mathematicians generally was motivated by astronomy and astrology. In fact, there were no separate mathematical texts; the mathematical material is presented in chapters of works on astronomy.

The Hindu methods of writing numbers up to A.D. 600 were numerous and even included using words or syllables for number symbols. By 600 they reverted to the older Brahmi symbols, though the precise form of these symbols changed throughout the period. Positional notation in base 10, which had been in limited use for about a hundred years, now became standard. Also, the zero, which the Alexandrian Greeks had earlier used only to denote the absence of a number, was treated as a complete number. Mahāvīra says that multiplication of a number by 0 gives 0 and that subtracting 0 does not diminish a number. He also says, however, that a number divided by 0 remains unchanged. Bhāskara, in talking about a fraction whose denominator is 0, says that this fraction remains the same though much be added and subtracted, just as no change takes place in the immutable Deity when worlds are created and destroyed. A number divided by zero, he adds, is termed an infinite quantity.

For fractions in astronomy, the Hindus used sexagesimal positional notation. For other purposes they used a ratio of integers but without the bar, thus: $\frac{3}{4}$.

The arithmetic operations of the Hindus were much like ours. For example, Mahāvīra gives our rule for division by a fraction: Invert and multiply.

The Hindus introduced negative numbers to represent debts; in such situations, positive numbers represented assets. The first known use is by Brahmagupta about 628; he also states the rules for the four operations with negative numbers. Bhāskara points out that the square root of a positive number is twofold, positive and negative. He brings up the matter of the square root of a negative number but says that there is no square root because a negative number is not a square. No definitions, axioms, or theorems are given.

The Hindus did not unreservedly accept negative numbers. Even Bhāskara, while giving 50 and −5 as two solutions of a problem, says, "The second value is in this case not to be taken, for it is inadequate; people do not approve of negative solutions." However, negative numbers gained acceptance slowly.

The Hindus took another great step in arithmetic by facing up to the problem of irrational numbers; that is, they started to operate with these numbers by correct procedures, which, though not proven generally by them, at least permitted useful conclusions to be drawn. For example, Bhāskara says, "Term the sum of two irrationals the greater surd; and twice their product the lesser one. The sum and difference of them reckoned like integers are so." He then shows how to add them, as follows: Given the irrationals $\sqrt{3}$ and $\sqrt{12}$,

$$\sqrt{3} + \sqrt{12} = \sqrt{(3 + 12) + 2\sqrt{3 \cdot 12}} = \sqrt{27} = 3\sqrt{3}.$$

The general principle, in our notation, is

(1) $$\sqrt{a} + \sqrt{b} = \sqrt{(a + b) + 2\sqrt{ab}}.$$

We should note the phrase "reckoned like integers" in the above quotation. Irrationals were treated as though they possessed the same properties as integers. Thus if we had integers c and d we would certainly write

$$(2) \qquad c + d = \sqrt{(c+d)^2} = \sqrt{c^2 + d^2 + 2cd}.$$

Now if $c = \sqrt{a}$ and $d = \sqrt{b}$, then (2) is exactly (1).

Bhāskara also gives the following rule for the sum of two irrationals: "The root of the quotient of the greater irrational divided by the lesser one being increased by one; the sum being squared and multiplied by the smaller irrational quantity is the sum of the two surd roots." This means, for example,

$$\sqrt{3} + \sqrt{12} = \sqrt{\left(\sqrt{\frac{12}{3}} + 1\right)^2 \cdot 3},$$

which yields $3\sqrt{3}$. He also gives rules for the multiplication, division, and square root of irrational expressions.

The Hindus were less sophisticated than the Greeks in that they failed to see the logical difficulties involved in the concept of irrational numbers. Their interest in calculation caused them to overlook philosophic distinctions, or distinctions based on principles that in Greek thought were fundamental. But in blithely applying to irrationals procedures like those used for rationals, they helped mathematics progress. Moreover, their entire arithmetic was completely independent of geometry.

The Hindus also made some progress in algebra. They used abbreviations of words and a few symbols to describe operations. As in Diophantus, there was no symbol for addition; a dot over the subtrahend indicated subtraction; other operations were called for by key words or abbreviations; thus *ka* from the word *karana* called for the square root of what followed. For the unknowns, when more than one was involved, they had words that denoted colors. The first one was called the unknown and the remaining ones black, blue, yellow, and so forth. The initial letter of each word was also used as a symbol. This symbolism, though not extensive, was enough to classify Hindu algebra as almost symbolic and certainly more so than Diophantus' syncopated algebra. Problems and solutions were written in this quasi-symbolic style. Only the steps were given; no reasons or proofs accompanied them.

The Hindus recognized that quadratic equations have two roots and included negative roots as well as irrational roots. The three types of quadratics $ax^2 + bx = c$, $ax^2 = bx + c$, $ax^2 + c = bx$, a, b, c positive, separately treated by Diophantus, were treated as one case $px^2 + qx + r = 0$, because the Hindus allowed some coefficients to be negative. They used the method of completing the square, which of course was not new with them. Since they did not recognize square roots for negative numbers, they could not solve all quadratics. Mahāvīra also solves $x/4 + 2\sqrt{x} + 15 = x$, which arises from a verbal problem.

In indeterminate equations the Hindus advanced beyond Diophantus. These equations arose in problems of astronomy; the solutions showed when certain constellations would appear in the heavens. The Hindus sought all integral solutions, whereas Diophantus sought one rational solution. The method of obtaining integral solutions of $ax \pm by = c$, where a, b, and c are positive integers, was introduced by Āryabhata and improved by his successors. It is the same as the modern one. Let us consider $ax + by = c$. If a and b have a common factor m that does not divide c, then no integral solutions are possible because the left side is divisible by m whereas the right side is not. If a, b, and c have a common factor it is removed, and, in the light of the preceding remark, one need consider only the case where a and b are relatively prime. Now Euclid's algorithm for finding the greatest common divisor of two integers a and b, with $a > b$, calls first for dividing b into a so that $a = a_1 b + r$, where a_1 is the quotient and r is the remainder. Hence $a/b = a_1 + r/b$. This can be expressed as

(3) $a/b = a_1 + 1/(b/r).$

The second step in Euclid's algorithm is to divide r into b. Then $b = a_2 r + r_1$ or $b/r = a_2 + r_1/r$. If we insert this value of b/r in (3), we can write

(4) $$\frac{a}{b} = a_1 + \cfrac{1}{a_2 + \cfrac{1}{r/r_1}}.$$

The continuation of Euclid's algorithm leads to what is called a continued fraction

$$\frac{a}{b} = a_1 + \cfrac{1}{a_2 + \cfrac{1}{a_3 + \cdots}},$$

which is also written as

$$\frac{a}{b} = a_1 + \frac{1}{a_2 +} \quad \frac{1}{a_3 +} \cdots.$$

The process applies also when $a < b$. In this case a_1 is zero, and then the process is continued as before. For the case of a and b integral, the continued fraction terminates.

The fractions obtained by stopping with the first, second, third, and generally nth quotient are called the first, second, third, and nth convergent, respectively. Since in the case of a and b integral the continued fraction terminates, there is a convergent that just precedes the exact expression for a/b. If p/q is the value of this convergent, then one can show that

$$aq - bp = \pm 1.$$

Let us consider the positive value. We can now return to our original indeterminate equation, and since $aq - bp = 1$ we can write

$$ax + by = c(aq - bp)$$

and by rearranging terms obtain

$$\frac{cq - x}{b} = \frac{y + cp}{a}.$$

If we let t represent each of these fractions we have

(5) $x = cq - bt$ and $y = at - cp.$

We may now assign integral values to t, and since all the other quantities are integers, we thereby obtain integral values of x and y. The minor modifications to take care of the cases where $aq - bp = -1$, or when the original equation is $ax - by = c$, are readily devised. Brahmagupta gave solution (5), though not of course in terms of general letters a, b, p, and q.

The Hindus also worked with indeterminate quadratic equations. They solved the type

$$y^2 = ax^2 + 1, \qquad a \text{ not a perfect square,}$$

and recognized that this type was fundamental in treating

$$cy^2 = ax^2 + b.$$

The methods involved are too specialized to warrant consideration here.

It is noteworthy that they found pleasure in many mathematical problems and stated them in fanciful or verse form, or in some historical context, to please and attract people. The original reason for doing so may have been to aid the memory, because the old Brahman practice was to trust to memory and avoid writing things down.

Algebra was applied to the usual commercial problems: the calculation of interest, discount, division of the profits of a partnership, and the allocation of shares in an estate; but astronomy was the main application.

3. *Hindu Geometry and Trigonometry of the Period* A.D. 200–1200

Geometry during this period showed no notable advances; it consisted of formulas (correct and incorrect) for areas and volumes. Many of these, such as Heron's formula for the area of a triangle and Ptolemy's theorem, came from the Alexandrian Greeks. Sometimes the Hindus were aware that a formula was only approximately correct and sometimes they were not. Their values of π were generally inaccurate; $\sqrt{10}$ was commonly used, though the better value of 3.1416 appears at times. For the area of any quadrilateral they gave the formula $\sqrt{(s - a)(s - b)(s - c)(s - d)}$, where s is half the perimeter and a, b, c, and d are the sides, a formula which is correct only for quadrilaterals inscribed in circles. They offered no geometric proofs; on the whole they cared little for geometry.

In trigonometry the Hindus made a few minor advances. Ptolemy had used the chords of arcs, calculated on the basis of 120 units in the diameter of a circle. Varāhamihira used 120 units for the radius. Hence Ptolemy's table of chords became for him a table of half chords, but still associated with the full arc. Āryabhata then made two changes. First he associated the half chord with half of the arc of the full chord; this Hindu notion of sine was used by all later Hindu mathematicians. Secondly, he introduced a radius of 3438 units. This number comes from assigning $360 \cdot 60$ units (the number of minutes) to the circumference of a circle and using $C = 2\pi r$, with π approximated by 3.14. Thus in Āryabhata's scheme the sine of an arc of 30°, that is, the length of the half chord corresponding to an arc of 30°, was 1719. While the Hindus used the equivalent of our cosine, they more often used the sine of the complementary arc. They also used the notion of versed sine, or $1 - $ cosine.

Since the radius of a circle now contained 3438 units, Ptolemy's values for the chords were no longer suitable, and the Hindus recomputed a table of half chords, starting from the fact that the half chord corresponding to an arc of 90° is 3438 and the half chord corresponding to an arc of 30° is 1719. Then, by using trigonometric identities such as Ptolemy had established, they were able to calculate the half chords of arcs at intervals of 3°45′. This angle resulted from dividing each quadrant of 90° into 24 parts. It is noteworthy that they used the identities in algebraic form, unlike the geometrical arguments of Ptolemy, and made arithmetic calculations on the basis of the algebraic relations. Their practice was, in principle, like ours.

The motivation for the trigonometry was astronomy, of which practically all of Hindu trigonometry was a by-product. Standard astronomical works included the *Sûrya Siddhânta* (System of the Sun, 4th cent.) and the *Āryabhatiya* of Āryabhata (6th cent.). The major work was the *Siddhânta Siromani* (Diadem of an Astronomical System) written by Bhāskara in 1150. Two chapters of this work are titled *Lîlāvatî* (The Beautiful) and *Vîja-ganita* (Root Extraction); these are devoted to arithmetic and algebra.

Though astronomy was a primary interest in the period after A.D. 200, the Hindus made no significant progress. They took over a minor Hellenistic activity in arithmetical astronomy (of Babylonian origin), which predicts planetary and lunar positions by extrapolation from observational data. Even the Hindu words for center, minute, and other terms were just the Greek words transliterated. The Hindus were only slightly concerned with the geometrical theory of deferent and epicycle, though they did teach the sphericity of the earth.

About the year 1200 scientific activity in India declined and progress in mathematics ceased. After the British conquered India in the eighteenth century, a few Indian scholars went to England to study and on their return did initiate some research. However, this modern activity is part of European mathematics.

As our survey indicates, the Hindus were interested in and contributed to the arithmetical and computational activities of mathematics rather than to the deductive patterns. Their name for mathematics was *ganita*, which means "the science of calculation." There is much good procedure and technical facility, but no evidence that they considered proof at all. They had rules, but apparently no logical scruples. Moreover, no general methods or new viewpoints were arrived at in any area of mathematics.

It is fairly certain that the Hindus did not appreciate the significance of their own contributions. The few good ideas they had, such as separate symbols for the numbers from 1 to 9, the conversion to base 10, and negative numbers, were introduced casually with no realization that they were valuable innovations. They were not sensitive to mathematical values. Along with the ideas they themselves advanced, they accepted and incorporated the crudest ideas of the Egyptians and Babylonians. The Persian historian al-Bîrûnî (973–1048) says of the Hindus, "I can only compare their mathematical and astronomical literature . . . to a mixture of pearl shells and sour dates, or of pearl and dung, or of costly crystals and common pebbles. Both kinds of things are equal in their eyes, since they cannot raise themselves to the methods of a strictly scientific deduction."

4. *The Arabs*

Thus far the Arab role in the history of mathematics has been to deliver the final blow to the Alexandrian civilization. Before starting their conquests, they had been nomads occupying the region of modern Arabia. They were stirred to activity and unity by Mohammed, and less than a century after his death in 632, had conquered lands from India to Spain, including North Africa and southern Italy. In 755 the Arab empire split into two independent kingdoms, the eastern part having its capital at Bagdad and the western one at Cordova in Spain.

Their conquests completed, the former nomads settled down to build a civilization and a culture. Rather quickly the Arabs became interested in the arts and sciences. Both centers attracted scientists and supported their work, though Bagdad proved to be the greater; an academy, a library, and an astronomical observatory were established there.

The cultural resources available to the Arabs were considerable. They invited Hindu scientists to settle in Bagdad. When Justinian closed Plato's Academy in A.D. 529, many of its Greek scholars went to Persia, and the Greek learning that flourished there became, a century later, part of the Arab world. The Arabs also established contacts with the Greeks of the independent Byzantine Empire; in fact the Arab caliphs bought Greek manuscripts from the Byzantines. Egypt, the center of Greek learning in the Alexandrian period, had been conquered by the Arabs, so the learning that survived there contributed to the activity in the Arab empire. The Syrian schools of Antioch, Emesa, and Damascus and the school of Nestorian Christians at Edessa, which had become the chief repositories in the Near East of Greek works after the destruction of Alexandria in 640, and Christian monasteries in the Near East, which also possessed these works, all came under Arab rule. Thus the Arabs had control over, or access to, the men and culture of the Byzantine Empire, Egypt, Syria, Persia, and the lands farther east, including India.

One speaks of Arabic mathematics, but it was Arabic in language primarily. Most of the scholars were Greeks, Christians, Persians, and Jews. However, it is to the credit of the Arabs that after the period of conquest, which was marked by religious fanaticism, they were liberal to other peoples and sects and the infidels were able to function freely.

Fundamentally, what the Arabs possessed was Greek knowledge obtained directly from Greek manuscripts or from Syrian and Hebrew versions. All the major works became accessible to them. They obtained a copy of Euclid's *Elements* from the Byzantines about 800 and translated it into Arabic. Ptolemy's *Mathematical Syntaxis* was translated into Arabic in 827 and to the Arabs became a preeminent, almost divine book; it became known as the *Almagest*, meaning the greatest work. They also translated Ptolemy's *Tetrabiblos* and this work on astrology was popular among them. In the course of time the works of Aristotle, Apollonius, Archimedes, Heron, Diophantus, and of the Hindus became accessible in Arabic. The Arabs then improved the translations and made commentaries. It is these translations, some still extant, that became available to Europe later, the Greek originals having been lost. Until 1300 the Arab civilization was dynamic and its learning became widespread.

5. *Arabic Arithmetic and Algebra*

When the Arabs were still nomads, they had words for numbers but no symbols. They took over and improved the Hindu number symbols and the idea of positional notation. They used these number symbols for whole numbers and common fractions (adding a bar to the Hindu scheme) in their mathematical texts and Arabic alphabetic numerals, on the Greek pattern, for astronomical texts. For astronomy they used the sexagesimal fractions in imitation of Ptolemy.

Like the Hindus, the Arabs worked freely with irrationals. In fact, Omar Khayyam (1048?–1122) and Nasîr-Eddin (1201–74) clearly state that every ratio of magnitudes, whether commensurable or incommensurable, may be called a number, a statement Newton felt obliged to reaffirm in his *Universal Arithmetic* of 1707. The Arabs took over the operations with irrational numbers that the Hindus had introduced, and such transformations as $\sqrt{a^2 b} = a\sqrt{b}$ and $\sqrt{ab} = \sqrt{a}\sqrt{b}$ became common.

In arithmetic the Arabs took one step backward. Though they were familiar with negative numbers and the rules for operating with them through the work of the Hindus, they rejected negative numbers.

To algebra the Arabs contributed first of all the name. The word "algebra" comes from a book written in 830 by the astronomer Mohammed ibn Musa al-Khowârizmî (*c.* 825), titled *Al-jabr w'al muqâbala*. The word *al-jabr* meant "restoring," in this context, restoring the balance in an equation by placing on one side of an equation a term that has been removed from the other; thus if -7 is removed from $x^2 - 7 = 3$, the balance is restored by writing $x^2 - 7 + 3$. *Al' muqâbala* meant "simplification," as by combining $3x$ and $4x$ into $7x$ or by subtracting equal terms from both sides of an equation. *Al-jabr* also came to mean "bonesetter," that is, a restorer of broken bones. When the Moors brought the word into Spain, it became *algebrista* and meant "bonesetter." At one time it was not uncommon in Spain to see a sign "Algebrista y Sangrador" (bonesetter and bloodletter) over the entrance to a barbershop because barbers at that time and even centuries later administered the simpler medical treatments. In sixteenth-century Italy, algebra meant the art of bonesetting. When al-Khowârizmî's book was first translated into Latin in the twelfth century, the title was rendered as *Ludus algebrae et almucgrabalaeque*, though other titles were also used. The name of the subject was finally shortened to algebra.

The algebra of al-Khowârizmî is founded on Brahmagupta's work but also shows Babylonian and Greek influences. Al-Khowârizmî performs some operations just as Diophantus does. For example, in equations involving several unknowns, he reduces to one unknown and then solves. Diophantus refers to his unknown s as the side when s^2 also occurs; so does al-Khowârizmî. Al-Khowârizmî calls the square of the unknown "power," which is Diophantus' word. He also uses, as did Diophantus, special names for the powers of the unknown. The unknown he refers to as the "thing" or the "root" (of a plant), whence our term root. Al-Karkhî of Bagdad (died *c.* 1029), who wrote a superior Arabic algebra text in the early part of the eleventh century, certainly follows the Greeks and especially Diophantus. However, the Arabs did not use symbolism. Their algebra is entirely rhetorical and, in this respect, a step backward, as compared with the Hindus and even Diophantus.

In his algebra al-Khowârizmî gives the product of $(x \pm a)$ and $(y \pm b)$. He shows how to add and subtract terms from an expression of the form $ax^2 + bx + c$. He solves linear and quadratic equations, but keeps the six separate forms, such as $ax^2 = bx$, $ax^2 = c$, $ax^2 + c = bx$, $ax^2 + bx = c$, and

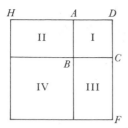

Figure 9.1

$ax^2 = bx + c$, so that a, b, and c are always positive. This avoids negative numbers standing alone and the subtraction of quantities that may be larger than the minuend. In this practice of using the separate forms, al-Khowârizmî follows Diophantus. Al-Khowârizmî recognizes that there can be two roots of quadratics, but gives only the real positive roots, which can be irrational. Some writers give both positive and negative roots.

One example of a quadratic treated by al-Khowârizmî reads as follows: "A square and ten of its roots are equal to nine and thirty dirhems, that is you add ten roots to one square, the sum is equal to nine and thirty." He gives the solution thus: "Take half the number of roots, that is, in this case five, then multiply this by itself and the result is five and twenty. Add this to the nine and thirty, which gives sixty-four; take the square root, or eight, and subtract from it half the number of roots, namely five, and there remains three. This is the root." The solution is exactly what the process of completing the square calls for.

Though the Arabs gave algebraic solutions of quadratic equations, they explained or justified their processes geometrically. Undoubtedly they were influenced by the Greek reliance upon geometrical algebra; while they arithmetized the processes, they must have believed that the proof had to be made geometrically. Thus to solve the equation, which is $x^2 + 10x = 39$, al-Khowârizmî gives the following geometrical method. Let AB (Fig. 9.1) represent the value of the unknown x. Construct the square $ABCD$. Produce DA to H and DC to F so that $AH = CF = 5$, which is one-half of the coefficient of x. Complete the square on DH and DF. Then the areas I, II, and III are x^2, $5x$, and $5x$, respectively. The sum of these is the left side of the equation. To both sides we now add area IV, which is 25. Hence the entire square is $39 + 25$ or 64 and its side must be 8. Then AB or AD is $8 - 5$ or 3. This is the value of x. The geometric argument rests on Proposition 4 of Book II of the *Elements*.

The Arabs solved some cubics algebraically and gave a geometrical explanation in the manner just illustrated for quadratics. This was done, for example, by Tâbit ibn Qorra (836–901), a pagan of Bagdad, who was also a physician, philosopher, and astronomer, and by the Egyptian al-Hasan ibn al-Haitham, known generally as Alhazen (c. 965–1039). As for the general

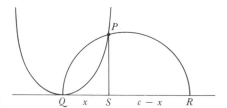

Figure 9.2

cubic, Omar Khayyam believed this could be solved only geometrically, by using conic sections. We shall illustrate the method he used in his *Algebra* (*c.* 1079) to solve some types of cubics by considering one of the simpler cases he treats, namely, $x^3 + Bx = C$, where B and C are positive.

Khayyam writes the equation as $x^3 + b^2x = b^2c$ wherein $b^2 = B$ and $b^2c = C$. He then constructs a parabola (Fig. 9.2), whose latus rectum is b. This quantity does indeed fix the parabola, and though the curve itself cannot be constructed with straightedge and compass, as many points as desired can be so constructed. He constructs next the semicircle on the diameter QR which has length c. Then the intersection P of the parabola and semicircle determines the perpendicular PS, and QS is the solution of the cubic equation.

Khayyam's proof is purely synthetic. From the geometric property of the parabola as given in Apollonius (or as we can see from the equation $x^2 = by$)

(6) $$x^2 = b \cdot PS$$

or

(7) $$\frac{b}{x} = \frac{x}{PS}.$$

Now consider the right triangle QPR. The altitude PS is the mean proportional between QS and SR. Hence

(8) $$\frac{x}{PS} = \frac{PS}{c - x}.$$

From (7) and (8) we have

(9) $$\frac{b}{x} = \frac{PS}{c - x}.$$

But from (7)

$$PS = \frac{x^2}{b}.$$

If we substitute this value of PS in (9) we see that x satisfies the equation $x^3 + b^2x = b^2c$.

Khayyam also solved the type $x^3 + ax^2 = c^3$, whose roots are determined by the intersection of a hyperbola and a parabola, and the type $x^3 \pm ax^2 + b^2x = b^2c$, whose roots are determined by the intersection of an ellipse and a hyperbola. He also solved one quartic, $(100 - x^2)(10 - x)^2 = 8100$, the roots of which are determined by the intersection of a hyperbola and a circle. He gives only positive roots.

The solution of cubic equations by using intersecting conics is the greatest Arab step in algebra. The mathematics is of precisely the same sort as the Greek geometrical algebra, though it uses conic sections. The goal should be an arithmetic answer, but the Arabs could obtain this only by measuring the final length representing x. In this work the influence of Greek geometry is clear.

The Arabs also solved indeterminate equations of the second and third degree. A couple of writers stated and tried to prove that $x^3 + y^3 = z^3$ cannot be solved integrally. They also gave the sums of the first, second, third, and fourth powers of the first n natural numbers.

6. *Arabic Geometry and Trigonometry*

Arabic geometry was influenced mainly by Euclid, Archimedes, and Heron. The Arabs did make critical commentaries of Euclid's *Elements*, which is surprising because it shows appreciation of rigor despite their usual indifference to it in algebra. These commentaries included work on the parallel axiom, which we shall consider later (Chap. 36). They are valuable less for what they have to offer in the way of new results or new proofs than for the information they supply about the Greek manuscripts the Arabs had at their disposal that have since been lost. One new problem, which became popular in Renaissance Europe, was explored by the Persian Abû'l-Wefâ or Albuzjani (940–98): constructions with a straightedge and a fixed circle (i.e., fixed compass opening).

The Arabs made a little progress in trigonometry. Theirs, like the Hindus', is arithmetical, rather than geometric as in Hipparchus and Ptolemy. Thus, to calculate some cosine values from sine values they used an identity such as $\sin^2 A + \cos^2 A = 1$ and algebraic steps. Like the Hindus, they used sines of arcs rather than chords of double the arcs, though (as in the Hindu work) the number of units in the sine or half chord depends upon the number of units taken for the radius. Tâbit ibn Qorra and the astronomer al-Battânî (*c.* 858–929) introduced this use of sines among the Arabs.

Arab astronomers introduced what we call the tangent and cotangent ratios, but as lines containing a number of units, just as the sine of an arc was a length containing a number of units. These two ratios are to be found in the work of al-Battânî. Abû'l-Wefâ introduced the secant and cosecant as lengths in a work on astronomy. He also computed tables of sines and tangents for every 10 minutes of angles. Al-Bîrûnî gave the law of sines for plane triangles and a proof.

The systematization of plane and spherical trigonometry in a work independent of astronomy was achieved by Nasîr-Eddin in his *Treatise on the Quadrilateral.* This work contains six fundamental formulas for the solution of right spherical triangles and shows how to solve more general triangles by what we now call the polar triangle. Unfortunately the Europeans did not know Nasîr's work until about 1450; until then trigonometry remained, as it had been from its inception, an appendage of astronomy, in the texts as well as in application.

The Arab scientific effort, though not original, was extensive; however, we cannot do more here than note the continuation of lines inaugurated by the Greeks. Unlike the Hindus, the Arabs did take over Ptolemy's astronomy. Astronomy was emphasized so that the hours for prayers could be accurately known and so that Arabs throughout the vast empire could face Mecca during prayers. Astronomical tables were enlarged; instruments were improved; and observatories were built and used. As in India, practically all mathematicians were primarily astronomers. Astrology also played a big role in stimulating work in astronomy and, therefore, mathematics.

Another science pursued by the Arabs was optics. Alhazen, who was a physicist as well as a mathematician, wrote the great treatise *Kitab al-manazer* or *Treasury of Optics,* which exercised an enormous influence. In it he stated the full law of reflection, including the fact that the incident ray, the reflected ray, and the normal to the reflecting surface all lie in one plane. But, like Ptolemy, he did not succeed in finding the law for the angle of refraction, despite much effort and experimentation. He discussed spherical and parabolic mirrors, lenses, the camera obscura, and vision. Optics was a favorite subject with the Arabs because it lent itself to occult and mystic thoughts. However, they made no important original contributions.

The Arab uses of mathematics were of the sort we have encountered before. Astronomy, astrology, optics, and medicine (through astrology) required it, though some of the algebra, as one of the Arab mathematicians put it, was "most needed in problems of distribution, inheritance, partnership, land measurement. . . ." Mathematics was studied by the Arabs mainly to further the few sciences they pursued, not for its own sake. Nor did they study science for its own sake. They were not interested in the Greek goal of understanding the mathematical design of nature or in comprehending God's ways, as were the medieval Europeans. The Arab objective, new in the history of science, was power over nature. They thought they could achieve that power through alchemy, magic, and astrology, which were an integral part of their scientific effort. This objective was taken up later by more critical minds, who could distinguish science from pseudoscience and were more profound in their approach.

The significant contribution to mathematics that we owe to the Arabs was to absorb Greek and Hindu mathematics, preserve it, and, ultimately, through events we have yet to look into, transmit it to Europe. Arab activity reached its peak about the year 1000. Between 1100 and 1300 the Christian attacks in the Crusades weakened the eastern Arabs. Subsequently, their territory was overrun and conquered by the Mongols; after 1258 the caliphate at Bagdad ceased to exist. Further destruction by the Tartars under Tamerlane just about wiped out this Arab civilization, though a bit of mathematical work was done there after the Tartar invasion. In Spain the Arabs were constantly attacked and finally conquered in 1492 by the Christians; this ended the mathematical and scientific activity in that region.

7. *Mathematics* circa 1300

Though the mathematical work of the Hindus and Arabs was not brilliant, it did bring about some changes in the content and character of mathematics that were material for the future of the subject. Positional notation in base 10 (using special number symbols for the quantities from 1 to 9 and zero as a number), the introduction of negative numbers, and the free use of irrationals as numbers not only extended arithmetic vastly but paved the way for a more significant algebra, an algebra in which letters and operations could apply to a far broader class of numbers.

Both peoples worked with equations, determinate and indeterminate, on an arithmetic rather than a geometric basis. Though algebra, as initiated by the Egyptians and Babylonians, was arithmetically grounded, the Greeks had subverted it by requiring a geometric basis. Hence the Hindu and Arabic work not only restored algebra to its proper foundation but even advanced the art in several ways. The Hindus used more symbolism and made progress in indeterminate equations, while the Arabs ventured further in the subject of third degree equations, though Khayyam's work still relied upon geometry.

Euclidean geometry was not furthered but trigonometry was. The introduction of the sine or half chord did prove to be a technical advantage. The arithmetical or algebraic technique for handling identities and for calculation in trigonometry was an expeditious step; and the segregation of trigonometric knowledge from astronomy revealed a more broadly applicable science.

Two developments occurred that were significant for the forthcoming recognition that algebra is coextensive with geometry. The acceptance of irrational numbers made it possible to assign numerical values to all line segments and two- and three-dimensional figures, that is, to express lengths, areas, and volumes by numbers. Moreover, the Arab practice of solving equations algebraically while justifying the processes by means of a geometric representation exhibited the parallelism of the two subjects. The fuller development of that parallelism was to lead to analytic geometry.

Perhaps most interesting is the Hindus' and Arabs' self-contradictory concept of mathematics. Both worked freely in arithmetic and algebra and yet did not concern themselves at all with the notion of proof. That the Egyptians and Babylonians were content to accept their few arithmetic and geometric rules on an empirical basis is not surprising; this is a natural basis for almost all human knowledge. But the Hindus and Arabs were aware of the totally new concept of mathematical proof promulgated by the Greeks. The Hindu behavior can be somewhat rationalized; though they did indeed possess some knowledge of the classic Greek works, they paid them little attention and followed primarily the Alexandrian Greek treatment of arithmetic and algebra. Even this preference for one kind of mathematics over another raises a question. But the Arabs were fully aware of Greek geometry and even made critical studies of Euclid and other Greek authors. Moreover, conditions for the pursuit of pure science were favorable over a period of several centuries, so that the pressure to produce practically useful results need not have caused mathematicians to sacrifice proof for immediate utility. How could both peoples have treated the two areas of mathematics so differently from the Greeks?

There are numerous possible answers. Both civilizations were on the whole uncritical, despite the Arabic commentaries on Euclid. Hence they may have been content to take mathematics as they found it; that is, geometry was to be deductive but arithmetic and algebra could be empirical or heuristic. A second possibility is that both of these peoples—more likely, the Arabs—recognized the widely different standards for geometry as opposed to arithmetic and algebra but did not see how to supply a logical foundation for arithmetic. One fact that seems to support such an explanation is that the Arabs did at least explain their solution of quadratics by giving the geometric basis.

There are other possible explanations. Both the Hindus and Arabs favored arithmetic, algebra, and the algebraic formulation of and operation with trigonometric relationships. This predisposition may bespeak a different mentality or it may reflect a response to the needs of the civilizations. Both of these civilizations were practically oriented and, as we have had occasion to note in connection with the Alexandrian Greeks, practical needs do call for quantitative results, which are supplied by arithmetic and algebra. One bit of evidence favoring the thesis that different mentalities were involved is the reaction of the Europeans to the very same mathematical heritage which the Hindus and Arabs received. As we shall see, the Europeans were far more troubled by the disparate states of arithmetic and geometry.

In the absence of any exhaustive and definitive study we must take the position that the Hindus and Arabs were conscious of the precarious basis for arithmetic and algebra but had the audacity, reinforced by practical needs, to develop these branches. Though they undoubtedly did not appreciate what they were accomplishing, they took the only course mathematical innovation can pursue. New ideas can come about only by the free and bold pursuit of heuristic and intuitive insights. The logical justification and corrective measures, should the latter be needed, can be brought into play only when there is something to logicize. Hindu and Arab venturesomeness brought arithmetic and algebra to the fore once again and placed it almost on a par with geometry.

Two independent traditions or concepts of mathematics had now become established: on the one hand, the logical deductive body of knowledge that the Greeks established, which served the larger purpose of understanding nature; and on the other, the empirically grounded, practically oriented mathematics founded by the Egyptians and Babylonians, resuscitated by some of the Alexandrian Greeks, and extended by the Hindus and Arabs. The one favored geometry and the other arithmetic and algebra. Both traditions and both goals were to continue to operate.

Bibliography

Ball, W. W. R.: *A Short Account of the History of Mathematics*, Dover (reprint), 1960, Chap. 9.

Berry, Arthur: *A Short History of Astronomy*, Dover (reprint), 1961, pp. 76–83.

Boyer, Carl B.: *A History of Mathematics*, John Wiley and Sons, 1968, Chaps. 12–13.

Cajori, Florian: *A History of Mathematics*, Macmillan, 1919, pp. 83–112.

Cantor, Moritz: *Vorlesungen über Geschichte der Mathematik*, 2nd ed., B. G. Teubner, 1894, Johnson Reprint Corp., 1965, Vol. 1, Chaps. 28–30, 32–37.

Coolidge, Julian L.: *The Mathematics of Great Amateurs*, Dover (reprint), 1963, Chap. 2.

Datta, B., and A. N. Singh: *History of Hindu Mathematics*, 2 vols., Asia Publishing House (reprint), 1962.

Dreyer, J. L. E.: *A History of Astronomy from Thales to Kepler*, Dover (reprint), 1953, Chap. 11.

Karpinski, L. C.: *Robert of Chester's Latin Translation of the Algebra of al-Khowarizmi*, Macmillan, 1915. English version also.

Kasir, D. S.: "The Algebra of Omar Khayyam," Columbia University Teachers College thesis, 1931.

O'Leary, De Lacy: *How Greek Science Passed to the Arabs*, Routledge and Kegan Paul, 1949.

Pannekoek, A.: *A History of Astronomy*, John Wiley and Sons, 1961, Chap. 15.

Scott, J. F.: *A History of Mathematics*, Taylor and Francis, 1958, Chap. 5.

Smith, David Eugene: *History of Mathematics*, Dover (reprint), 1958, Vol. 1, pp. 138–47, 152–92, 283–90.

Struik, D. J.: "Omar Khayyam, Mathematician," *The Mathematics Teacher*, 51, 1958, 280–85.

10

The Medieval Period in Europe

> In most sciences one generation tears down what another has
> built and what one has established another undoes. In mathe-
> matics alone each generation adds a new story to the old
> structure.   HERMANN HANKEL

1. *The Beginnings of a European Civilization*

Western and central Europe entered into the development of mathematics
as the Arab civilization began to decline. However, to familiarize ourselves
with the state of affairs in medieval Europe, to learn how the European
civilization got its start, and to understand the directions it took, we must
go back, briefly at least, to its beginnings.

During the ages when the Babylonians, Egyptians, Greeks, and Romans
flourished, the area now called Europe (except for Italy and Greece)
possessed a primitive civilization. The Germanic tribes who lived there
mastered neither writing nor learning. The Roman historian Tacitus (1st
cent. A.D.) describes these tribes of about the time of Christ as honest,
hospitable, hard-drinking, hating peace, and proud of the loyalty of their
wives. Cattle-raising, hunting, and the cultivation of grain were the main
occupations. Beginning in the fourth century of the Christian era the Huns
drove the Goths and the Germanic tribes occupying central Europe farther
west. In the fifth century the Goths took over the western Roman Empire
itself.

Though parts of France and England had previously acquired some
learning while under the domination of the Roman Empire, by A.D. 500
new civilizing influences began to operate in Europe. Even before the empire
collapsed the Catholic Church was organized and powerful. The Church
gradually converted the Germanic and Gothic barbarians to Christianity
and began to found schools; these were attached to already existing monasteries that possessed fragments of Greek and Roman learning and had been
teaching people how to read the church services and the sacred books. A
little later, the need to train men for ecclesiastical posts motivated the
development of higher schools.

200

In the latter half of the eighth century some secular rulers founded additional schools. In Charlemagne's empire schools were organized by Alcuin of York (730–804), an Englishman who came to Europe at the invitation of Charlemagne himself. These schools too were attached to cathedrals or monasteries and emphasized Christian theology and music. Ultimately the universities of Europe grew out of the church schools, with teachers supplied by church orders such as the Franciscans and Dominicans. Bologna, the first university, was founded in 1088. The universities of Paris, Salerno, Oxford, and Cambridge were established about the year 1200. Of course at the outset these were hardly universities in the modern sense. Also, though formally independent, they were essentially devoted to the interests of the Church.

2. The Materials Available for Learning

As the Church extended its influence it imposed the culture it favored. Latin was the official language of the Church and so Latin became the international language of Europe and the language of mathematics and science. It was also the language of instruction in European schools until well into the eighteenth century. It became inevitable that Europeans would seek their knowledge largely from Latin—that is, Roman—books. Since Roman mathematics was insignificant, all the Europeans learned was a very primitive number system and a few facts of arithmetic. They also acquired a bit of Greek mathematics through a few translators.

The principal translator, whose works were widely used until the twelfth century, was Anicius Manlius Severinus Boethius (c. 480–524), a descendant of one of the oldest Roman families. Using Greek sources, he compiled in Latin selections from elementary treatises on arithmetic, geometry, and astronomy. Of Euclid's *Elements* he may have translated as many as five books or as few as two, and these constituted part of his *Geometry*. In this subject he gave definitions and theorems but no proofs. He also included in this work some material on the geometry of mensuration. Some results are incorrect and others only approximations. Peculiarly, the *Geometry* also contained material on the abacus and on fractions, the latter preliminary to material on astronomy (which we do not have). Boethius also wrote *Institutis arithmetica*, a translation of Nichomachus' *Introductio arithmetica*, though he omitted some of Nichomachus' results. This book was the source of all arithmetic taught in the schools for almost a thousand years. Finally, Boethius translated some works of Aristotle and wrote an astronomy based on Ptolemy and a book on music based on Euclid's, Ptolemy's, and Nichomachus' works. It is very likely that Boethius did not understand all he translated. It was he who introduced the word "quadrivium" for arithmetic, geometry, music, and astronomy. His best-known work, still read today, is the *Consolations of Philosophy*, which he wrote while imprisoned for alleged treason (for which he was ultimately beheaded).

Other translators were the Roman Aurelius Cassiodorus (*c.* 475–570), who rendered a few parts of the Greek works on mathematics and astronomy in his own poor version; Isidore of Seville (*c.* 560–636), who wrote the *Etymologies*, a work in twenty books on material ranging from mathematics to medicine; and the Englishman, the Venerable Bede (674–735). These men were the main links between Greek mathematics and the early medieval world.

In all of the problems appearing in books written by early medieval mathematicians, only the four operations with integers were involved. Since in practice calculation was done on various forms of the abacus, the rules of operation were specially adapted to it. Fractions were rarely employed, and where they were, the Roman fractions with names rather than special symbolism were used; for example, *uncia* for 1/12, *quincunx* for 5/12, *dodrans* for 9/12. Irrational numbers did not appear at all. Good calculators were known in the Middle Ages as practitioners of the "Black Art," magic.

In the tenth century the study of mathematics was improved somewhat by Gerbert (d. 1003), a native of Auvergne, later Pope Sylvester II. His writings, however, were confined to elementary arithmetic and elementary geometry.

3. *The Role of Mathematics in Early Medieval Europe*

Though the mathematical material available for instruction was scanty, mathematics was relatively important in the curriculum of even the early medieval schools. The curriculum was divided into the quadrivium and the trivium. The quadrivium included arithmetic, considered as the science of pure numbers; music, regarded as an application of numbers; geometry, or the study of magnitudes such as length, area, and volumes at rest; and astronomy, the study of magnitudes in motion. The trivium covered rhetoric, dialectic, and grammar.

Learning even the little mathematics we have described served several purposes. After the time of Gerbert it was applied to finding heights and distances, the astrolabe and mirror being the field instruments. The clergy was expected to defend the theology and rebut arguments by reasoning, and mathematics was regarded as good training for theological reasoning, just as Plato had regarded it as good training for philosophy. The Church advocated teaching mathematics for its application in keeping the calendar and predicting holidays. Each monastery had at least one man who could do the necessary calculations, and various improvements in arithmetic and in the method of keeping the calendar were devised in the course of this work.

Another motivation for the study of some mathematics was astrology.

This pseudoscience, which had had some vogue in Babylonia, in Hellenistic Greece, and among the Arabs, was almost universally accepted in medieval Europe. The basic doctrine of astrology was, of course, that the heavenly bodies influenced and controlled human bodies and fortunes. To understand the influences of the heavenly bodies and to predict what special heavenly events, such as conjunctions and eclipses, portended, some astronomical knowledge was needed; and thus some mathematics was indispensable.

Astrology was especially important in the late medieval centuries. Every court had astrologers and the universities had professors of astrology and courses in the subject. Astrologers advised princes and kings on political decisions, military campaigns, and personal matters. The curious thing is that even rulers who had become learned in and attached to Greek thought relied upon astrologers. In the late medieval period and in the Renaissance astrology not only became a major activity but was regarded as a branch of mathematics.

Through astrology mathematics was linked to medicine (Chap. 7, sec. 8). Although the Church dismissed the physical body as relatively unimportant, the physicians could not subscribe to this belief. Since the heavenly bodies presumably influenced health, physicians sought the relations between heavenly events and particular constellations, on the one hand, and the health of individuals, on the other. Records were kept of the constellations that appeared at the births, marriages, sicknesses, and deaths of thousands of people and used to predict the success of medical treatment. Such a wide knowledge of mathematics was required for this purpose that physicians had to become learned in the field. In fact they were astrologers and mathematicians far more than students of the human body.

The application of mathematics to medicine through astrology became more widespread in the latter part of the medieval period. Bologna had a school of medicine and mathematics in the twelfth century. When the astronomer Tycho Brahe attended the University of Rostock in 1566, there were no astronomers there but there were astrologers, alchemists, mathematicians, and physicians. In many universities, professors of astrology were more common than professors of medicine and astronomy proper. Galileo did lecture to medical students on astronomy, but for the sake of astrology.

4. *The Stagnation in Mathematics*

The early medieval period extends from about 400 to about 1100: seven hundred years during which the European civilization might have developed some mathematics. It could have derived immense help from the Greek works had it pursued the few available leads on the vast knowledge embodied in them. Yet during this period mathematics made no progress, nor were there any serious attempts to build mathematics. The reasons are of interest to those who seek to understand under what circumstances mathematics can flourish.

The primary reason for the low level of mathematics was lack of interest in the physical world. Christianity, which dominated Europe, prescribed its own goals, values, and way of life. The important concerns were spiritual, so much so that inquiries into nature stimulated by curiosity or practical ends were regarded as frivolous or unworthy. Christianity and even the later Greek philosophers, the Stoics, Epicureans, and neo-Platonists, emphasized lifting mind above flesh and matter and preparing the soul for an after-life in heaven. The ultimate reality was the everlasting life of the soul; and the soul's health was strengthened by learning moral and spiritual truths. The doctrines of sin, fear of hell, salvation, and aspiration to heaven were dominant. Since the study of nature did not help achieve such goals or prepare for the afterlife, it was opposed as worthless and even heretical.

Where, then, did the Europeans secure any knowledge about the nature and design of the universe and of man? The answer is that all knowledge was derived from the study of the Scriptures. The creeds and dogmas of the Church Fathers, which were amplifications and interpretations of the Scriptures, were taken as the supreme authority. St. Augustine (354–430), a very learned man and most influential in spreading neo-Platonism, said, "Whatever knowledge man has acquired outside of Holy Writ, if it be harmful it is there condemned; if it be wholesome it is there contained." This quotation, though not representative of Augustine, is representative of the early medieval attitude toward nature.

This brief sketch of the early medieval civilization, rather one-sided as it is, because we have been concerned largely with its relation to mathematics, may nevertheless give some idea of what was indigenous to Europe and what Europe, building on a meager legacy from Rome, produced under the leadership of the Church. Until 1100 the medieval period did not produce any great culture in intellectual spheres. The characteristics of its intellectual state were an indistinctness of ideas, dogmatism, mysticism, and a reliance upon authorities, who were constantly consulted, analyzed, and commented on. The mystical leanings caused people to elevate vague ideas into realities and even accept them as religious truths. What little theoretical science existed was static. Theology embraced all knowledge and the Fathers of the Church authored systems of universal knowledge. But they did not conceive or search for principles other than those contained in Christian doctrines.

The Roman civilization was unproductive in mathematics because it was too much concerned with practical and immediately applicable results. The civilization of medieval Europe was unproductive in mathematics for exactly the opposite reason. It was not at all concerned with the physical world. Mundane matters and problems were unimportant. Christianity put its emphasis on life after death and on preparation for that life.

Apparently mathematics cannot flourish in either an earthbound or a heavenbound civilization. We shall see that it has been most successful in a free intellectual atmosphere which couples an interest in the problems presented by the physical world with a willingness to think about ideas suggested by these problems in an abstract form that makes no promise of immediate or practical return. Nature is the matrix from which ideas are born. The ideas must then be studied for themselves. Then, paradoxically, a new insight into nature, a richer, broader, more powerful understanding, is achieved, which in turn generates deeper mathematical activities.

5. *The First Revival of the Greek Works*

By 1100 the civilization of Europe was somewhat stabilized. Though the society was largely feudal, there were already numerous independent merchants, the beginnings of industry, arts and crafts carried on by free people, large-scale farming, manufacturing, mining, banking, and cattle-raising. Foreign trade, notably with the Arabs and the Near East, had been established. Finally, the wealth necessary to support scholarship and the arts was acquired by princes, church officials, and merchants.

Though there was a stable society there is little indication that the Europeans, if left to pursue their own way of life, would ever have abandoned the outlook and emphasis already sketched and turned to a serious study of mathematics. Western Europe was the successor of Christianized Rome and neither Rome nor Christianity was inclined toward mathematics. But about 1100, new influences began to affect the intellectual atmosphere. Through trade and travel the Europeans had come into contact with the Arabs of the Mediterranean area and the Near East and with the Byzantines of the Eastern Roman Empire. The Crusades (c. 1100–c. 1300), military campaigns to conquer territory, brought Europeans into Arab lands. The Crusaders were men of action rather than learning; it may be that the importance of the contact through the Crusades has been overestimated. At any rate the Europeans began to learn about the Greek works from the Arabs and Byzantine Greeks.

Awareness of the Greek learning created great excitement; Europeans energetically sought out copies of the Greek works, their Arabic versions, and texts written by Arabs. Princes and church leaders supported scholars in the hunt for these treasures. The scholars went to Arab centers in Africa, Spain, southern France, Sicily, and the Near East to study the works and bring back what they could purchase. Adelard of Bath (c. 1090–c. 1150) went to Syria, which was under Arab control, to Cordova, disguised as a Mohammedan student, and to southern Italy. Leonardo of Pisa learned arithmetic in North Africa. The republics of northern Italy and the papacy sent missions and ambassadors to the Byzantine Empire and to Sicily, which was the original home of famous Greek centers and which up to 878 had been under Byzantine rule. In 1085 Toledo was captured by the Christians and thus a major center for the Arabic works was opened up to European scholars. Sicily was taken from the Arabs by the Christians in 1091 and the works there became freely accessible. A search in Rome, which possessed Greek works from the days of the Empire, unearthed more manuscripts.

As they secured these works the Europeans undertook more and more to translate them into Latin. The twelfth-century translations from Greek were, on the whole, not good because Greek was not well known. They were *de verbo ad verbum*; but they were better than the translations of Greek works that had passed through Arabic, a language quite dissimilar to Greek. Hence until well into the seventeenth century there was a steady output of new, improved translations.

Thus Europe got to know the works of Euclid and Ptolemy, al-Khowârizmî's *Arithmetic* and *Algebra*, the *Sphaerica* of Theodosius, many works of Aristotle and Heron, and a couple of Archimedes' works, particularly his *Measurement of a Circle*. (The rest of his work was translated into Latin in 1544 by Hervagius of Basle.) Neither Apollonius nor Diophantus was translated during the twelfth and thirteenth centuries. Works on philosophy, medicine, science, theology, and astrology were also translated. Since the Arabs did have almost all the Greek works, the Europeans acquired a tremendous literature. They admired these works so much and were so fascinated by the novel ideas that they became disciples of Greek thought. They valued these works far more than their own creations.

6. *The Revival of Rationalism and Interest in Nature*

A rational approach to natural phenomena and explanation in terms of natural causes, as opposed to moral or purposive explanations, began to show signs of life almost immediately after the first translations of the Greek and Arabic works reached Europe. A group at Chartres in France, Gilbert de la Porée (*c.* 1076–1154), Thierry of Chartres (d. *c.* 1155), and Bernard Sylvester (*c.* 1150), had begun to seek rational explanations even of biblical passages and spoke, at least, of the need to use mathematics in the study of nature. Their doctrines followed Plato's *Timaeus* but were more rational than that dialogue. However, their pronouncements on physical phenomena, though noteworthy in medieval thought, were neither significant nor influential enough to warrant attention here.

With the influx of Greek works, the trend to rational explanations, study of the physical world, and interest in the enjoyment of the real world through food, a physical life, and the pleasures of nature became marked. Some men even began to pit their own reason against the authority of the Church. Thus Adelard of Bath said he would not listen to those who are "led in a halter; . . . Wherefore if you want to hear anything from me, give and take reason."

Peculiarly enough, the introduction of some of the Greek works retarded the awakening of Europe for a couple of centuries. By 1200 or so the extensive writings of Aristotle became reasonably well known. The European intellectuals were pleased and impressed by his vast store of facts, his acute distinctions, his cogent arguments, and his logical arrangement of knowledge. The defect in Aristotle's doctrines was that he accepted those that appealed to the mind almost without regard for their correspondence with experience. He offered concepts, theories, and explanations, such as the doctrine of basic substances, the distinction between earthly and heavenly bodies (Chap. 7, sec. 3), and the emphasis on final cause, which had little basis in reality, or were not fruitful. Since all of these doctrines were accepted uncritically, new ideas were either not entertained or failed to gain a hearing, and progress was delayed. It was perhaps also a hindrance that Aristotle assigned a minor role to mathematics, certainly a role subordinate to physical explanation, which, for Aristotle, was qualitative.

The scientific work of the period from about 1100 to 1450 was done by the Scholastics, who espoused doctrines based on the authority of the Christian Fathers and of Aristotle; the work suffered accordingly. Some of the Scholastics revolted against the prevailing dogmatism and against the insistence on the absolute correctness of Aristotle's science. One who felt the need for obtaining general principles from experimentation and for deductions in which mathematics would play a part and which could then be tested against facts was the natural philosopher Robert Grosseteste (c. 1168–1253), Bishop of Lincoln.

The most eloquent protester against authority, and one who had genuine ideas to offer, was Roger Bacon (1214–94), the *Doctor Mirabilis*. He declared, "If I had the power over the works of Aristotle, I would have them all burned; for it is only a loss of time to study in them and a cause of error, and a multiplication of ignorance beyond expression." Bacon's enormous knowledge covered the sciences of his time and many languages, including Arabic. Long before they became widely known he was informed on the latest inventions and scientific advances: gunpowder, the action of lenses, mechanical clocks, the construction of the calendar, and the formation of the rainbow. He even discussed ideas on submarines, airplanes, and automobiles. His writings on mathematics, mechanics, optics, vision, astronomy, geography, chronology, chemistry, perspective, music, medicine, grammar, logic, metaphysics, ethics, and theology were sound.

What is especially striking about Bacon is that he understood how reliable knowledge is obtained. He inquired into the causes that produce or prevent the advance of science and speculated on the reform of the methods of inquiry. Though he did recommend study of the Scriptures, he emphasized mathematics and experimentation and foresaw great prospects that could be realized through science.

Mathematical ideas, he affirms, are innate in us and identical with things as they are in nature, for nature is written in the language of geometry. Hence mathematics offers truth. It is prior to the other sciences because it takes cognizance of quantity, which is apprehended by the intuition. He "proves" in one chapter of his *Opus Majus* that all science requires mathematics, and his arguments show a just appreciation of the office of mathematics in science. Though he stresses mathematics, he also has a full appreciation of the role and importance of experimentation as a means of discovery and as a test of results obtained theoretically or in any other way. "Argument concludes a question; but it does not make us feel certain, or acquiesce in the contemplation of truth, except the truth also be found to be so by experience."

Bacon's *Opus Majus* has much on the usefulness of mathematics for geography, chronology, music, the explanation of the rainbow, calendar-reckoning, and the certification of faith. He also treats the role of mathematics in state administration, meteorology, hydrography, astrology, perspective, optics, and vision.

However, even Bacon was a product of his times. He believed in magic and astrology and maintained that theology is the goal of all learning. He was also a victim of his times: he ended up in prison, as did many other intellectual leaders who had begun to assert the priority and independence of human reason and the importance of observation and experimentation. His influence on his age was not great.

William of Ockham (*c.* 1300–49) continued the weighty attacks on Aristotle, criticizing Aristotle's views on causation. Final cause, he said, is pure metaphor. All causes are immediate, and the total cause is the aggregate of all the antecedents that suffice to bring about an event. This knowledge of connections has a universal validity because of the uniformity of nature. The primary function of science is to establish sequences of observations. As for substance, Ockham said, we know only properties, not a fundamental substantial form.

He also attacked contemporary physics and metaphysics, saying that knowledge gained from experience is real, whereas rational constructs are not; they are merely invented to explain the observed facts. His famous principle is "Ockham's razor" (already stated by Grosseteste and Duns Scotus [1266–1308]): It is futile to work with more entities when fewer suffice. He divorced theology from natural philosophy (science) on the ground that theology derives knowledge from revelations whereas natural philosophy should derive it from experience.

These dissenters did not suggest new scientific ideas. But they did press for freedom of speculation, thought, and inquiry and urged experience as the source of scientific knowledge.

7. *Progress in Mathematics Proper*

Despite the rigid bounds to thought in the period from 1100 to about 1450, some mathematical activity did take place, the main centers being the universities of Oxford, Paris, Vienna (founded in 1365), and Erfurt (founded in 1392). The initial work was a direct response to the Greek and Arabic literature.

The first European worthy of mention is Leonardo of Pisa (*c.* 1170–1250), also called Fibonacci. He was educated in Africa, traveled extensively in Europe and Asia Minor, and was famous for his sovereign possession of the entire mathematical knowledge of his own and preceding generations. He resided at Pisa and was well known to Frederick II of Sicily and the philosophers of the court, to whom most of his extant works are dedicated.

In 1202 Leonardo wrote the epoch-making and long-used *Liber Abaci*, a free rendition of Arabic and Greek materials into Latin. Arabic notation for numbers and Hindu methods of calculation were already known to some extent in Europe, but only in the monasteries. People in general used Roman numerals and avoided zero because they did not understand it. Leonardo's book exerted great influence and changed the picture; it taught the Hindu methods of calculation with integers and fractions, square roots, and cube roots. These methods were subsequently improved by the Florentine merchants.

In both the *Liber Abaci* and a later work, the *Liber Quadratorum* (1225), Leonardo treated algebra. He followed the Arabs in using words rather than symbols and in basing the algebra on arithmetical methods. He presented the solution of determinate and indeterminate equations of the first and second degree and some cubic equations. Like Khayyam, he believed that general cubic equations could not be solved algebraically.

On the geometric side, Leonardo in his *Practica Geometriae* (1220) reproduced much of Euclid's *Elements* and Greek trigonometry. His teaching of surveying by trigonometric methods rather than by the Roman geometric methods represented a slight advance.

The most significant new feature of Leonardo's work is the observation that Euclid's classification of irrationals in Book X of the *Elements* did not include all irrationals. Leonardo showed that the roots of $x^3 + 2x^2 + 10x = 20$ are not constructible with straightedge and compass. This was the first indication that the number system contained more than the Greek criterion of existence by construction allowed. Leonardo also introduced the notion of what are still called Fibonacci sequences, wherein each term is the sum of the preceding two.

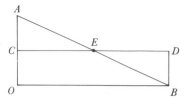

Figure 10.1

Beyond the observation on irrationals made by Leonardo, there were some germinal ideas in the work of Nicole Oresme (*c.* 1323–82), Bishop of Lisieux and a teacher in the Parisian College of Navarre. In his unpublished *Algorismus Proportionum* (*c.* 1360), he introduced both a notation and some computation with fractional exponents. His thinking was that since (in our notation) $4^3 = 64$ and $(4^3)^{1/2} = 8$, then $4^{3/2} = 8$. The notion of fractional exponents reappeared in the work of several sixteenth-century writers, but was not widely used until the seventeenth century.

Another contribution of Oresme's involves the study of change. We may recall that Aristotle distinguished sharply between quality and quantity. Intensity of heat was a quality to Aristotle. To change intensity some substance, a species of heat, had to be lost and another added. Oresme affirmed that there were not different kinds of heat but only more or less of the same kind. Several fourteenth-century Scholastics at Oxford and Paris, including Oresme, began to think about change and rate of change quantitatively. These men studied uniform motion (motion with constant velocity), difform motion (motion with varying velocity), and uniformly difform motion (motion with constant acceleration).

This line of thought culminated in that period with Oresme's doctrine of the latitude of forms. On this subject he wrote *De Uniformitate et Difformitate Intensionum* (*c.* 1350) and *Tractatus de Latitudinibus Formarum* (n.d.). To study change and rate of change, Oresme followed the Greek tradition in asserting that measurable quantities other than numbers can be represented by points, lines, and surfaces. Thus, to represent velocity changing with time, he represents time along a horizontal line, which he called the longitude, and the velocities at various times by vertical lines, which he called the latitudes. To represent a velocity decreasing uniformly from the value *OA* (Fig. 10.1) at *O* to zero at *B*, he gives a triangular figure. He also points out that the rectangle *OBDC*, determined by *E*, the midpoint of *AB*, has the same area as the triangle *OAB* and represents a uniform motion over the same time interval. Oresme associated physical change with the entire geometrical figure. The whole area represented the variation in question; numerical values were not involved.

It is often said that Oresme contributed to the formation of the function concept, to the functional representation of physical laws, and to the classifi-

cation of functions. He is also credited with the creation of coordinate
geometry and the graphical representation of functions. Actually the latitude
of forms is a dim idea, at best a kind of graph. Though Oresme's representation of intensity under the name *latitudines formarum* was a major technique in
the Scholastic attempt to study physical change, was taught in the universities, and was applied in efforts to revise Aristotle's theory of motion, its
influence on subsequent thought was minor. Galileo did use this figure, but
with far more clarity and pointedness. Since Descartes carefully avoided any
reference to his predecessors, we do not know if he was influenced by
Oresme's ideas.

8. *Progress in Physical Science*

Because the progress of mathematics depended vitally on the renewal of an
interest in science, we shall note briefly the efforts made in this area by the
medieval men.

In mechanics, they took over the highly acceptable Greek works on the
lever and centers of gravity and Archimedes' work on hydrostatics. They did
little more than comprehend the theory of the lever, though slight additions
were made by Jordanus Nemorarius (d. 1237). It was the theory of motion
that received most attention.

Because Aristotle's science had acquired ascendancy, it was his theory
that became the starting point for investigations of motion. As we pointed
out in Chapter 7, there were several apparent difficulties in Aristotle's theory.
These the earlier medieval scientists sought to resolve within the basic
Aristotelian framework. Thus, to account for the acceleration of falling bodies,
several thirteenth-century men interpreted Aristotle's vague notions about
gravity to mean that the weight of a body increases as it nears the center of
the earth. Then, since the force increases, so does the velocity. Others questioned the correctness of Aristotle's fundamental law that velocity is force
divided by resistance.

The school at Chartres was succeeded in the fourteenth century by a
school at Paris whose leaders were Oresme and Jean Buridan (*c.* 1300–
c. 1360). Here at the university Aristotelian views had been dominant. To
explain the continuing motion of objects launched by a force, Buridan put
forth a new theory, the theory of impetus. Following the sixth-century
Christian scholar Philoponus, Buridan said that the motive power impressed
on an arrow or projectile was impressed on the object itself and not on the air.
This impetus, rather than the propelling action of the air, would maintain
the object's uniform velocity indefinitely, were it not for the action of
external forces. In the case of falling bodies, impetus was gradually increased by natural gravity acting to add successive increments of impetus

to that which the object already had. In rising bodies—for example, projectiles—the impetus given to the object was gradually decreased by the air resistance and natural gravity. Impetus was what God gave to the celestial spheres and these needed no heavenly agents to keep them running. Buridan defined impetus as the quantity of matter multiplied by the velocity; hence, in modern terms, it is momentum.

This new theory was remarkable for several reasons. By applying it to heavenly and earthly motions, Buridan linked the two in one theory. Furthermore, the theory implied that, contrary to Aristotle's law, a force would alter motion and not just sustain it. Third, the concept of impetus itself was a great advance; it transferred the motive power from the medium to the moving object and also made possible the consideration of a void. Buridan is one of the founders of modern dynamics. His theory gained wide acceptance in his own century and for two centuries thereafter.

Perhaps projectile motion received this attention because by the thirteenth century improved weaponry, such as catapults, crossbows, and longbows, could throw projectiles over long and highly arched paths; a century later cannonballs were in use. Aristotle had said that a body can move only under one type of force at a time; if two were acting, one would destroy the other. Hence a body thrown up and out would move along a straight line until the "violent" motion was spent and then the body would drop straight to earth under a natural motion. Of the various revisions of this theory, the idea of Jordanus Nemorarius proved most helpful; he showed that the force under which a body thrown straight out moved at any instant could be resolved into two components, natural gravity acting downward and a "violent" horizontal force of projection. This idea was taken up by da Vinci, Stevin, Galileo, and Descartes.

The Parisian school of Buridan and Oresme went on to consider not only uniform but difform and uniformly difform motion and proved to its own satisfaction that the effective velocity in uniformly difform motion was the average of the initial and final velocities. Perhaps most significant about the efforts in thirteenth- and fourteenth-century mechanics was the attempt to introduce quantitative considerations and to replace qualitative by quantitative arguments.

The major interest of the medieval scientists lay in the field of optics. One reason for this is that the Greeks had laid a firmer foundation in what we now call geometrical optics than in other physical fields; and by the late medieval period their numerous works in optics were known in Europe. In addition, the Arabs had made some advances over the Greeks. By 1200 some of the basic laws of light were well known, including the straight-line motion of light in a uniform medium; the law of reflection; and Ptolemy's incorrect law of refraction (he believed the angle of refraction was proportional to the angle of incidence). Also the knowledge of spherical and parabolic mirrors, spherical aberration, the pinhole camera, the uses of lenses, the functioning of the eye, atmospheric refraction, and the possibility of magnification had passed on to Europe from the Greeks and the Arabs.

The scientists Grosseteste, Roger Bacon, Vitello (13th cent.), John Peckham (d. 1292), and Theodoric of Freiberg (d. c. 1311) made advances in light. Having learned of the refraction of light by lenses, they determined the focal lengths of some lenses, studied combinations of lenses, suggested magnification by combinations of lenses, and made improvements in the theory of the rainbow. Glass mirrors were perfected during the thirteenth century; spectacles date from 1299. Vitello observed the dispersion of light under refraction; that is, he produced colors from white light passed through a hexagonal crystal. He also directed light through a bowl of water to study the rainbow, since he had previously noted that when sunlight passes through a bowl of water the colors of the rainbow appear in the emerging light. Optics continued to be a major science; we shall find Kepler, Galileo, Descartes, Fermat, Huygens, and Newton active in it.

9. Summary

In science, as in other fields, the Middle Ages devoted itself to time-tested and authoritative works. The schools produced diligent excerpting from older manuscripts, summarizing, and commentaries. The spirit of the times forced minds to follow trusted, prescribed, and rigid ways. The search for a universal philosophy covering all phenomena of man, nature, and God is characteristic of the late medieval period. But the contributions suffered from an indistinctness of ideas, mysticism, dogmatism, and a commentatorial spirit directed toward the analysis of authorities.

Nevertheless, as the world gradually changed, awareness of discrepancy and conflict between beliefs and overt facts grew stronger, and the need for a revision of learning and beliefs grew clearer. Before Galileo demonstrated the value of experience, before Descartes taught people to look within themselves, and before Pascal formulated the idea of progress, it was the unconventional thinkers, largely the dissident Scholastics, who attempted to advance along new lines, to challenge established outlooks, and to place stronger reliance upon observation of nature than had the Greeks.

Experimentation, motivated in part by the search for magical powers, and the use of induction to obtain general principles or scientific laws began to be important sources of knowledge, despite the fact that the major scientific method of the late Middle Ages was rational explanation, presented in a formal or geometrical demonstration based on a priori principles.

The value of mathematics in the investigation of nature also received some recognition. Though on the whole the medieval scientists followed

Aristotle in looking for material or physical explanations, these were hard to obtain and not too helpful; they found more and more that it was easier to correlate observations and experimental results mathematically and then check the mathematical law. Thus in astronomy it was not Aristotle's physical modification of Eudoxus' theory that was used by the scientists concerned with astronomical theory proper, navigation, and the calendar, but Ptolemy's. As a consequence, mathematics began to play a larger role than Aristotle had assigned to it.

Despite the new trends and activities, it is doubtful that medieval Europe, if permitted to pursue an unchanging course, would have ever developed any real science and mathematics. Free inquiry was not permitted. The few universities already in existence by 1400 were controlled by the Church and the professors were not free to teach what they thought correct. If no scientific doctrines were condemned in the medieval period, it was only because no new major ones were promulgated. Any real dissent from Christian thought that appeared in any sphere was suppressed with dispatch, cruelty, and viciousness unequaled in history, largely through the Inquisitions initiated by Pope Innocent III in the thirteenth century.

Other factors, relatively minor, delayed changes in Europe. The revived Greek knowledge reached only the few scholars who had the leisure and opportunity to study it. Manuscripts were expensive; many who wanted to acquire them could not. Also, Europe in the period from 1100 to 1500 was broken up into a number of independent dukedoms, principalities, more or less democratic or oligarchic city-states, and the Papal States. The wars among all these political units were continuous and absorbed the energies of the people. The Crusades, which began about 1100, wasted a fantastic number of lives. The Black Death in the second half of the fourteenth century took about a third of the population of Europe and set back the entire civilization. Fortunately, forces of revolutionary strength did begin to exert their effects on the European intellectual, political, and social scene.

Bibliography

Ball, W. W. R.: *A Short Account of the History of Mathematics*, Dover (reprint), 1960, Chaps. 8, 10, 11.

Boyer, Carl B.: *A History of Mathematics*, John Wiley and Sons, 1968, Chap. 14.

Cajori, Florian: *A History of Mathematics*, Macmillan, 1919, pp. 113–29.

Clagett, Marshall: *The Science of Mechanics in the Middle Ages*, University of Wisconsin Press, 1959.

———: *Nicole Oresme and the Geometry of Qualities and Motions*, University of Wisconsin Press, 1968.

Crombie, A. C.: *Augustine to Galileo*, Falcon Press, 1952, Chaps. 1–5.

———: *Robert Grosseteste and the Origins of Experimental Science*, Oxford University Press, 1953.

Easton, Stewart: *Roger Bacon and His Search for a Universal Science*, Columbia University Press, 1952.

Hofmann, J. E.: *The History of Mathematics*, Philosophical Library, 1957, Chaps. 3–4.

Smith, David Eugene: *History of Mathematics*, Dover (reprint), 1958, Vol. 1, pp. 177–265.

11

The Renaissance

> It appears to me that if one wants to make progress in mathematics, one should study the masters and not the pupils.
>
> N. H. ABEL

1. *Revolutionary Influences in Europe*

In the period from about 1400 to about 1600, which we shall adopt as the period of the Renaissance (though this term is used to describe different chronological periods by various writers), Europe was profoundly shaken by a number of events that ultimately altered drastically the intellectual outlook and stirred up mathematical activity of unprecedented scale and depth.

The revolutionary influences were widespread. Political changes resulted from almost incessant wars involving every city and state of Europe. Italy, the mother of the Renaissance, is a prime example. Though the history of the Italian states in the fifteenth and sixteenth centuries is marred by constant intrigues, mass murders, and destruction caused by wars, the very fluid nature of political conditions and the establishment of some democratic governments were favorable to the rise of the individual. Wars against the papacy, a leading political and military power in that era, not only freed people from the domination of the Church but encouraged intellectual opposition as well.

Italy acquired great wealth during the latter part of the Middle Ages. This was largely due to its geographical position. Its seaports were most favorably located to import goods from Africa and Asia for shipment to the rest of Europe. Great banking houses made Italy the financial center. This wealth was essential to the support of learning. And it was in Italy, the most bedeviled of all countries and seething with turmoil, that the thoughts that were to mold Western civilization were first conceived and expressed.

During the fifteenth century the Greek works reached Europe in enormously greater numbers. In the early part of the century the connections between Rome and the Byzantine Empire, which possessed the largest collection of Greek documents but had been isolated, became stronger. The Byzantines, at war with the Turks, sought the help of the Italian states.

Under the improved relationships teachers of Greek were brought to Italy and Italians went to the Byzantine Empire to learn Greek. When the Turks conquered Constantinople in 1453, the Greek scholars fled to Italy, bringing more manuscripts with them. Thus not only were many more Greek works made available to Europe, but many of the newly acquired manuscripts were far better than the ones previously acquired in the twelfth and thirteenth centuries. The subsequent translations, made into Latin directly from the Greek, were more reliable than those made from Arabic.

Johann Gutenberg's invention, about 1450, of printing with movable type speeded up the spread of knowledge. Linen and cotton paper, which the Europeans learned about from the Chinese through the Arabs, replaced parchment and papyrus from the twelfth century on. Commencing in 1474, mathematical, astronomical, and astrological works appeared in printed form. For example, the first printed edition of Euclid's *Elements* in a Latin translation by Johannes Campanus (13th cent.) appeared in Venice in 1482. During the next century the first four books of Apollonius' *Conic Sections*, Pappus' works, and Diophantus' *Arithmetica*, as well as other treatises, appeared in printed editions.

The introduction of the compass and gunpowder had significant effects. The compass made possible navigation out of sight of land. Gunpowder, introduced in the thirteenth century, changed the method of warfare and the design of fortifications and made the study of the motion of projectiles important.

A new economic era was initiated by a tremendous growth in manufacturing, mining, large-scale agriculture and a great variety of trades. Each of these enterprises encountered technical problems that were tackled more vigorously than in any previous civilization. In contrast to the slave societies of Egypt, Greece, and Rome, and the serfdom of medieval feudalism, the new society possessed an expanding class of free artisans and laborers. Independent mechanics and wage-paying employers had an incentive to search for labor-saving devices. The competition of a capitalist economy also stimulated direct studies of physical phenomena and causal connections to improve materials and methods of production. Since the Church had offered explanations of many of these phenomena, conflicts arose. We may be sure that whenever the physical explanation proved more useful than the theological, the latter was ignored.

The merchant class contributed to a new order in Europe by promoting the geographical explorations of the fifteenth and sixteenth centuries. Prompted by the need for better trade routes and for sources of commodities, the explorations brought to Europe much knowledge of strange lands, plants, animals, climates, ways of life, beliefs, and customs. This knowledge challenged medieval dogmas and stimulated imaginations.

Doubts as to the soundness of Church science and cosmology raised by direct observation or by information filtering into Europe from explorers and traders, objections to the Church's suppression of experimentation and thought on the problems created by the new social order, the moral degeneration of some Church leaders, corrupt Church practices such as the sale of salvation, and, finally, serious doctrinal differences, all culminated in the Protestant Reformation. The reformers were supported by the merchants and princes, who were anxious to break the power of the Church.

The Reformation as such did not liberalize thought or free men's minds. The Protestant leaders aimed only to set up their own brand of dogmatism. However, in raising questions concerning the nature of the sacraments, the authority of Church government, and the meaning of passages in the Scriptures, Luther, Calvin, and Zwingli unintentionally encouraged many people to think and do what they would not otherwise have dared. Thought was stimulated and argumentation provoked. Further, to win adherents the Protestants professed that individual judgment, rather than papal authority, was the basis for belief. Thus, variations in belief were sanctioned. Called upon to choose between Catholic and Protestant claims, many declared "a plague on both your houses" and turned from the two faiths to nature, observation, and experimentation as the sources of knowledge.

2. The New Intellectual Outlook

The Church rested on authority, revered Aristotle, and branded doubt as criminal. It also deprecated material satisfactions and emphasized salvation of the soul for an afterlife. These tenets contrasted sharply with values the Europeans learned from the Greeks, though they were not pronounced in Aristotle: the study of nature, the enjoyment of the physical world, the perfecting of mind and body, freedom of inquiry and expression, and confidence in human reason. Irked by Church authority, restrictions on the physical life, and reliance on Scripture as the source of all knowledge and the authority for all assertions, the intellectuals grasped eagerly at the new values. Instead of endless disputations and wrangling as to the meaning of biblical passages to determine truth, men turned to nature itself.

Almost as a corollary to the revival of Greek knowledge and values came the revival of interest in mathematics. From the works of Plato especially, which became known in the fifteenth century, the Europeans learned that nature is mathematically designed and that this design is harmonious, aesthetically pleasing, and the inner truth. Nature is rational, simple and orderly, and it acts in accordance with immutable laws. Platonic and Pythagorean works also emphasized number as the essence of reality, a doctrine that had already received some attention from the deviating Scholastics of the thirteenth and fourteenth centuries. The revival of Platonism clarified and crystallized the ideas and methods with which these men had been struggling. The Pythagorean-Platonic emphasis on quantitative relations as the essence of reality gradually became dominant. Copernicus, Kepler, Galileo, Descartes, Huygens, and Newton were in this respect essentially Pythagoreans and by their work established the principle that the goal of scientific activity should be quantitative mathematical laws.

To the intellectuals of the Renaissance, mathematics appealed for still another reason. The Renaissance was a period in which the medieval civilization and culture were discredited as new influences, information, and revolutionary movements swept Europe. These men sought new and sound bases for the erection of knowledge, and mathematics offered such a foundation. Mathematics remained the one accepted body of truths amid crumbling philosophical systems, disputed theological beliefs, and changing ethical values. Mathematical knowledge was certain and offered a secure foothold in a morass; the search for truth was redirected toward it.

Mathematicians and scientists did receive some inspiration from the theological bias of the Middle Ages, which had inculcated the view that all the phenomena of nature are not only interconnected but operate in accordance with an overall plan: all the actions of nature follow the plan laid down by a single first cause. How, then, was the theological view of God's universe to be reconciled with the search for the mathematical laws of nature? The answer was a new doctrine, namely, that God designed the universe mathematically. In other words, by crediting God with being a supreme mathematician, it became possible to regard the search for the mathematical laws of nature as a religious quest.

This doctrine inspired the work of the sixteenth-, seventeenth-, and even some eighteenth-century mathematicians. The search for the mathematical laws of nature was an act of devotion; it was the study of the ways and nature of God and of His plan of the universe. The Renaissance scientist was a theologian, with nature instead of the Bible for his subject. Copernicus, Brahe, Kepler, Galileo, Pascal, Descartes, Newton, and Leibniz speak repeatedly of the harmony that God imparted to the universe through His *mathematical* design. Mathematical knowledge, since it is in itself truth about the universe, is as sacrosanct as any line of Scripture, indeed superior, because it is clear, undisputed knowledge. Galileo said, "Nor does God less admirably reveal Himself to us in Nature's actions than in the Scriptures' sacred dictions." To which Leibniz added, "*Cum Deus calculat, fit mundus*" (As God calculates, so the world is made). These men sought mathematical relations that would reveal the grandeur and glory of God's handiwork. Man could not hope to learn the divine plan as clearly as God himself understood it, but with humility and modesty, man could seek at least to approach the mind of God.

The scientists persisted in the search for mathematical laws underlying natural phenomena because they were convinced that God had incorporated these laws in His construction of the universe. Each discovery of a law of nature was hailed more as evidence of God's brilliance than of the brilliance of the investigator. Kepler in particular wrote paeans to God on the occasion of each discovery. The beliefs and attitudes of the mathematicians and scientists exemplify the larger cultural phenomenon that swept Renaissance Europe. The Greek works impinged on a deeply devout Christian world, and the intellectual leaders born in one and attracted by the other fused the doctrines of both.

3. *The Spread of Learning*

For several reasons the diffusion of the new values took place slowly. The Greek works were, at first, to be found only at the courts of secular and ecclesiastical princes and were not accessible to the people at large. Printing aided enormously in making books generally available, but the effect was gradual because even printed books were expensive. The problem of spreading knowledge was complicated by two additional factors. First, many of those who would gladly have put mathematics and science to use in industry, crafts, navigation, architecture, and other pursuits were uneducated; schooling was by no means common. The second factor was language. The learned—scholars, professors, and theologians—were at home with Latin and, to some extent, with Greek. But artists, artisans, and engineers knew only the vernacular—French, German, and the several Italian languages— and were not benefited by the Latin translations of the Greek works.

Beginning in the sixteenth century, many Greek classics were translated into the popular languages. Mathematicians themselves took a hand in this activity. For example, Tartaglia translated Euclid's *Elements* from Latin into Italian in 1543. The translation movement continued well into the seventeenth century; but it proceeded slowly because many scholars were hostile to the common people. The former were contemptuous of the latter; they preferred Latin because they believed that the weight of tradition attached to it would give authority to their pronouncements. To counter such scholars and to reach the public and obtain its support for his ideas, Galileo deliberately wrote in Italian. Descartes wrote in French for the same reason; he hoped that those who used their natural reason might be better judges of his teachings than those who slavishly respected the ancient writings.

Another measure to educate the public, adopted in a few cities of Italy, was the establishment of libraries. The Medici family financed libraries in Florence and some of the popes did the same in Rome. Partly to spread education and partly to provide a meeting place for communication among scholars, several liberal rulers founded academies. Most notable among these was the Florentine Academy of Design, founded by Cosimo I de' Medici (1519–74) in 1563, which became a center for mathematical studies, and the

Roman Accademia dei Lincei (Lynx-like) founded in 1603. Members of these academies translated Latin works into the popular languages, gave lectures to the public, and extended and deepened their own knowledge by communication with each other. These academies were the forerunners of the more famous ones founded later in England, France, Italy, and Germany that were so enormously helpful in spreading knowledge.

Regrettably the universities of the fifteenth and sixteenth centuries played almost no role in these developments. Theology ruled the universities and its study was the purpose of learning. Knowledge was considered complete and final. Hence experimentation was unnecessary, and the new discoveries of outsiders were ignored. The conservative university professors did their best to cling to medieval learning as formulated by the Scholastics since the thirteenth century. The universities did teach arithmetic, geometry, astronomy, and music, but the astronomy was based on Ptolemy and was not observational. Natural philosophy meant studying Aristotle's *Physics*.

4. *Humanistic Activity in Mathematics*

While the Scholastics adhered firmly to late medieval doctrines, a new group of humanists devoted themselves to collecting, organizing, and critically studying Greek and Roman learning. These men made assiduous studies, and with on the whole questionable sagacity strove to cleanse the texts of errors and to restore lost material. They slavishly accepted, repeated, and endlessly interpreted what they found in the ancient and medieval manuscripts, even undertaking philological studies to determine precise meanings. They also wrote many books that redid the ancient works in a Scholastic reinterpretation. Though this activity may have aroused an interest in learning, it nevertheless fostered the deception that learning consisted in deepening and confirming the accepted body of knowledge.

Typical of the humanists of the sixteenth century was the algebraist Jerome Cardan (Gerolamo Cardano), who was born in Pavia in 1501. His career as rogue and scholar is one of the most fascinating among the fantastic careers of Renaissance men. He gives an account of his life in his *De Vita Propria* (Book of My Life), written in his old age, in which he praises and abases himself. He says that his parents endowed him only with misery and scorn; he passed a wretched boyhood and was so poor for the first forty years of his life that he ceased to regard himself as poor because, as he put it, he had nothing left to lose. He was high-tempered, devoted to erotic pleasures, vindictive, quarrelsome, conceited, humorless, incapable of compunction, and purposely cruel in speech. Though he had no passion for gambling he played dice every day for twenty-five years and chess for forty years as an escape from poverty, chronic illnesses, calumnies, and injustices. In his *Liber de Ludo Aleae* (Book on Games of Chance), published posthumously in

1663, he says one should gamble for stakes to compensate for the time lost and he gives advice on how to cheat to insure that compensation.

After devoting his youth to mathematics, physics, and gambling, he graduated from the University of Pavia in medicine. He practiced this art, later taught it in Milan and Bologna, and became celebrated all over Europe as a physician. He also served as professor of mathematics in several Italian universities. In 1570 he was sent to prison for the heresy of casting the horoscope of Jesus. Astonishingly, the pope subsequently hired him as an astrologer. At seventy-five, shortly before his death in 1576, he boasted of fame, a grandson, wealth, learning, powerful friends, belief in God, and fourteen good teeth.

His writings embraced mathematics, astronomy, astrology, physics, medicine, and an enormous variety of other subjects, including moral aphorisms (to atone for his cheating at cards). Despite much training in the sciences, Cardan was a man of his times; he firmly believed in astrology, dreams, charms, palmistry, portents, and superstitions, and wrote many volumes on these subjects. He is the rational apologist for these occult arts, which, he contended, permit as much certitude as do navigation and medicine. He also wrote encyclopedic treatises on the inhabitants of the universe, that is, angels, demons, and various intelligences, including in these books material undoubtedly stolen from his father's distinguished friend, Leonardo da Vinci. The extant material fills about 7000 pages.

The syncretic tendency, exhibited by the natural philosophers in their writings, which united all of reality in gigantic works, is represented in mathematics by Cardan. With uncritical diligence he brought together from ancient, medieval, and contemporary sources the available mathematical knowledge in an encyclopedic mass, using theoretical and empirical sources equally. With his cherished magical and mystical number theory was associated his predilection for algebraic speculation, which he carried further than his contemporaries. In addition to being a celebrated doctor, Cardan is distinguished from the other learned natural philosophers of the sixteenth century by his deeper interest in mathematics. But mathematics was not method to him; it was a special magical talent and an emotionally charged form of speculation.

One of the lesser-known humanists, Ignazio Danti (1537–86), a professor of mathematics at Bologna, wrote a book on mathematics for laymen that reduced all pure and applied mathematics to a series of synoptic tables. *Le scienze matematiche ridotte in tavole* (Bologna, 1577) is characteristic of the classifying spirit of the times; it served to direct the course of mathematical instruction in the schools of the late sixteenth century. Danti was one of the few mathematicians and astronomers who advocated applied mathematics as a branch of learning (as did Galileo later). The subjects covered are significant because they show the range of mathematics at that time: arithmetic, geometry, music, astrology, goniometry (especially measurement of volumes), meteorology, dioptrics, geography, hydrography, mechanics, architecture, military architecture, painting, and sculpture. The first four topics represented pure mathematics; the remainder, applied mathematics.

Characteristic humanist efforts are also evident in the investigations into mechanics by such learned mathematicians as Guidobaldo del Monte (1545–1607), Bernadino Baldi (1553–1617), and Giovanni Battista Benedetti (1530–90). These men hardly grasped the theorems of Archimedes; the works of Pappus made more sense to them and had greater appeal because Pappus elaborated on the proofs given in the earlier Greek classics. They deviated little from the Scholastics in tackling the standard problems and limited themselves to the correction of individual statements and theorems. They accepted much that was false and, in addition, lacked the ability to separate the living, vital ideas from the dead ones. Their humanistic schooling inclined them to incorporate in Euclidean deductions all old and new knowledge, regardless of how this squared with experiments. Consequently their critical faculties were weakened and their own experience lost its value. Their experimentation was free of magical ingredients and their erudition was indeed primarily humanistic, but in principle and in essence they were the last of the medievalists rather than the founders of new methods of thought and research. The Italian mathematicians and physicists Francesco Maurolycus (1494–1575), Benedetti, Baldi, and del Monte, the men whom Galileo later generously called his teachers, and who in some respects prepared the way for him, did not, because of their dependence on ancient modes of thought, make ground-breaking contributions in formulating or solving mathematical or physical problems.

5. *The Clamor for the Reform of Science*

As in the previous centuries of its existence, mathematics was to derive its major inspirations and themes from physical science. However, for science to flourish it was essential that the Europeans break away from slavish adherence to authority. A number of men realized that the methodology of science must be changed; they initiated a real break with Scholasticism and the uncritical acceptance of Greek knowledge.

One of the earliest to call clearly for a new approach to knowledge was the famous Renaissance artist Leonardo da Vinci (1452–1519). Incredibly endowed both physically and mentally, he achieved greatness as a linguist, botanist, zoologist, anatomist, geologist, musician, sculptor, painter, architect, inventor, and engineer. Leonardo made quite a point of distrusting the knowledge that scholars professed so dogmatically. These men of book learning he described as strutting about puffed up and pompous, adorned not by their own labors but by the labors of others whose work they merely repeated; they were but the reciters and trumpeters of other people's learning. He also criticized the concepts, methods, and goals of the bookish scholars because they did not deal with the real world. He almost boasted that he was not a man of letters and could do bigger and better things by learning from experience. And indeed he learned for himself many facts of mathematics, some principles of mechanics, and the laws of equilibrium of the lever. He made remarkable observations on the flight of birds, the flow of water, the structure of rocks, and the structure of the human body. He studied light, color, plants, and animals. Famous are his words, "If you do not rest on the good foundations of nature you will labor with little honor and less profit." Experience, he says, is never deceptive though our judgment may be. "In the study of the sciences which depend upon mathematics, those who do not consult nature but authors, are not the children of nature but only her grandchildren."

Leonardo did believe in the combination of theory and practice. He says, "He who loves practice without theory is like the sailor who boards ship without a rudder and compass and never knows where he may be cast." On the other hand, he said, theory without practice cannot survive and dies as quickly as it lives. "Theory is the general; experiments are the soldiers." He wished to use theory to direct experiments.

Nevertheless, Leonardo did not fully grasp the true method of science. In fact, he had no methodology, nor any underlying philosophy. His work was that of a practical investigator of nature, motivated by aesthetic drives but otherwise undirected. He was interested in and sought quantitative relationships and in this respect was a forerunner of modern science. However, he was not as consciously quantitative as Galileo. While his writings on mechanics and science were used by such sixteenth-century men as Cardan, Baldi, Tartaglia, and Benedetti, they were not the stimulus for Galileo, Descartes, Stevin, and Roberval.

Leonardo's views on mathematics and his working knowledge and use of it were peculiar to his time and illustrate its spirit and approach. Reading Leonardo one finds many statements suggesting that he was a learned mathematician and profound philosopher who worked on the level of the professional mathematician. He says, for example, "The man who discredits the supreme certainty of mathematics is feeding on confusion, and can never silence the contradictions of sophistical sciences, which lead to eternal quackery . . . for no human inquiry can be called science unless it pursues its path through mathematical exposition and demonstration." To pass beyond observation and experience there was for him only one trustworthy road through deceptions and mirages—mathematics. Only by holding fast to mathematics could the mind safely penetrate the labyrinth of intangible and insubstantial thought. Nature works through mathematical laws and nature's forces and operations must be studied through quantitative investigations. These mathematical laws, which one must approach through experience, are the goal of the study of nature. On the basis of such pronouncements, no doubt, Leonardo is often credited with being a greater mathematician than he actually was. When one examines Leonardo's notebooks one realizes how little he knew of mathematics and that his approach was empirical and intuitive.

More influential in urging a reform of the methods of science was Francis Bacon (1561–1626). Bacon sought methods of obtaining truths in intellectual, moral, political, and physical spheres. Though changes were already occurring in the methods of physical science during the sixteenth century, the public at large and even many men of letters were not aware of this. Bacon's lofty eloquence, wide learning, comprehensive views, and bold pronouncements as to the future made men turn more than a passing gaze upon what was going on and note the "Great Instauration" he depicted. He formulated trenchant aphorisms that caught their attention. When people finally noted that science was beginning to make the advances Bacon advocated, they hailed him as the author and leader of the revolution he had merely perceived earlier. Actually he did understand the changes taking place better than his contemporaries.

The salient feature of his philosophy was the confident and emphatic announcement of a new era in the progress of science. In 1605 he published his treatise *Advancement of Learning*; it was followed by the *Novum Organum* (New Method) of 1620. In the latter book he is more explicit. He points out the feebleness and scanty results of prior efforts to study nature. Science, he says, has served only medicine and mathematics or been used to train immature youths. Progress lies in a change of method. All knowledge begins with observation. Then he makes his extraordinary contribution: an insistence on a "graduate and successive induction" instead of hasty generalization. Bacon says, "There are two ways, and can only be two, of seeking and finding truth. The one, from senses and particulars, takes a flight to the most general axioms, and from those principles and their truths, settled once for all, invents and judges of intermediate axioms. The other method collects axioms from senses and particulars, ascending continuously and by degrees, so that in the end it arrives at the more general axioms; this latter way is the true one, but hitherto untried." By "axioms" he means general propositions arrived at by induction and suited to be the starting point of deductive reasoning.

Bacon attacked the Scholastic approach to natural phenomena in these words: "It cannot be that the axioms discovered by argumentations should avail for the discovery of new works; since the subtlety of nature is greater many times over than the subtlety of arguments. . . . Radical errors in the first concoction of the mind are not to be cured by the excellence of functions and remedies subsequent. . . . We must lead men to the particulars themselves, while men on their side must force themselves for a while to lay their notions by and begin to familiarize themselves with the facts."

Bacon did not realize that science must measure so as to obtain quantitative laws. He did not see, in other words, what kinds of gradual inquiries were needed and the order in which they must be taken. Nor did he appreciate the inventive genius that all discovery requires. In fact, he says that "there is not much left to acuteness and strength of genius, but all degrees of genius and intellect are brought to the same level."

Though he did not create it, Bacon issued the manifesto for the experimental method. He attacked preconceived philosophical systems, brain creations, and idle displays of learning. Scientific work, he said, should not be entangled in a search for final causes, which belongs to philosophy. Logic and rhetoric are useful only in organizing what we already know. Let us close in on nature and come to grips with her. Let us not have desultory, haphazard experimentation; let it be systematic, thorough, and directed. Mathematics is to be a handmaiden to physics. In all, Bacon offered a fascinating program for future generations.

Another doctrine and program is associated with Francis Bacon, though it antedates him. The Greeks had, on the whole, been content to derive from their mathematics and science an understanding of nature's ways. The few early medieval scientists and the Scholastics studied nature largely to determine the final cause or purpose served by phenomena. However, the Arabs, a more practical people, studied nature to acquire power over it. Their astrologers, seers, and alchemists sought the elixir of life, the philosopher's stone, methods of converting less useful to more useful metals, and magic properties of plants and animals, in order to prolong man's life, heal his sicknesses, and enrich him materially. While these pseudosciences continued to flourish in medieval times, some of the more rational Scholastics—for example, Robert Grosseteste and Roger Bacon—began to envision the same goal, but by more proper scientific investigations. Thanks to Francis Bacon's exhortations, the mastery of nature became a positive doctrine and a pervasive motivation.

Bacon wished to put knowledge to use. He wanted to command nature for the service and welfare of man, not to please and delight scholars. As he put it, science was to ascend to axioms and then descend to works. In *The New Atlantis* Bacon describes a society of scholars provided with space and equipment for the acquisition of useful knowledge. He foresaw that science could provide man with "infinite commodities," "endow human life with inventions and riches, and minister to the conveniences and comforts of man." These, he says, are the true and lawful goals of science.

Descartes, in his *Discourse on Method*, echoed this thought:

> It is possible to attain knowledge which is very useful in life, and instead of that Speculative Philosophy which is taught in the Schools, we may

find a practical philosophy by means of which, knowing the force and the action of fire, water, air, the stars, the heavens, and all other bodies that environ us, as distinctly as we know the different crafts of our artisans, we can in the same way employ them in all those uses to which they are adapted, and thus render ourselves the masters and possessors of nature.

The chemist Robert Boyle said, "The good of mankind may be much increased by the naturalist's insight into the trades."

The challenge thrown out by Bacon and Descartes was quickly accepted, and scientists plunged optimistically into the task of mastering as well as understanding nature. These two motivations are still the major driving forces, and indeed the interconnections between science and engineering grew rapidly from the seventeenth century onward.

This program was taken up most seriously even by governments. The French Academy of Sciences, founded by Colbert in 1666, and the Royal Society of London, founded in 1662, were dedicated to the cultivation of "such knowledge as has a tendency to use" and to making science "useful as well as attractive."

6. *The Rise of Empiricism*

While reformers of science urged the return to nature and the need for experimental facts, practically oriented artisans, engineers, and painters were actually obtaining the hard facts of experience. Using the natural intuitive approach of ordinary men and seeking not ultimate meanings but merely helpful explanations of phenomena they encountered in their work, these technicians obtained knowledge that in effect mocked the sophistical distinctions, the long etymological derivations of meanings, the involved logical arguments, and the pompous citation of Greek and Roman authorities advanced by the learned scholars and even the humanists. Because the technical accomplishments of Renaissance Europe were superior to and more numerous than those achieved by any other civilization, the empirical knowledge acquired in their pursuit was immense.

The artisans, engineers, and artists had to grapple with real and workable mechanical ideas and properties of materials. Nevertheless, the new physical insights gained in this way were impressive. It was the spectacle makers who, without discovering a single law of optics, invented the telescope and the microscope. The technicians arrived at laws by attending to phenomena. By measured, gradual steps, such as no speculative view of scientific method has suggested, they obtained truths as profound and comprehensive as any conjecture had dared to anticipate. Whereas the theoretical reformers were bold, self-confident, hasty, ambitious, and contemptuous of antiquity, the practical reformers were cautious, modest, slow, and receptive to all knowledge, whether derived from tradition or from observation. They worked rather than speculated, dealt with particulars rather than generalities, and added to science instead of defining it or proposing how to obtain it. In physics, the plastic arts, and technical fields generally, experience rather than theory and speculation became the new source of knowledge.

Coupled with the pure empiricism of the artisans and, indeed, in part suggested by the problems they presented, was the gradual rise of systematic observation and experimentation, carried out largely by a more learned group. Greeks such as Aristotle and Galen had observed a great deal and had discussed the inductions that might be made on the basis of observations; but one cannot say that the Greeks ever possessed an experimental science. The Renaissance activity, very modest in extent, marks the beginning of the now vast scientific enterprise. The most significant groups of Renaissance experimentalists were the physiologists led by Andreas Vesalius (1514–64), the zoologists led by Ulysses Aldrovandi (1522–1605), and the botanists led by Andrea Cesalpino (1519–1603).

In the area of the physical sciences, the experimental work of William Gilbert (1540–1603) on magnetism was by far the most outstanding. In his famous *De Magnete* (1600), he stated explicitly that we must start from experiments. Though he respected the ancients because a stream of wisdom came from them, he scorned those who quoted others as authorities and did not experiment or verify what they were told. His series of carefully conducted, detailed, and simple experiments is a classic in the experimental method. He notes, incidentally, that Cardan, in *De Rerum Varietate*, had described a perpetual-motion machine and comments, "May the gods damn all such sham, pilfered, distorted works, which do but muddle the minds of the students."

We have been pointing out the variegated practical interests that led to a vast expansion of the study of nature and the consequent impulse to systematic experimentation. Side by side with this practical work, largely independent but not oblivious of it, some men pursued the larger goal of science—the understanding of nature. The work of the later Scholastics on falling bodies, described in the preceding chapter, was continued in the sixteenth century. Their predominant goal was to secure the basic laws of motion. The work on projectile motion, often described as a response to practical needs, was motivated far more by broad scientific interests in mechanics. The work of Copernicus and Kepler in astronomy (Chap. 12, sec. 5) was certainly motivated by the desire to improve astronomical theory. Even the artists of the Renaissance sought to penetrate to the essence of reality.

Fortunately technicians and scientists began to recognize common interests and to appreciate the assistance each could derive from the other. The technicians of the fifteenth and sixteenth centuries, the early engineers who had relied on manual dexterity, mechanical ingenuity, and sheer inventiveness and cared little for principles, became aware of the aid to practice they could secure from theory. On their side scientists became aware that the artisans were obtaining facts of nature that correct theory had to comprehend, and that they could secure pregnant suggestions for investigation from the work of the artisans. In the opening paragraph of his *Dialogues Concerning Two New Sciences* (the sciences are strength of materials and the theory of motion), Galileo acknowledges this inspiration for his investigations. "The constant activity which you Venetians display in your famous arsenal suggests to the studious mind a large field for investigation, especially that part of the work which involves mechanics; for in this department all types of instruments and machines are constantly being constructed by many artisans, among whom there must be some who, partly by inherited experiences and partly by their own observations, have become highly expert and clever in explanation."

The practical and purely scientific interests were fused in the seventeenth century. When the larger principles and problems had emerged from the empirical needs, and the mathematical knowledge of the Greeks had become fully available to the scientists, the latter were able to proceed more effectively with pure science. Without losing sight of the goal of understanding the design of the universe, they also willingly sought to aid practice. The outcome was an expansion in scientific activity on an unprecedented scale, plus far-reaching and weighty technical improvements that culminated in the Industrial Revolution.

The great importance of the beginnings of modern science for us is, of course, that it paved the way for the major developments in mathematics. Its immediate effect was involvement with concrete problems. Since the Renaissance mathematicians worked for the republics and princes and collaborated with architects and handworkers—Maurolycus was an engineer for the city of Messina, Baldi was a mathematician for the Duke of Urbino, Benedetti was the chief engineer for the Duke of Savoy, and Galileo was court mathematician for the Grand Duke of Tuscany—they took up the observations and experiences of the practical people. Up to the time of Galileo, the impact of the technicians and architects can be seen largely in the work of Nicolò Tartaglia (1499–1557), a genius who was self-taught in the science of his time. Tartaglia made the transition from the practical to the learned mathematician, singling out with discernment useful problems and observations from empirical knowledge. His uniqueness lies in this achievement and in his complete independence from the magical influences that characterize the work of his rival Cardan. Tartaglia's position is midway between Leonardo and Galileo—not merely chronologically, but because his work on the mathematics of dynamical problems raised that subject to a new science and influenced the forerunners of Galileo.

The long-range effect was that modern mathematics, guided by the

Platonic doctrine that it is the essence of reality, grew almost entirely out of the problems of science. Under the new directive to study nature and to obtain laws embracing observations and experimental results, mathematics broke away from philosophy and became tied to physical science. The consequence for mathematics was a burst of activity and original creation that was the most prolific in its history.

Bibliography

Ball, W. W. R.: *A Short Account of the History of Mathematics*, Dover (reprint), 1960, Chaps. 12–13.

Burtt, E. A.: *The Metaphysical Foundations of Modern Physical Science*, Routledge and Kegan Paul, 1932.

Butterfield, Herbert: *The Origins of Modern Science*, Macmillan, 1951, pp. 1–87.

Cajori, Florian: *A History of Mathematics*, 2nd ed., Macmillan, 1919, pp. 128–45.

Cardano, Gerolamo: *Opera Omnia*, Johnson Reprint Corp., 1964.

————: *The Book of My Life*, Dover (reprint), 1962.

————: *The Book on Games of Chance*, Holt, Rinehart and Winston, 1961.

Clagett, Marshall, ed.: *Critical Problems in the History of Science*, University of Wisconsin Press, 1959, pp. 3–196.

Crombie, A. C.: *Augustine to Galileo*, Falcon Press, 1952, Chaps. 5–6.

————: *Robert Grosseteste and the Origins of Experimental Science*, Oxford University Press, 1953, Chap. 11.

Dampier-Whetham, W. C. D.: *A History of Science*, Cambridge University Press, 1929, Chap. 3.

Farrington, B.: *Francis Bacon*, Henry Schuman, 1949.

Mason, S. F.: *A History of the Sciences*, Routledge and Kegan Paul, 1953, Chaps. 13, 16, 19, and 20.

Ore, O.: *Cardano: The Gambling Scholar*, Princeton University Press, 1953.

Randall, John H., Jr.: *The Making of the Modern Mind*, Houghton Mifflin, 1940, Chaps. 6–9.

Russell, Bertrand: *A History of Western Philosophy*, Simon and Schuster, 1945, pp. 491–557.

Smith, David Eugene: *History of Mathematics*, Dover (reprint), 1958, Vol. 1, pp. 242–65, and Chap. 8.

Smith, Preserved: *A History of Modern Culture*, Holt, Rinehart and Winston, 1940, Vol. 1, Chaps. 5–6.

Strong, Edward W.: *Procedures and Metaphysics*, University of California Press, 1936; reprinted by Georg Olms, 1966, pp. 1–134.

Taton, René, ed.: *The Beginnings of Modern Science*, Basic Books, 1964, pp. 3–51, pp. 82–177.

Vallentin, Antonina: *Leonardo da Vinci*, Viking Press, 1938.

White, Andrew D.: *A History of the Warfare of Science with Theology*, George Braziller (reprint), 1955.

12

Mathematical Contributions in the Renaissance

> The chief aim of all investigations of the external world should be to discover the rational order and harmony which has been imposed on it by God and which He revealed to us in the language of mathematics.   JOHANNES KEPLER

1. *Perspective*

Though the Renaissance men grasped only dimly the outlooks, values, and goals of the Greek works, they did take some original steps in mathematics; and they made advances in other fields that paved the way for the tremendous seventeenth-century upsurge in our subject.

The artists were the first to manifest the renewal of interest in nature and to apply seriously the Greek doctrine that mathematics is the essence of nature's reality. The artists were self-taught and learned through practice. Fragments of Greek knowledge filtered down to them, but on the whole they sensed rather than grasped the Greek ideas and intellectual outlook. To an extent this was an advantage because, lacking formal schooling, they were free of indoctrination. Also, they enjoyed freedom of expression because their work was deemed "harmless."

The Renaissance artists were by profession universal men—that is, they were hired by princes to perform all sorts of tasks from the creation of great paintings to the design of fortifications, canals, bridges, war machines, palaces, public buildings, and churches. Hence they were obliged to learn mathematics, physics, architecture, engineering, stonecutting, metalworking, anatomy, woodworking, optics, statics, and hydraulics. They performed manual work and yet tackled the most abstract problems. In the fifteenth century, at least, they were the best mathematical physicists.

To appreciate their contribution to geometry we must note their new goals in painting. In the medieval period the glorification of God and the illustration of biblical themes were the purposes of painting. Gilt backgrounds suggested that the people and objects portrayed existed in some heavenly region. Also the figures were intended to be symbolic rather than realistic. The painters produced forms that were flat and unnatural and did not deviate from the pattern. In the Renaissance the depiction of the real world became the goal. Hence the artists undertook to study nature in order to reproduce it faithfully on their canvases and were confronted with the mathematical problem of representing the three-dimensional real world on a two-dimensional canvas.

Filippo Brunelleschi (1377–1446) was the first artist to study and employ mathematics intensively. Giorgio Vasari (1511–74), the Italian artist and biographer, says that Brunelleschi's interest in mathematics led him to study perspective and that he undertook painting just to apply geometry. He read Euclid, Hipparchus, and Vitello on mathematics and optics and learned mathematics from the Florentine mathematician Paolo del Pozzo Toscanelli (1397–1482). The painters Paolo Uccello (1397–1475) and Masaccio (1401–28) also sought mathematical principles for a system of realistic perspective.

The theoretical genius in mathematical perspective was Leone Battista Alberti (1404–72), who presented his ideas in *Della pittura* (1435), printed in 1511. This book, thoroughly mathematical in character, also includes some work on optics. His other important mathematical work is *Ludi mathematici* (1450), which contains applications to mechanics, surveying, time-reckoning, and artillery fire. Alberti conceived the principle that became the basis for the mathematical system of perspective adopted and perfected by his artist successors. He proposed to paint what one eye sees, though he was well aware that in normal vision both eyes see the same scene from slightly different positions and that only through the brain's reconciliation of the two images is depth perceived. His plan was to further the illusion of depth by such devices as light and shade and the diminution of color with distance. His basic principle can be explained in the following terms. Between the eye and the scene he interposed a glass screen standing upright. He then imagined lines of light running from the eye or station point to each point in the scene itself. These lines he called a pyramid of rays or a projection. Where these rays pierced the glass screen (the picture plane), he imagined points marked out; the collection of points he called a section. The significant fact about it is that it creates the same impression on the eye as the scene itself, because the same lines of light come from the section as from the original scene. Hence the problem of painting realistically is to get a true section onto the glass screen or, in practice, on a canvas. Of course the section depends upon the position of the eye and the position of the screen. This means merely that different paintings of the same scene can be made.

Since the painter does not look through his canvas to determine the section, he must have rules based on mathematical theorems, which tell him how to draw it. Alberti furnished some correct rules[1] in his *Della pittura*, but did not give all the details. He intended his book as a summary to be supplemented by discussions with his fellow painters and, in fact, apologizes for his brevity. He tried to make his material concrete, rather than formal and rigorous, and so gave theorems and constructions without proof.

Beyond introducing the concepts of projection and section, Alberti raised a very significant question. If two glass screens are interposed between the eye and the scene itself, the sections on them will be different. Further, if the eye looks at the same scene from two different positions and in each case a glass screen is interposed between the eye and the scene, again the sections will be different. Yet all these sections convey the original figure. Hence they must have some properties in common. The question is, What is the mathematical relation between any two of these sections, or what mathematical properties do they have in common? This question became the starting point of the development of projective geometry.

Though a number of artists wrote books on mathematical perspective and shared Alberti's philosophy of art, we can mention here only one or two leaders. Leonardo believed that painting must be an exact reproduction of reality and that mathematical perspective would permit this. It was "the rudder and guide rope of painting" and amounted to applied optics and geometry. Painting for him was a science because it reveals the reality in nature; for this reason it is superior to poetry, music, and architecture. Leonardo's writings on perspective are contained in his *Trattato della pittura* (1651), compiled by some unknown author who used the most valuable of Leonardo's notes on the subject.

The painter who set forth the mathematical principles of perspective in fairly complete form is Piero della Francesca (*c.* 1410–92). He, too, regarded perspective as the science of painting and sought to correct and extend empirical knowledge through mathematics. His main work, *De prospettiva pingendi* (1482–87), made advances on Alberti's idea of projection and section. In general, he gives procedures useful to artists and his directions employ strips of paper, wood, and the like. To help the artist he, like Alberti, gives intuitively understandable definitions. He then offers theorems which he "demonstrates" by constructions or by an arithmetical calculation of ratios. He was the painter-mathematician and the scientific artist par excellence, and his contemporaries so regarded him. He was also the best geometer of his time.

However, of all the Renaissance artists, the best mathematician was the German Albrecht Dürer (1471–1528). His *Underweysung der Messung mid dem*

---

1. For some of these rules, and paintings constructed in accordance with them, see the author's *Mathematics in Western Culture*, Oxford University Press, 1953.

Zyrkel und Rychtscheyd (Instruction in Measuring with Compass and Straight-edge, 1525), a work primarily on geometry, was intended to pass on to the Germans knowledge Dürer had acquired in Italy and in particular to help the artists with perspective. His book, more concerned with practice than theory, was very influential.

The theory of perspective was taught in painting schools from the sixteenth century onward according to the principles laid down by the masters we have been discussing. However, their treatises on perspective had on the whole been precept, rule, and ad hoc procedure; they lacked a solid mathematical basis. In the period from 1500 to 1600 artists and subsequently mathematicians put the subject on a satisfactory deductive basis, and it passed from a quasi-empirical art to a true science. Definitive works on perspective were written much later by the eighteenth-century mathematicians Brook Taylor and J. H. Lambert.

2. *Geometry Proper*

The developments in geometry apart from perspective during the fifteenth and sixteenth centuries were not impressive. One of the geometric topics discussed by Dürer, Leonardo, and Luca Pacioli (*c.* 1445–*c.* 1514), an Italian monk who was a pupil of Piero della Francesca and a friend and teacher of Leonardo, was the inscription of regular polygons in circles. These men attempted such constructions with a straightedge and a compass of fixed opening, a limitation already considered by the Arab Abû'l-Wefâ, but they gave only approximate methods.

The construction of the regular pentagon was a problem of great interest because it arose in the design of fortifications. In the *Elements*, Book IV, Proposition 11, Euclid had given a construction not limited by a compass of fixed opening. The problem of giving an exact construction with this limitation was tackled by Tartaglia, Ferrari, Cardan, del Monte, Benedetti, and many other sixteenth-century mathematicians. Benedetti then broadened the problem and sought to solve all Euclidean constructions with a straightedge and a compass of fixed opening. The general problem was solved by the Dane George Mohr (1640–97) in his *Compendium Euclidis Curiosi* (1673).

Mohr also showed, in his *Euclides Danicus* (1672), that the constructions that can be performed with straightedge and compass can be performed with only a compass. Of course without a straightedge, one cannot draw the straight line joining two points; but given the two points, one can construct the points of intersection of the line and the circle, and given two pairs of points, one can construct the point of intersection of the two lines determined by the two pairs. The fact that a compass alone suffices to perform the Euclidean constructions was rediscovered by Lorenzo Mascheroni (1750–1800) and published in his *La geometria del compasso* (1797).

Another of the Greek interests, centers of gravity of bodies, was also taken up by the Renaissance geometers. Leonardo, for example, gave a correct and an incorrect method of finding the center of gravity of an isosceles trapezoid. He then gave, without proof, the location of the center of gravity of a tetrahedron, namely, the center is one-quarter of the way up the line joining the center of gravity of the triangular base to the opposite vertex.

Two novel geometric ideas appear in minor works by Dürer. The first of these is space curves. He starts with helical space curves and considers the projection of these curves on the plane. The projections are various types of spirals and Dürer shows how to construct them. He also introduces the epicycloid, which is the locus of a point on a circle that rolls on the outside of a fixed circle. The second idea is the orthogonal projection of curves and of human figures on two and three mutually perpendicular planes. This idea, which Dürer merely touched on, was developed in the late eighteenth century into the subject of descriptive geometry by Gaspard Monge.

The work of Leonardo, Piero, Pacioli, and Dürer in pure geometry is certainly not significant from the standpoint of new results. Its chief value was that it spread widely some knowledge of geometry, crude as that knowledge was by Greek standards. The fourth part of Dürer's *Underweysung*, together with Piero's *De Corporibus Regularibus* (1487) and Pacioli's *De Divina Proportione* (1509), renewed interest in stereometry (mensuration of solid figures), which flourished in Kepler's time.

Another geometrical activity, map-making, served to stimulate further geometrical investigations. The geographical explorations had revealed the inadequacies of the existing maps; at the same time new geographical knowledge was being uncovered. The making and printing of maps was begun in the second half of the fifteenth century at centers such as Antwerp and Amsterdam.

The problem of map-making arises from the fact that a sphere cannot be slit open and laid flat without distorting distance. In addition, directions (angles) or area, or both, may be distorted. The most significant new method of map-making is due to Gerhard Kremer, known also as Mercator (1512–94), who devoted his life to the science. In 1569 he put out a map using the famous Mercator projection. In this scheme the lines of latitude and longitude are straight. The longitude lines are equally spaced, but the spacing between latitude lines is increased. The purpose of this increase is to keep the ratio of a length of one minute of longitude to one minute of latitude correct. On the earth a change of 1′ in latitude equals 6087 feet; but only on the equator is a change of 1′ in longitude equal to 6087 feet. For example, at latitude 20° a change of 1′ in longitude equals 5722 feet, producing the ratio

$$\frac{1' \text{ change in longitude}}{1' \text{ change in latitude}} = \frac{5722}{6087}.$$

For this true ratio to hold on Mercator's straight-line map, where longitude lines are equally spaced and each minute of change equals 6087 feet, he increases the spaces between latitude lines by the factor $1/\cos L$ as latitude L increases. At 20° latitude on his map a 1′ change in latitude equals a distance of 6087 $(1/\cos 20°)$, or 6450 feet. Thus at latitude 20°

$$\frac{1' \text{ change in longitude}}{1' \text{ change in latitude}} = \frac{6087}{6450},$$

and this ratio equals the true ratio of 5722/6087.

The Mercator map has several advantages. Only on this projection are two points on the map at correct compass course from one another. Then a course of constant compass bearing on the sphere, that is, a curve called a loxodrome or rhumb line, which cuts all meridians at the same angle, becomes a straight line on the map. Distance and area are not preserved; in fact, the map distorts badly around the poles. However, since direction is preserved, so is the angle between two directions at a point, and the map is said to be conformal.

Though no large new mathematical ideas emerged from the sixteenth-century work on map-making, the problem was taken over later by mathematicians and led to work in differential geometry.

3. Algebra

Up to the appearance of Cardan's *Ars Magna* (1545), which we shall treat in the next chapter, there were no Renaissance developments of any consequence in algebra. However, the work of Pacioli is worth noting. Like most others of his century he believed that mathematics is the broadest systematic learning and that it applies to the practical and spiritual life of all people. He also realized the advantages of theoretical knowledge for practical work. Theory must be master and guide, he tells the mathematicians and technicians. Like Cardan he belonged to the humanist circle. Pacioli's major publication is the *Summa de Arithmetica, Geometria, Proportione et Proportionalita* (1494). The *Summa* was a compendium of the available knowledge and was representative of the times because it linked mathematics with a great variety of practical applications.

The contents covered the Hindu-Arabic number symbols, which were already in use in Europe, business arithmetic, including bookkeeping, the algebra thus far created, a poor summary of Euclid's *Elements*, and some trigonometry taken from Ptolemy. The application of the concept of proportion to reveal design in all phases of nature and in the universe itself was a major theme. Pacioli called proportion "mother" and "queen" and applied it to the sizes of the parts of the human body, to perspective, and even to color mixing. His algebra is rhetorical; he follows Leonardo and the Arabs in calling the unknown quantity the "thing." The square of the unknown Pacioli called *census*, which he sometimes abbreviates as *ce* or *Z*; the cube of the unknown, *cuba*, is sometimes represented by *cu* or *C*. Other abbreviations for words, such as *p* for plus and *æ* for *æqualis*, also occur. In writing equations, whose coefficients are always numerical, he puts terms on the side that permits the use of positive coefficients. Though an occasional subtraction of a term, for example, $-3x$, does appear, no purely negative numbers are used; only the positive roots of equations are given. He used algebra to calculate geometrical quantities much as we would use an arithmetic proportion to relate the lengths of sides of similar triangles and perhaps to find one unknown length, though Pacioli's uses are often more involved. He closes his book with the remark that the solution of $x^3 + mx = n$ and of $x^3 + n = mx$ (we use modern notation) are as impossible as the quadrature of the circle.

Though there was nothing original in the *Summa*, this book and his *De Divina Proportione* were valuable because they contained much more than was taught in the universities. Pacioli served as the intermediary between what existed in the scholarly works and the knowledge acquired by artists and technicians. He tried to help these men learn and use mathematics. Nevertheless it is a significant commentary on the mathematical development of arithmetic and algebra between 1200 and 1500 that Pacioli's *Summa*, appearing in 1494, contained hardly anything more than Leonardo of Pisa's *Liber Abaci* of 1202. In fact, the arithmetic and algebra in the *Summa* were based on Leonardo's book.

4. *Trigonometry*

Until 1450, trigonometry was largely spherical trigonometry; surveying continued to use the geometric methods of the Romans. About that date plane trigonometry became important in surveying, though Leonardo of Pisa in his *Practica Geometriae* (1220) had already initiated the method.

New work in trigonometry was done by Germans of the late fifteenth and early sixteenth centuries, who usually studied in Italy and then returned to their native cities. At the time Germany had become prosperous, some of the wealth having been acquired by the Hanseatic League of North Germany, which controlled much trade; hence merchant patrons were able to support the work of many of the men we shall mention. The trigonometric work was motivated by navigation, calendar-reckoning, and astronomy, interest in the last-mentioned field having been heightened by the creation of the heliocentric theory, about which we shall say more later.

George Peurbach (1423–61) of Vienna started to correct the Latin translations of the *Almagest*, which had been made from Arabic versions, but which he proposed to make from the original Greek. He also began to make more accurate trigonometric tables. However, Peurbach died young and his work was taken up by his pupil Johannes Müller (1436–76), known as Regiomontanus, who vitalized trigonometry in Europe. Having studied astronomy and trigonometry at Vienna under Peurbach Regiomontanus went to Rome, studied Greek under Cardinal Bessarion (*c.* 1400–72), and collected Greek manuscripts from the Greek scholars who had fled the Turks. In 1471 he settled in Nuremberg under the patronage of Bernard Walther. Regiomontanus made translations of several Greek works—the *Conic Sections* of Apollonius and parts of Archimedes and Heron—and founded his own press to print them.

Following Peurbach, he adopted the Hindu sine, that is, the half chord of the half arc, and then constructed a table of sines based on a radius of 600,000 units and another based on a radius of 10,000,000 units. He also calculated a table of tangents. In the *Tabulae Directionum* (written in 1464–67), he gave five-place tangent tables and decimal subdivision of the angles, a very unusual procedure for those times.

Many men of the fifteenth and sixteenth centuries constructed tables, among them George Joachim Rhaeticus (1514–76), Copernicus, François Vieta (1540–1603), and Bartholomäus Pitiscus (1561–1613). Characteristic of this work was the use of a larger and larger number of units in the radius so that the values of the trigonometric quantities could be obtained more accurately without the use of fractions or decimals. For example, Rhaeticus calculated a table of sines based on a radius of 10^{10} units and another based on 10^{15} units, and gave values for every 10 seconds of arc. Pitiscus in his *Thesaurus* (1613) corrected and published the second table of Rhaeticus. The word "trigonometry" is his.

More fundamental was the work on the solution of plane and spherical triangles. Until about 1450 spherical trigonometry consisted of loose rules based on Greek, Hindu, and Arabic versions, the last of which came from Spain. The works of the Eastern Arabs Abû'l-Wefâ and Nâsir-Eddin were not known to Europe until this time. Regiomontanus was able to take advantage of Nâsir-Eddin's work, and in *De Triangulis*, written in 1462–63, put together in more effective fashion the available knowledge of plane trigonometry, spherical geometry, and spherical trigonometry. He gave the law of sines for spherical triangles, namely,

$$\frac{\sin a}{\sin A} = \frac{\sin b}{\sin B} = \frac{\sin c}{\sin C}$$

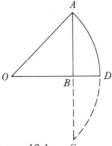

Figure 12.1

and the law of cosines involving the sides, that is,

$$\cos a = \cos b \cos c + \sin b \sin c \cos A.$$

The *De Triangulis* was not published until 1533; in the meantime, Johann Werner (1468–1528) improved and published Regiomontanus's ideas in *De Triangulis Sphaericis* (1514).

For many years after the work of Regiomontanus spherical trigonometry continued to be troubled by the need for a multitude of formulas, partly because Regiomontanus in his *De Triangulis*, and even Copernicus a century later, used only the sine and cosine functions. Also the negative values of the cosine and tangent functions for obtuse angles were not recognized as numbers.

Rhaeticus, who was a pupil of Copernicus, changed the meaning of sine. Instead of speaking of *AB* (Fig. 12.1) as the sine of *AD*, he spoke of *AB* as the sine of angle *AOB*. However, the length of *AB* was still expressed in a number of units dependent on the number of units chosen for the length of the radius. As a consequence of Rhaeticus's change, the triangle *OAB* became the basic structure and the circle with radius *OA* incidental. Rhaeticus used all six functions.

Plane and spherical trigonometry were further systematized and extended slightly by François Vieta, a lawyer by profession but recognized far more as the foremost mathematician of the sixteenth century. His *Canon Mathematicus* (1579) was the first of his many works on trigonometry. Here he gathered together the formulas for the solution of right and oblique plane triangles, including his own contribution, the law of tangents:

$$\frac{a - b}{a + b} = \frac{\tan \left(\dfrac{A - B}{2} \right)}{\tan \left(\dfrac{A + B}{2} \right)}.$$

For spherical right triangles he gave the complete set of formulas needed to calculate any one part in terms of two other known parts, and the rule for remembering this collection of formulas, which we now call Napier's rule. He also contributed the law of cosines involving the angles of an oblique spherical triangle:

$$\cos A = -\cos B \cos C + \sin B \sin C \cos a.$$

Many trigonometric identities had been established by Ptolemy; Vieta added to these. For example, he gave the identity

$$\sin A - \sin B = 2 \cos \frac{A + B}{2} \sin \frac{A - B}{2}$$

and identities for $\sin n\theta$ and $\cos n\theta$ in terms of $\sin \theta$ and $\cos \theta$. The latter identities are contained in his *Sectiones Angulares*, published posthumously in 1615.[2] Vieta expressed and worked with the identities in algebraic form, though the notation was by no means modern.

He used the formula for $\sin n\theta$ to solve the problem posed by the Belgian mathematician Adrianus Romanus (1561–1615) in his book *Ideae Mathematicae* (1593) as a challenge to all Frenchmen. The problem was to solve an equation of the forty-fifth degree in x. Henry IV of France sent for Vieta, who recognized that the problem amounted to this: Given the chord of an arc, to find the chord of the forty-fifth part of that arc. This is equivalent to expressing $\sin 45A$ in terms of $\sin A$ and finding $\sin A$. If $x = \sin A$ then the algebraic equation is of the forty-fifth degree in x. Vieta knew that problem could be solved by breaking this equation up into a fifth-degree equation and two third-degree equations; these he solved quickly. He gave the 23 positive roots but ignored the negative ones. In his *Responsum* (1595)[3] he explained his method of solution.

In the sixteenth century, trigonometry began to break away from astronomy and acquire status as a branch of mathematics. The application to astronomy continued to be extensive, but other applications—as, for example, surveying—warranted the study of the subject from a more detached viewpoint.

5. *The Major Scientific Progress in the Renaissance*

The Renaissance mathematicians prepared the ground for the upsurge of mathematical study in Europe by translating the Greek and Arabic works and by compiling encyclopedic works on the existing knowledge. But the motivations and directions of the subsequent mathematical creations of the Europeans stemmed primarily from scientific and technological problems,

2. *Opera*, Leyden, 1646, 287–304.
3. *Opera*, 305–24.

though there were some exceptions. The rise of algebra was, at least at the outset, a continuation of Arab lines of activity; and some new work in geometry was suggested by problems raised by artists.

By far the most significant Renaissance development in motivating the mathematics of the next two centuries was the revolution in astronomy led by Copernicus and Kepler. When the Greek works became available, after about 1200, both Aristotle's astronomical theory (a modification of Eudoxus') and Ptolemy's theory became widely known and were pitted against each other. Strictly speaking, various additions to both schemes had been introduced by the Arabs and the late medieval astronomers to improve the accuracy of both systems or to accommodate Aristotle's scheme to Christian theology. Ptolemy's scheme, reasonably accurate for the time, was purely mathematical and hence regarded only as a hypothesis, not a description of real structures. Aristotle's theory was the one most men accepted, though Ptolemy's was more useful for astronomical predictions, navigation, and calendar-reckoning.

A few Arab, late medieval, and Renaissance figures, including al-Bîrûnî (973–1048), Oresme, and Cardinal Nicholas of Cusa (1401–64), perhaps responding to Greek ideas, seriously considered that the earth might be rotating and that it might be equally possible to build an astronomical theory on the basis of the earth moving around the sun, but none worked out a new theory.

Among astronomers, Nicholas Copernicus suddenly appeared as a colossus. Born in Thorn, Poland, in 1473, Copernicus studied mathematics and science at the University of Cracow. At the age of twenty-three he went to Bologna to further his studies and there became familiar with Pythagorean and other Greek doctrines, including astronomical theory. He also studied medicine and church law. In 1512 he returned to Poland to become canon (an administrator) at the Cathedral of Frauenberg where he remained until his death in 1543. While performing his duties he devoted himself to intensive studies and observations that culminated in a revolutionary astronomical theory. This achievement in the domain of thought outclasses in significance, boldness, and grandeur the conquest of the seas.

It is difficult to determine what caused Copernicus to overthrow the fourteen-hundred-year-old Ptolemaic theory. The indications in the preface of his classic work, *De Revolutionibus Orbium Coelestium* (On the Revolutions of the Heavenly Spheres, 1543) are incomplete and somewhat enigmatic. Copernicus states that he was aroused by divergent views on the accuracy of the Ptolemaic system, by the view that the Ptolemaic theory was just a convenient hypothesis, and by the conflict between adherents of the Aristotelian and Ptolemaic theories.

Copernicus retained some principles of Ptolemaic astronomy. He used the circle as the basic curve on which his explanation of the motions of the heavenly bodies was to be constructed. Moreover, like Ptolemy, he used the fact that the motion of the planets must be built up through a sequence of motions with constant speed. His reason was that a change in speed could be caused only by a change in motive power, and since God, the cause of motion, was constant, the effect had to be, as well. He also took over the Greek scheme of epicyclic motion on a deferent. But Copernicus objected to the uniform equant motion utilized by Ptolemy because this motion did not call for uniform linear velocity.

By using Aristarchus' idea of putting the sun rather than the earth at the center of each deferent, Copernicus was able to replace the complicated diagrams formerly required to portray the motion of each heavenly body by much simpler ones. Instead of 77 circles he needed only 34 to explain the motion of the moon and the six known planets. Later he refined this scheme somewhat by putting the sun near but not quite at the center of the system.

Copernicus's theory was not in better accord with observations than the current modifications of Ptolemy's theory. The merit of his system was rather that it made the motion of the earth around the sun account for the major irregularities in the motions of the planets without employing as many epicycles. Moreover, his scheme treated all the planets in the same general way, whereas Ptolemy had used somewhat different schemes for the inner planets, Mercury and Venus, than for the outer planets, Mars, Jupiter, and Saturn. Finally, the calculation of the positions of the heavenly bodies was simpler in Copernicus's scheme, so much so that even in 1542 astronomers, using his theory, began preparation of new tables of celestial positions.

Copernicus's theory met both sound and prejudiced opposition. The discrepancies between Copernican theory and observations caused Tycho Brahe (1546–1601) to abandon the theory and seek a compromise. Vieta, for the same reason, rejected it altogether and turned to improving Ptolemaic theory. Most intellectuals rejected the theory either because they failed to understand it or because they would not entertain revolutionary ideas. The mathematics certainly was difficult to understand; as Copernicus himself said in his preface, the book was addressed to mathematicians. The observation of a new star by Brahe and German astronomers in 1572 did help; the sudden appearance and disappearance of stars contradicted the Aristotelian and Scholastic dogma of the invariability of the heavens.

The fate of the heliocentric theory would have been very uncertain were it not for the work of Johannes Kepler (1571–1630). He was born in Weil, a city in the Duchy of Württemberg. His father was a drunkard who turned from being a soldier of fortune to keeping a tavern. While attending elementary school, Kepler had to help in the tavern. He was then withdrawn from school and sent to work as a laborer in the fields. When still a boy he contracted smallpox, which left him with crippled hands and impaired eyesight. However, he managed to get a bachelor's degree at the College of Maulbronn in 1588; then, headed for the ministry, he studied at the University of Tübingen. There a friendly professor of mathematics and astronomy, Michael Mästlin (Möstlin, 1550–1631) taught him Copernican theory privately. Kepler's superiors at the university questioned his devoutness and in 1594 offered him a professorship of mathematics and morals at Grätz in Austria, which Kepler accepted. To fulfill his duties he was required to know astrology; this turned him further toward astronomy.

Kepler was expelled from Grätz when the city came under Catholic control, and he became an assistant to Tycho Brahe in the latter's observatory at Prague. When Brahe died, Kepler was appointed to his place. Part of his job was to cast horoscopes for his employer, Emperor Rudolph II. Kepler consoled himself with the thought that astrology enabled astronomers to make a living.

During his entire life Kepler was harassed by all sorts of difficulties. His first wife and several children died. As a Protestant, he suffered in various ways from persecution by Catholics. He was often in desperate financial state. His mother was accused of witchcraft and Kepler had to defend her. Yet throughout all his misfortunes he pursued his scientific work with perseverance, extraordinary labor, and fertile imagination.

In his approach to scientific problems Kepler is a transitional figure. Like Copernicus and the medieval thinkers he was attracted by a pretty and rational theory. He accepted the Platonic doctrine that the universe is ordered according to a preestablished mathematical plan. But unlike his predecessors, he had an immense regard for facts. His more mature work was based entirely on facts and advanced from facts to laws. In the search for laws he displayed inventiveness in hypotheses, a love of truth, and a lively fancy playing with but not obstructing reason. Though he devised a great number of hypotheses he did not hesitate to discard them when they did not fit the facts.

Moved by the beauty and harmony of Copernicus's system, he decided to devote himself to the search for whatever additional geometrical harmonies the much more accurate observations supplied by Tycho Brahe might permit. His search for the mathematical relationships of whose existence he was convinced led him to spend years following up false trails. In the preface to his *Mysterium Cosmographicum* (1596), we find him saying: "I undertake to prove that God, in creating the universe and regulating the order of the cosmos, had in view the five regular bodies of geometry as known since the days of Pythagoras and Plato, and that he has fixed according to those dimensions, the number of heavens, their proportions, and the relations of their movements."

And so he postulated that the radii of the orbits of the six planets were the radii of spheres related to the five regular solids in the following way. The

Figure 12.2   Figure 12.3

largest radius was that of the orbit of Saturn. In a sphere of this radius he
supposed a cube to be inscribed. In this cube a sphere was inscribed whose
radius should be that of the orbit of Jupiter. In this sphere he supposed a
tetrahedron to be inscribed and in this, in turn, another sphere, whose
radius was to be that of the orbit of Mars, and so on through the five regular
solids. This allowed for six spheres, just enough for the number of planets
then known. However, the deductions from this hypothesis were not in
accord with observations and he abandoned the idea, but not before he
had made extraordinary efforts to apply it in modified form.

Although the attempt to use the five regular solids to ferret out nature's
secrets did not succeed, Kepler was eminently successful in later efforts to
find harmonious mathematical relations. His most famous and important
results are known today as Kepler's three laws of planetary motion. The
first two were published in a book of 1609 bearing a long title, which is
sometimes shortened to *Astronomia Nova* and sometimes to *Commentaries on the
Motions of Mars*.

The first law states that the path of each planet is not the resultant of
a combination of moving circles but is an ellipse with the sun at one focus
(Fig. 12.2). Kepler's second law is best understood in terms of a diagram
(Fig. 12.3). The Greeks, we saw, believed that the motion of a planet must
be explained in terms of constant linear speeds. Kepler, like Copernicus, at
first held firmly to the doctrine of constant speeds. But his observations
compelled him to abandon this cherished belief too. His joy was great when
he was able to replace it by something equally attractive, for his conviction
that nature follows mathematical laws was reaffirmed. If MM' and NN'
are distances traversed by a planet in equal intervals of time, then, according
to the principle of constant speed, MM' and NN' would have to be equal
distances. However, according to Kepler's second law, MM' and NN' are
generally not equal but the *areas SMM'* and *SNN'* are equal. Thus Kepler
replaced equal distances by equal areas. To wrest such a secret from the
planets was indeed a triumph, for the relationship described is by no means
as easily discernible as it may appear to be on paper.

Kepler made even more extraordinary efforts to obtain the third law of
motion. This law says the square of the period of revolution of any planet is
equal to the cube of its average distance from the sun, provided the period
of the earth's revolution and its distance from the sun are the units of time

and distance.[4] Kepler published this result in *The Harmony of the World* (1619).

Kepler's work is far more revolutionary than Copernicus's; equally daring in adopting heliocentrism, Kepler broke radically from authority and tradition by utilizing the ellipse (as opposed to a composition of circular motions) and nonuniform velocities. He hewed firmly to the position that scientific investigations are independent of all philosophical and theological doctrines, that mathematical considerations alone should determine the wisdom of any hypothesis, and that the hypotheses and deductions from them must stand the test of empirical confirmation.

The work of Copernicus and Kepler is remarkable on many accounts, but we must confine ourselves to its relevance to the history of mathematics. In view of the many serious counter-arguments against a heliocentric theory, their work demonstrates how strongly the Greek view that the truths of nature lie in mathematical laws had already taken hold in Europe.

There were weighty scientific objections, many of which had already been advanced by Ptolemy against Aristarchus' suggestion. How could such a heavy body as the earth be set into and kept in motion? The other planets were in motion, even according to Ptolemaic theory, but the Greeks and medieval thinkers had maintained that these were composed of some special light substance. There were other objections. Why, if the earth rotates from west to east, does an object thrown up into the air not fall back to the west of its original position? Why doesn't the earth fly apart in its rotation? Copernicus's very weak answer to the latter question was that the sphere is a natural shape and moves naturally, and hence the earth would not destroy itself. Further, it was asked, why do objects on the earth and the air itself stay with an earth rotating at about 3/10 of a mile per second and revolving around the sun at the rate of about 18 miles per second? If, as Ptolemy and Copernicus believed, the speed of a body in natural motion is proportional to its weight, the earth should leave behind it objects of lesser weight. Copernicus replied that the air possessed "earthiness" and would therefore stay with the earth.

There were additional scientific objections from astronomers. If the earth moved, why did the direction of the "fixed" stars not change? An angle of parallax of 2' required that the distance of the stars be at least four million times the radius of the earth; such a distance was inconceivable at that time. Not detecting any parallax of the stars (which implied that they had to be even farther away), Copernicus declared that "the heavens are immense by comparison with the Earth and appear to be of infinite size.... The bounds of the universe are unknown and unknowable." Then, realizing the inadequacy of this answer, he assigned the problem to the philosophers

4. Though Kepler stated it this way, the correct statement calls for replacing average distance by the semimajor axis.

and thereby evaded it. Not until 1838 did the mathematician Bessel measure the parallax of one of the nearest stars and find it to be 0.31″.

If Copernicus and Kepler had been "sensible" men, they would never have defied their senses. We do not feel either the rotation or the revolution of the earth despite the high speeds involved. On the other hand, we do see the motion of the sun.

Copernicus and Kepler were highly religious; yet both denied one of the central doctrines of Christianity, that man, the chief concern of God, was at the center of the universe, and everything in the universe revolved around him. In contrast, the heliocentric theory, by putting the sun at the center of the universe, undermined this comforting dogma of the Church, because it made man appear to be just one of a possible host of wanderers drifting through a cold sky. It seemed less likely that he was born to live gloriously and to attain paradise upon his death. Less likely, too, was it that he was the object of God's ministrations. Thus, by displacing the earth, Copernicus and Kepler removed a cornerstone of Catholic theology and imperiled its structure. Copernicus pointed out that the universe is so immense, compared to the earth, that to speak of a center is meaningless. However, this argument put him all the more into opposition with religion.

Against all the objections Copernicus and Kepler had only one retort, but a weighty one. Each had achieved mathematical simplification and, indeed, overwhelmingly harmonious and aesthetically superior theory. If mathematical relationships were to be the goal of scientific work, and if a better mathematical account could be given, then this fact, reinforced by the belief that God had designed the world and would clearly have used the superior theory, was sufficient to outweigh all objections. Each felt, and clearly stated, that his work revealed the harmony, symmetry, and design of the divine workshop and overpowering evidence of God's presence. Copernicus could not restrain his gratification: "We find, therefore, under this orderly arrangement, a wonderful symmetry in the universe, and a definite relation of harmony in the motion and magnitude of the orbs, of a kind that it is not possible to obtain in any other way." The very title of Kepler's work of 1619, *The Harmony of the World*, and endless paeans to God, expressing satisfaction with the grandeur of God's mathematical design, attest to his beliefs.

It is not surprising that at first only mathematicians supported the heliocentric theory. Only a mathematician, and one convinced that the universe was mathematically designed, would have had the mental fortitude to disregard the prevailing philosophical, religious, and physical beliefs. Not until Galileo focused his telescope on the heavens did astronomical evidence favor the mathematical argument. Galileo's observations, made in the early 1600s, revealed four moons circling around Jupiter, showing that planets could have moons. It followed that the earth, too, need not be more than a planet just because it had a moon. Galileo also observed that the moon had a rough surface and mountains and valleys like the earth. Hence the earth was also likely to be just one more heavenly body and not necessarily the center of the universe.

The heliocentric theory finally won acceptance because it was simpler for calculations, because of its superior mathematics, and because observations supported it. This meant that the science of motion had to be recast in the light of a rotating and revolving earth. In brief, a new science of mechanics was needed.

Investigations into light and optics continued in an unbroken line from what we have already noted in the medieval period. In the sixteenth century the astronomers became more interested in these subjects because the refractive effect of air on light changes the direction of light rays as they come from the planets and stars and so gives misleading information as to the directions of these bodies. Toward the end of the sixteenth century, the telescope and microscope were invented. The scientific uses of these instruments are obvious; they opened up new worlds, and interest in optics, already extensive, surged still higher. Almost all of the seventeenth-century mathematicians worked on light and lenses.

6. *Remarks on the Renaissance*

The Renaissance did not produce any brilliant new results in mathematics. The minor progress in this area contrasts with the achievements in literature, painting, and architecture, where masterpieces that still form part of our culture were created, and in science, where the heliocentric theory eclipsed the best of Greek astronomy and dwarfed any Arabic or medieval contributions. For mathematics the period was primarily one of absorption of the Greek works. It was not so much a rebirth as a recovery of an older culture.

Equally important for the health and growth of mathematics was that it had once more, as in Alexandrian times, reestablished its intimate connections with science and technology. In science the realization that mathematical laws are the goal, the be-all and the end-all, and in technology, the appreciation that the mathematical formulation of the results of investigations was the soundest and most useful form of knowledge and the surest guide to design and construction, guaranteed that mathematics was to be a major force in modern times and gave promise of new developments.

Bibliography

Armitage, Angus: *Copernicus*, W. W. Norton, 1938.
————: *John Kepler*, Faber and Faber, 1966.
————: *Sun, Stand Then Still*, Henry Schuman, 1947; in paperback as *The World of Copernicus*, New American Library, 1951.

Ball, W. W. Rouse: *A Short Account of the History of Mathematics*, Dover (reprint), 1960, Chap. 12.

Baumgardt, Carola: *Johannes Kepler, Life and Letters*, Victor Gollancz, 1952.

Berry, Arthur: *A Short History of Astronomy*, Dover (reprint), 1961, pp. 86–197.

Braunmühl, A. von: *Vorlesungen über die Geschichte der Trigonometrie*, 2 vols., B. G. Teubner, 1900 and 1903, reprinted by M. Sändig, 1970.

Burtt, E. A.: *The Metaphysical Foundations of Modern Physical Science*, 2nd ed., Routledge and Kegan Paul, 1932, Chaps. 1–2.

Butterfield, Herbert: *The Origins of Modern Science*, Macmillan, 1951, Chaps. 1–7.

Cantor, Moritz: *Vorlesungen über Geschichte der Mathematik*, 2nd ed., B. G. Teubner, 1900, Vol. 2, pp. 1–344.

Caspar, Max: *Johannes Kepler*, trans. Doris Hellman, Abelard-Schuman, 1960.

Cohen, I. Bernard: *The Birth of a New Physics*, Doubleday, 1960.

Coolidge, Julian L.: *The Mathematics of Great Amateurs*, Dover (reprint), 1963, Chaps. 3–5.

Copernicus, Nicolaus: *De Revolutionibus Orbium Coelestium* (1543), Johnson Reprint Corp., 1965.

Crombie, A. C.: *Augustine to Galileo*, Falcon Press, 1952, Chap. 4.

Da Vinci, Leonardo: *Philosophical Diary*, Philosophical Library, 1959.

————: *Treatise on Painting*, Princeton University Press, 1956.

Dampier-Whetham, William C. D.: *A History of Science*, Cambridge University Press, 1929, Chap. 3.

Dijksterhuis, E. J.: *The Mechanization of the World Picture*, Oxford University Press, 1961, Parts 3 and 4.

Drake, Stillman and I. E. Drabkin: *Mechanics in Sixteenth-Century Italy*, University of Wisconsin Press, 1969.

Dreyer, J. L. E.: *A History of Astronomy from Thales to Kepler*, Dover (reprint), 1953, Chaps. 12–16.

————: *Tycho Brahe, A Picture of Scientific Life and Work in the Sixteenth Century*, Dover (reprint), 1963.

Gade, John A.: *The Life and Times of Tycho Brahe*, Princeton University Press, 1947.

Galilei, Galileo: *Dialogue on the Great World Systems* (1632), University of Chicago Press, 1953.

Hall, A. R.: *The Scientific Revolution*, Longmans Green, 1954, Chaps. 1–6.

Hallerberg, Arthur E.: "George Mohr and *Euclidis Curiosi*," *The Mathematics Teacher*, 53, 1960, 127–32.

————: "The Geometry of the Fixed-Compass," *The Mathematics Teacher*, 52, 1959, 230–44.

Hart, Ivor B.: *The World of Leonardo da Vinci*, Viking Press, 1962.

Hofmann, Joseph E.: *The History of Mathematics*, Philosophical Library, 1957.

Hughes, Barnabas: *Regiomontanus on Triangles*, University of Wisconsin Press, 1967. A translation of *De Triangulis*.

Ivins, W. M., Jr.: *Art and Geometry* (1946), Dover (reprint), 1965.

Kepler, Johannes: *Gesammelte Werke*, C. H. Beck'sche Verlagsbuchhandlung, 1938–59.

————: *Concerning the More Certain Foundations of Astrology*, Cancy Publications, 1942. Many of Kepler's books have been reprinted and a few translated.

Koyré, Alexandre: *From the Closed World to the Infinite Universe*, Johns Hopkins Press, 1957.

———: *La Révolution astronomique*, Hermann, 1961.

Kuhn, Thomas S.: *The Copernican Revolution*, Harvard University Press, 1957.

MacCurdy, Edward: *The Notebooks of Leonardo da Vinci*, George Braziller (reprint), 1954.

Pannekoek, A.: *A History of Astronomy*, John Wiley and Sons, 1961, Chaps. 16–25.

Panofsky, Erwin: *"Dürer as a Mathematician,"* in James R. Newman, *The World of Mathematics*, Simon and Schuster, 1956, pp. 603–21.

Santillana, G. de: *The Crime of Galileo*, University of Chicago Press, 1955.

Sarton, George: *The Appreciation of Ancient and Medieval Science During the Renaissance*, University of Pennsylvania Press, 1955.

———: *Six Wings: Men of Science in the Renaissance*, Indiana University Press, 1957.

Smith, David Eugene: *History of Mathematics*, Dover (reprint), 1958, Vol. 1, Chap. 8; Vol. 2, Chap. 8.

Smith, Preserved: *A History of Modern Culture*, Holt, Rinehart and Winston, 1930, Vol. 1, Chaps. 2–3.

Taylor, Henry Osborn: *Thought and Expression in the Sixteenth Century*, Crowell-Collier (reprint), 1962, Part V.

Taylor, R. Emmet: *No Royal Road: Luca Pacioli and His Times*, University of North Carolina Press, 1942.

Tropfke, Johannes: *Geschichte der Elementarmathematik*, 7 vols., 2nd. ed., W. De Gruyter, 1921–24.

Vasari, Giorgio: *Lives of the Most Eminent Painters, Sculptors, and Architects* (many editions).

Wolf, Abraham: *A History of Science, Technology and Philosophy in the Sixteenth and Seventeenth Centuries*, George Allen and Unwin, 1950, Chaps. 1–6.

Zeller, Sister Mary Claudia: "The Development of Trigonometry from Regiomontanus to Pitiscus", Ph.D. dissertation, University of Michigan, 1944; Edwards Brothers, 1946.

13
Arithmetic and Algebra in the Sixteenth and Seventeenth Centuries

> Algebra is the intellectual instrument for rendering clear the
> quantitative aspects of the world.
>
> ALFRED NORTH WHITEHEAD

1. Introduction

The first major new European mathematical developments took place in arithmetic and algebra. The Hindu and Arabic work had put practical arithmetical calculations in the forefront of mathematics and had placed algebra on an arithmetic instead of a geometric basis. This work also attracted attention to the problem of solving equations.

In the first half of the sixteenth century there was hardly any change from the Arab attitude or spirit, but merely an increase in the kind of activity the Europeans had learned about from the Arab works. By the middle of the century the practical and scientific needs of European civilization were prompting further steps in arithmetic and algebra. Technological applications of scientific work and practical needs require, as we have pointed out, quantitative results. For example, the far-ranging geographical explorations required more accurate astronomical knowledge. At the same time the interest in correlating the new astronomical theory with increasingly accurate observations demanded better astronomical tables, which in turn meant more accurate trigonometric tables. In fact, a good deal of the sixteenth-century interest in algebra was motivated by the need to solve equations and work with identities in making trigonometric tables. Developing banking and commercial activities called for an improved arithmetic. The response to these interests is evident in the writings of Pacioli, Tartaglia, and Stevin, among others. Pacioli's *Summa* and Tartaglia's *General trattato de' numeri e misure* (1556) contain an immense number of problems in mercantile arithmetic. Finally, the technical work of the artisans, especially in architecture, cannon-making and projectile motion, called for quantitative thinking. Beyond these applications, a totally new use for algebra—the representation of curves—motivated an immense amount of work. Under the pressure of these needs, progress in algebra accelerated.

We shall consider the new developments under four headings: arithmetic, symbolism, the theory of equations, and the theory of numbers.

2. The Status of the Number System and Arithmetic

By 1500 or so, zero was accepted as a number and irrational numbers were used more freely. Pacioli, the German mathematician Michael Stifel (1486?– 1567), the military engineer Simon Stevin (1548–1620), and Cardan used irrational numbers in the tradition of the Hindus and Arabs and introduced more and more types. Thus Stifel worked with irrationals of the form $\sqrt[m]{a + \sqrt[n]{b}}$. Cardan rationalized fractions with cube roots in them. The extent to which irrationals were used is exemplified by Vieta's expression for π.[1] By considering regular polygons of 4, 8, 16, ... sides inscribed in a circle of unit radius, Vieta found that the value of π is given by

$$\frac{2}{\pi} = \cos\frac{90°}{2} \cos\frac{90°}{4} \cos\frac{90°}{8} \cdots$$

$$= \sqrt{\frac{1}{2}} \sqrt{\frac{1}{2} + \frac{1}{2}\sqrt{\frac{1}{2}}} \sqrt{\frac{1}{2} + \frac{1}{2}\sqrt{\frac{1}{2} + \frac{1}{2}\sqrt{\frac{1}{2}}}} \cdots.$$

Though calculations with irrationals were carried on freely, the problem of whether irrationals were really numbers still troubled people. In his major work, the *Arithmetica Integra* (1544), which deals with arithmetic, the irrationals in the tenth book of Euclid, and algebra, Stifel considers expressing irrationals in the decimal notation. On the one hand, he argues:

> Since, in proving geometrical figures, when rational numbers fail us irrational numbers take their place and prove exactly those things which rational numbers could not prove ... we are moved and compelled to assert that they truly are numbers, compelled that is, by the results which follow from their use—results which we perceive to be real, certain, and constant. On the other hand, other considerations compel us to deny that irrational numbers are numbers at all. To wit, when we seek to subject them to numeration [decimal representation] ... we find that they flee away perpetually, so that not one of them can be apprehended precisely in itself. ... Now that cannot be called a true number which is of such a nature that it lacks precision.... Therefore, just as an infinite number is not a number, so an irrational number is not a true number, but lies hidden in a kind of cloud of infinity.

He then argues that real numbers are either whole numbers or fractions; obviously irrationals are neither, and so are not real numbers. A

1. The symbol π was first used by William Jones (1706).

century later, Pascal and Barrow said that a number such as $\sqrt{3}$ can be understood only as a geometric magnitude; irrational numbers are mere symbols that have no existence independent of continuous geometrical magnitude, and the logic of operations with irrationals must be justified by the Eudoxian theory of magnitudes. This was also the view of Newton in his *Arithmetica Universalis* (published in 1707 but based on lectures of thirty years earlier).

Others made positive assertions that the irrational numbers were independent entities. Stevin recognized irrationals as numbers and approximated them more and more closely by rationals; John Wallis in *Algebra* (1685), also accepted irrationals as numbers in the full sense. He regarded the fifth book of Euclid's *Elements* as essentially arithmetical in nature. Descartes, too, in *Rules for the Direction of the Mind* (*c.* 1628), admitted irrationals as abstract numbers that can represent continuous magnitudes.

As for negative numbers, though they had become known in Europe through the Arab texts, most mathematicians of the sixteenth and seventeenth centuries did not accept them as numbers, or if they did, would not accept them as roots of equations. In the fifteenth century Nicolas Chuquet (1445?–1500?) and, in the sixteenth, Stifel (1553) both spoke of negative numbers as absurd numbers. Cardan gave negative numbers as roots of equations but considered them impossible solutions, mere symbols; he called them fictitious, whereas he called positive roots real. Vieta discarded negative numbers entirely. Descartes accepted them, in part. He called negative roots of equations false, on the ground that they claim to represent numbers less than nothing. However, he had shown (see sec. 5) that, given an equation, one can obtain another whose roots are larger than the original one by any given quantity. Thus an equation with negative roots could be transformed into one with positive roots. Since we can turn false roots into real roots, Descartes was willing to accept negative numbers. Pascal regarded the subtraction of 4 from 0 as utter nonsense.

An interesting argument against negative numbers was given by Antoine Arnauld (1612–94), a theologian and mathematician who was a close friend of Pascal. Arnauld questioned that $-1:1 = 1:-1$ because, he said, -1 is less than $+1$; hence, How could a smaller be to a greater as a greater is to a smaller? The problem was discussed by many men. In 1712 Leibniz agreed[2] that there was a valid objection but argued that one can calculate with such proportions because their form is correct, just as one calculates with imaginary quantities.

One of the first algebraists to accept negative numbers was Thomas Harriot (1560–1621), who occasionally placed a negative number by itself on one side of an equation. But he did not accept negative roots. Raphael

2. *Acta Erud.*, 1712, 167–69 = *Math. Schriften*, 5, 387–89.

Bombelli (16th cent.) gave clear definitions for negative numbers. Stevin used positive and negative coefficients in equations and also accepted negative roots. In his *L'Invention nouvelle en l'algèbre* (1629), Albert Girard (1595–1632) placed negative numbers on a par with positive numbers and gave both roots of a quadratic equation, even when both were negative. Both Girard and Harriot used the minus sign for the operation of subtraction and for negative numbers.

On the whole not many sixteenth- and seventeenth-century mathematicians felt at ease with or accepted negative numbers as such, let alone recognizing them as true roots of equations. There were some curious beliefs about them. Though Wallis was advanced for his times and accepted negative numbers, he thought they were larger than infinity but not less than zero. In his *Arithmetica Infinitorum* (1655), he argued that since the ratio $a/0$, when a is positive, is infinite, then, when the denominator is changed to a negative number, as in a/b with b negative, the ratio must be greater than infinity.

Without having fully overcome their difficulties with irrational and negative numbers the Europeans added to their problems by blundering into what we now call complex numbers. They obtained these new numbers by extending the arithmetic operation of square root to whatever numbers appeared in solving quadratic equations by the usual method of completing the square. Thus, Cardan, in Chapter 37 of *Ars Magna* (1545), sets up and solves the problem of dividing 10 into two parts whose product is 40. The equation is $x(10 - x) = 40$. He obtains the roots $5 + \sqrt{-15}$ and $5 - \sqrt{-15}$ and then says, "Putting aside the mental tortures involved," multiply $5 + \sqrt{-15}$ and $5 - \sqrt{-15}$; the product is $25 - (-15)$ or 40. He then states, "So progresses arithmetic subtlety the end of which, as is said, is as refined as it is useless." As we shall soon see, Cardan became further involved with complex numbers in his solution of cubic equations (sec. 4). Bombelli too considered complex numbers in the solution of cubic equations and formulated in practically modern form the four operations with complex numbers; but he still regarded them as useless and "sophistic." Albert Girard did recognize complex numbers as at least formal solutions of equations. In *L'Invention nouvelle en l'algèbre* he says, "One could say: Of what use are these impossible solutions [complex roots]? I answer: For three things—for the certitude of the general rules, for their utility, and because there are no other solutions." However, Girard's advanced views were not influential.

Descartes also rejected complex roots and coined the term "imaginary." He says in *La Géométrie*, "Neither the true nor the false [negative] roots are always real; sometimes they are imaginary." He argued that, whereas negative roots can at least be made "real" by transforming the equation in which they occur into another equation whose roots are positive, this cannot be done for complex roots. These, therefore, are not real but imaginary; they are not numbers. Descartes did make a clearer distinction than his predecessors between real and imaginary roots of equations.

Even Newton did not regard complex roots as significant, most likely because in his day they lacked physical meaning. In fact he says, in *Universal Arithmetic*,[3] "But it is just that the Roots of Equations should be often impossible [complex], lest they should exhibit the cases of Problems that are impossible as if they were possible." That is, problems that do not have a physically or geometrically real solution should have complex roots.

The lack of clarity about complex numbers is illustrated by the oft-quoted statement by Leibniz, "The Divine Spirit found a sublime outlet in that wonder of analysis, that portent of the ideal world, that amphibian between being and not-being, which we call the imaginary root of negative unity."[4] Though Leibniz worked formally with complex numbers, he had no understanding of their nature.

During the sixteenth and seventeenth centuries, the operational procedures with real numbers were improved and extended. In Belgium (then part of the Netherlands), we find Stevin advocating in *La Disme* (Decimal Arithmetic, 1585) the use of decimals, as opposed to the sexagesimal system, for writing and operating with fractions. Others—Christoff Rudolff (*c.* 1500–*c.* 1545), Vieta, and the Arab al-Kashî (d. *c.* 1436)—had used them previously. Stevin advocated a decimal system of weights and measures; he was concerned to save the time and labor of bookkeepers (he himself had started his career as a clerk). He writes 5.912 as 5 ⓪ 9 ① 1 ② 2 ③, or as 5, 9′ 1″ 2‴. Vieta improved and extended the methods of performing square and cube roots.

The use of continued fractions in arithmetic is another development of the period. We may recall that the Hindus—Āryabhata in particular—had used continued fractions to solve linear indeterminate equations. Bombelli, in his *Algebra* (1572), was the first to use them in approximating square roots. To approximate $\sqrt{2}$, he writes

(1)
$$\sqrt{2} = 1 + \frac{1}{y}.$$

From this one finds

(2)
$$y = 1 + \sqrt{2}.$$

By adding 1 to both sides of (1) and using (2), it follows that

(3)
$$y = 2 + \frac{1}{y}.$$

3. Second ed., 1728, p. 193.
4. *Acta Erud.*, 1702 = *Math. Schriften*, 5, 350–61.

Hence, again by (1) and (3),

$$\sqrt{2} = 1 + \cfrac{1}{2 + \cfrac{1}{y}}.$$

And, since y is given by (3),

$$\sqrt{2} = 1 + \cfrac{1}{2 + \cfrac{1}{2 + \cfrac{1}{y}}}.$$

By repeated substitution of the value of y, Bombelli obtained

$$\sqrt{2} = 1 + \cfrac{1}{2 + \cfrac{1}{2 + \cfrac{1}{2 + \cfrac{1}{2} + \cdots}}}.$$

The right-hand side is also written as

$$\sqrt{2} = 1 + \frac{1}{2+}\ \frac{1}{2+}\ \frac{1}{2+}\ \cdots.$$

This continued fraction is simple because the numerators are all 1; it is periodic because the denominators repeat. Bombelli gave other examples of how to obtain continued fractions. He did not, however, consider the question of whether the expansion converged to the number it was supposed to represent.

The English mathematician John Wallis, in his *Arithmetica Infinitorum* (1655), represented $\frac{4}{\pi}$ as the infinite product $\frac{3\cdot3\cdot5\cdot5\cdot7\cdot7\cdots}{2\cdot4\cdot4\cdot6\cdot6\cdot8\cdots}$. In this book he also stated that Lord William Brouncker (1620–84), the first president of the Royal Society, had transformed this product into the continued fraction

$$\frac{4}{\pi} = 1 + \frac{1}{2+}\ \frac{9}{2+}\ \frac{25}{2+}\ \frac{49}{2+}\ \cdots.$$

Brouncker made no further use of this form. Wallis, however, took up the work. In his *Opera Mathematica*, I (1695), wherein he introduced the term "continued fraction," he gave the general rule for calculating the convergents of a continued fraction. That is, if p_n/q_n is the nth convergent of the continued fraction

$$\frac{b_1}{a_1+}\ \frac{b_2}{a_2+}\ \frac{b_3}{a_3+}\ \cdots, \quad \text{then} \quad \frac{p_n}{q_n} = \frac{a_n p_{n-1} + b_n p_{n-2}}{a_n q_{n-1} + b_n q_{n-2}}.$$

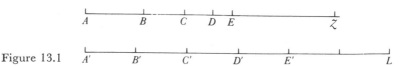

Figure 13.1

No definitive result on the convergence of p_n/q_n to the number that the continued fraction represents was obtained at this time.

The biggest improvement in arithmetic during the sixteenth and seventeenth centuries was the invention of logarithms. The basic idea was noted by Stifel. In his *Arithmetica Integra* he observed that the terms of the geometric progression

$$1, r, r^2, r^3, \ldots$$

correspond to the terms in the arithmetic progression formed by the exponents

$$0, 1, 2, 3, \ldots.$$

Multiplication of two terms in the geometric progression yields a term whose exponent is the sum of the corresponding terms in the arithmetic progression. Division of two terms in the geometric progression yields a term whose exponent is the difference of the corresponding terms in the arithmetic progression. This observation had also been made by Chuquet in *Le Triparty en la science des nombres* (1484). Stifel extended this connection between the two progressions to negative and fractional exponents. Thus the division of r^2 by r^3 yields r^{-1}, which corresponds to the term -1 in the extended arithmetic progression. Stifel, however, did not make use of this connection between the two progressions to introduce logarithms.

John Napier (1550–1617), the Scotsman who did develop logarithms about 1594, was guided by this correspondence between the terms of a geometric progression and those of the corresponding arithmetic progression. Napier was interested in facilitating calculations in spherical trigonometry that were being made on behalf of astronomical problems. In fact, he sent his preliminary results to Tycho Brahe for approval.

Napier explained his ideas in *Mirifici Logarithmorum Canonis Descriptio* (1614) and in the posthumously published *Mirifici Logarithmorum Canonis Constructio* (1619). Since he was concerned with spherical trigonometry, he dealt with the logarithms of sines. Following Regiomontanus, who used half chords and a circle whose radius contained 10^7 units, Napier started with 10^7 as the largest number to be considered. He let the sine values from 10^7 to 0 be represented on the line AZ (Fig. 13.1) and supposed that A moves toward Z with a velocity proportional to its distance from Z.

Strictly speaking, the velocity of the moving point varies continually with the distance from A and the proper expression of it calls for the calculus. However, if we consider any small interval of time t and let the lengths AB, BC, CD, ... be traversed in this interval, and if we assume that the velocity during the interval t is constant and the one possessed by the moving point at the beginning of the interval, then the lengths AZ, BZ, CZ, ... are in geometric progression. For, consider DZ, and let k be the proportionality constant relating the velocity and the distance of the moving point from Z. Now

$$DZ = CZ - CD.$$

Moreover, the velocity of the moving point at C is $k(CZ)$. Then the distance $CD = k(CZ)t$. Hence

$$DZ = CZ - k(CZ)t = CZ(1 - kt).$$

Thus any length in the sequence AZ, BZ, CZ, ... is $1 - kt$ times the preceding length.

Next Napier supposed that another point starting at the same time as A moves at a *constant* velocity on the line $A'L$ (Fig. 13.1), which extends indefinitely to the right, so that this point reaches B', C', D', ... as the first point reaches B, C, D, ..., respectively. The distances $A'B'$, $A'C'$, $A'D'$, ... are obviously in arithmetic progression. These distances $A'B'$, $A'C'$, $A'D'$, ... were taken by Napier to be the logarithms of BZ, CZ, DZ, ..., respectively. The logarithm of AZ or 10^7 was taken to be 0. Thus the logarithms increase in arithmetic progression while the numbers (sine values) decrease in geometric progression. The quantity here denoted by $1 - kt$ is $1 - \dfrac{1}{10^7}$ in Napier's original work. However, he changed this ratio as he proceeded in his calculations of logarithms.

Note that the smaller we take t the less the decrease in the sine values from AZ to BZ to CZ, etc. Hence the numbers of the logarithm table are closer together.

Napier took the distances $A'B'$, $A'C'$, ... to be 1, 2, 3, ..., though there was no need to do that. They could have been 1/2, 1, 1 1/2, 2, ... and the scheme would have worked just as well. Moreover, the original numbers were meaningful just as quantities, independent of the fact that they were sine values; so Napier's scheme really gave the logarithms of numbers. Napier himself applied logarithms to computations of spherical trigonometry.

The word "logarithm," coined by Napier, means "number of the ratio." "Ratio" refers to the common ratio of the sequence of numbers AZ, BZ, CZ, ... He also referred to logarithms as "artificial numbers."

Henry Briggs (1561–1631), a professor of mathematics and astronomy, suggested to Napier in 1615 that 10 be used as a base and that the logarithm of a number be the exponent in the power of 10 equal to that number. Here, unlike Napier's scheme, a base is chosen first. Briggs calculated his logarithms by taking successive square roots of 10, that is, $\sqrt{10}, \sqrt{\sqrt{10}}, \ldots$, until he reached, after 54 such root extractions, a number slightly greater than 1. That is, he obtained a number $A = 10^{[(1/2)^{54}]}$. Then he took $\log_{10} A$ to be $(1/2)^{54}$. By using the fact that the logarithm of a product of two numbers is the sum of their logarithms, he built up a table of logarithms for numbers that were closely spaced. The tables of common logarithms in current use are derived from those of Briggs.

Joost Bürgi (1552–1632), a Swiss watch and instrument maker and an assistant to Kepler in Prague, was also interested in facilitating astronomical calculations; he invented logarithms independently of Napier about 1600 but did not publish his work, *Progress Tabulen*, until 1620. Bürgi too was stimulated by Stifel's remarks that multiplication and division of terms in a geometric progression can be performed by adding and subtracting the exponents. His arithmetical work was similar to Napier's.

Variations of Napier's idea were gradually introduced. Also, many different tables of logarithms were calculated by algebraic means. The calculation of logarithms by the use of infinite series was made later by James Gregory, Lord Brouncker, Nicholas Mercator (born Kaufman, 1620–87), Wallis, and Edmond Halley (see Chap. 20, sec. 2).

Though the definition of logarithms as exponents of the powers that represent the numbers in a fixed base, as in Briggs's scheme, became the common approach, they were not defined as exponents in the early seventeenth century because fractional and irrational exponents were not in use. By the end of the century a number of men recognized that logarithms could be so defined, but the first systematic exposition of this approach was not made until 1742, when William Jones (1675–1749) gave such a presentation in the introduction to William Gardiner's *Table of Logarithms*. Euler had already defined logarithms as exponents and in 1728, in an unpublished manuscript (*Opera Posthuma*, II, 800–804), introduced e for the base of the natural logarithms.

The next development in arithmetic (whose further realization has proved momentous in recent times) was the invention of mechanical devices and machines to speed up the execution of arithmetic processes. The slide rule comes from the work of Edmund Gunter (1581–1626), who utilized Napier's logarithms. William Oughtred (1574–1660) introduced circular slide rules.

In 1642 Pascal invented a computing machine that handled addition by carrying from units to tens, tens to hundreds, etc., automatically. Leibniz saw it in Paris and then invented a machine which could perform multiplication. He showed the Royal Society of London his idea in 1677; a description was published by the Berlin Academy in 1710. In the late seventeenth century Samuel Morland (1625–95), independently invented one machine for addition and subtraction and another for multiplication.

We shall not pursue the history of calculating machines because until at least 1940, they were simply mechanical devices that performed only arithmetic and had no influence on the course of mathematics. We shall, however, note that the most significant step between those already described and modern electronic computers was made by Charles Babbage (1792–1871), who introduced a machine intended for astronomical and navigational computations. His "analytic engine" was designed to perform a whole series of arithmetic operations on the basis of instructions fed into the machine at the start. It would then work automatically under steam power. With support from the British government, he built demonstration models. Unfortunately the machine made excessive demands on the engineering competence of his time.

3. Symbolism

The advance in algebra that proved far more significant for its development and for analysis than the technical progress of the sixteenth century was the introduction of better symbolism. Indeed, this step made possible a science of algebra. Prior to the sixteenth century, the only man who had consciously introduced symbolism to make algebraic thinking and writing more compact and more effective was Diophantus. All other changes in notation were essentially abbreviations of normal words, rather casually introduced. In the Renaissance the common style was still rhetorical, that is, the use of special words, abbreviations, and of course number symbols.

The pressure in the sixteenth century to introduce symbolism undoubtedly came from the rapidly expanding scientific demands on mathematicians, just as the improvements in methods of calculation were a response to the increasing uses of that art. However, improvement was intermittent. Many changes were made by accident; and it is clear that the men of the sixteenth century certainly did not have an appreciation of what symbolism could do for algebra. Even after a decided advance in symbolism was made, it was not taken up at once by mathematicians.

Perhaps the first abbreviations, used from the fifteenth century on, were p for plus and m for minus. However, in the Renaissance and especially during the sixteenth and seventeenth centuries, special symbols were introduced. The symbols + and − were introduced by Germans of the fifteenth century to denote excess and defective weights of chests and were taken over by mathematicians; they appear in manuscripts after 1481. The symbol × for "times" is due to Oughtred; but Leibniz rightly objected to it because it can be confused with the letter x.

The sign = was introduced in 1557 by Robert Recorde (1510–58) of Cambridge, who wrote the first English treatise on algebra, *The Whetstone of Witte* (1557). He said he knew no two things more nearly alike than parallel lines and so two such lines should denote equality. Vieta, who at first wrote out "aequalis," later used $\sim$ for equality. Descartes used $\propto$. The symbols $>$ and $<$ are due to Thomas Harriot. Parentheses appear in 1544. Square brackets and braces, introduced by Vieta, date from about 1593. The square root symbol, $\sqrt{}$, was used by Descartes; but for cube root he wrote $\sqrt{c.}$.

As examples of some of the writing, we might note the following. Using R for square root, p for plus, and m for minus, Cardan wrote

$$(5 + \sqrt{-15}) \cdot (5 - \sqrt{-15}) = 25 - (-15) = 40$$

as

5p: R m:15
5m: R m:15
25m:m:15 qd. est 40.

He also wrote

$$\sqrt{7 + \sqrt{14}} \quad \text{as} \quad \text{R.V. 7.p: R14.}$$

The V indicated that all that followed was under the radical sign.

The use of symbols for the unknown and powers of the unknown had a rather surprisingly slow rise, in view of the simplicity and yet extraordinary value of the practice. (Of course Diophantus had used such symbols.) Early sixteenth-century authors, like Pacioli, referred to the unknown as *radix* (Latin for "root") or *res* (Latin for "thing"), *cosa* ("thing" in Italian), and *coss* ("thing" in German), for which reason algebra became known as the "cossic" art. In his *Ars Magna* Cardan referred to the unknown as *rem ignotam*. He wrote $x^2 = 4x + 32$ as *qdratu aeqtur* 4 *rebus p*:32.[5] The constant term, 32, was called the *numero*. The terms and notation varied a great deal; many symbols were derived from abbreviations. For example, one symbol for the unknown was the letter R, an abbreviation of *res*. The second power, represented by Z (from *zensus*), was called the *quadratum* or *censo*. C, taken from *cubus*, denoted x^3.

Gradually exponents were introduced to denote the powers of x. Exponents attached to numbers, we may recall, were used by Oresme in the fourteenth century. In 1484 Chuquet in *Triparty* wrote 12^3, 10^5, and 120^8 for $12x^3$, $10x^5$, and $120x^8$. He also wrote 12^0 for $12x^0$ and 7^{1m} for $7x^{-1}$. Thus 8^3, 7^{1m} equals 56^2 stood for $8x^3 \cdot 7x^{-1} = 56x^2$.

In his *Algebra* Bombelli used the word *tanto*, instead of *cosa*. For x, x^2, and x^3, he wrote $\underline{1}$, $\underline{2}$, and $\underline{3}$. Thus $1 + 3x + 6x^3 + x^2$ is $1\,p.\,3^{\underline{1}}\,p.\,6^{\underline{2}}\,p.\,1^{\underline{3}}$. In 1585 Stevin wrote for this expression $1^{\circledcirc} + 3^{\textcircled{1}} + 6^{\textcircled{2}} + \textcircled{3}$. Stevin also used fractional exponents, $\left(\dfrac{1}{2}\right)$ for square root, $\left(\dfrac{1}{3}\right)$ for cube root, and so on.

---

5. *Rem* and *rebus* are forms in the declension of *res*.

Claude Bachet de Méziriac (1581–1638) preferred to write $x^3 + 13x^2 + 5x + 2$ as $1C + 13Q + 5N + 2$. Vieta used the same notation for equations with numerical coefficients.

Descartes made rather systematic use of positive integral exponents. He expressed

$$1 + 3x + 6x^2 + x^3 \quad \text{as} \quad 1 + 3x + 6xx + x^3.$$

Occasionally he and others used x^2 also. For higher powers he used x^4, x^5, ... but not x^n. Newton used positive, negative, integral, and fractional exponents, as in $x^{5/3}$ and x^{-3}. When, in 1801, Gauss adopted x^2 for xx, the former became standard.

The most significant change in the character of algebra was introduced in connection with symbolism by François Vieta. Trained as a lawyer, he served in that capacity for the parliament in Brittany; he was later privy councillor to Henry of Navarre. When he was out of office from 1584 to 1589, as a result of political opposition, he devoted himself entirely to mathematics. Generally, he pursued mathematics as a hobby and printed and circulated his work at his own expense—a guarantee, as one writer put it, of oblivion.

Vieta was a humanist in spirit and intention; he wished to be the preserver, rediscoverer, and continuator of ancient mathematics. Innovation was for him renovation. He describes his *In Artem Analyticam Isagoge*[6] as the "work of the restored mathematical analysis." For this book he drew on the seventh book of Pappus' *Mathematical Collection* and on Diophantus' *Arithmetica*. He believed that the ancients had used a general algebraic type of calculation, which he reintroduced in his algebra, thus merely reactivating an art known and approved of in antiquity.

During the hiatus in his political career Vieta studied the works of Cardan, Tartaglia, Bombelli, Stevin, and Diophantus. From them and particularly from Diophantus he got the idea of using letters. Though a number of men, including Euclid and Aristotle, had used letters in place of specific numbers, these uses were infrequent, sporadic, and incidental. Vieta was the first to use letters purposefully and systematically, not just to represent an unknown or powers of the unknown but as general coefficients. Usually he used consonants for the known quantities and vowels for the unknown quantities. He called his symbolic algebra *logistica speciosa*, as opposed to *logistica numerosa*. Vieta was fully aware that when he studied the general quadratic equation $ax^2 + bx + c = 0$ (in our notation), he was studying an entire class of expressions. In making the distinction between *logistica numerosa* and *logistica speciosa* in his *Isagoge*, Vieta drew the line between arithmetic and algebra. Algebra, the *logistica speciosa*, he said, was a method of operating on species or forms of things. Arithmetic, the *numerosa*,

6. Introduction to the Analytic Art, 1591 = *Opera*, 1–12.

dealt with numbers. Thus in this one step, algebra became a study of general types of forms and equations, since what is done for the general case covers an infinity of special cases. Vieta used literal coefficients for positive numbers only.

Vieta sought to establish the algebraic identities concealed in geometrical form in the old Greek works, but to his mind clearly recognizable in Diophantus. Indeed, as we noted in Chapter 6, the latter had made many transformations of algebraic expressions by using identities that he did not cite explicitly. In his *Zeteticorum Libri Quinque*[7] Vieta sought to recapture these identities. He completed the square of a general quadratic expression and expressed general identities such as

$$a^3 + 3a^2b + 3ab^2 + b^3 = (a + b)^3,$$

though he wrote

a cubus + b in a quadr. 3 + a in b quad. 3 + b cubo aequalia

$$\overline{a + b} \text{ cubo.}$$

It would seem that Vieta's successors would immediately have been struck by the idea of general coefficients. But as far as one can judge, the introduction of letters for classes of numbers was accepted as a minor move in the development of symbolism. The idea of literal coefficients slipped almost casually into mathematics. However Vieta's ideas on symbolism were appreciated and made more flexible by Harriot, Girard, and Oughtred.

Improvements in Vieta's use of letters are due to Descartes. He used first letters of the alphabet for known quantities and last letters for unknowns, which is the modern practice. However, like Vieta, Descartes used letters for positive numbers only, though he did not hesitate to subtract terms with literal coefficients. Not until John Hudde (1633–1704) did so in 1657 was a letter used for positive and negative numbers. Newton did so freely.

Leibniz must be mentioned in the history of symbolism, though he postdates the significant steps in algebra. He made prolonged studies of various notations, experimented with symbols, asked the opinions of his contemporaries, and then chose the best. We shall encounter some of his symbolism in our survey of the calculus. He certainly appreciated the great saving of thought that good symbols make possible.

Thus, by the end of the seventeenth century, the deliberate use of symbolism—as opposed to incidental and accidental use—and the awareness of the power and generality it confers entered mathematics. Unfortunately, far too many symbols introduced in a hit-or-miss and thoughtless

7. Five Books of Analysis, 1593 = *Opera*, 41–81.

manner by men who did not appreciate the importance of this device became standard. In noting this, the historian Florian Cajori was impelled to say that "our symbols today are a mosaic of individual signs of rejected systems."

4. The Solution of Third and Fourth Degree Equations

The solution of quadratic equations by the method of completing the square had been known since Babylonian times, and about the only progress in that subject until 1500 was made by the Hindus, who treated quadratics such as $x^2 + 3x + 2 = 0$ and $x^2 - 3x - 2 = 0$ as one type, whereas their predecessors and even most of their Renaissance successors preferred to treat the latter equation in the form $x^2 = 3x + 2$. Cardan, as we noted, did solve one quadratic having complex roots but dismissed the solutions as useless. The cubic equation, except for isolated cases, had thus far defied the mathematicians; as late as 1494 Pacioli had asserted that the solution of general cubic equations was impossible.

Scipione dal Ferro (1465–1526), a professor of mathematics at Bologna, solved cubics of the type $x^3 + mx = n$ about 1500. He did not publish his method because in the sixteenth and seventeenth centuries discoveries were often kept secret and rivals were challenged to solve the same problem. However, about 1510, he did confide his method to Antonio Maria Fior (first half of 16th cent.) and to his son-in-law and successor Annibale della Nave (1500?–58).

Nothing more happened until Niccolò Fontana of Brescia (1499–1557) entered the scene. As a boy he had received a saber cut in the face from a French soldier, which caused him to stammer; as a consequence he was called Tartaglia, "Stammerer." Brought up in poverty, he taught himself Latin, Greek, and mathematics. He earned his living by teaching science in various Italian cities. In 1535 Fior challenged Tartaglia to solve thirty cubic equations. Tartaglia, who said he had already solved cubics of the form $x^3 + mx^2 = n$, m and n positive, solved all thirty, including the type $x^3 + mx = n$.

Pressed by Cardan to reveal his method, Tartaglia gave it to him in an obscure verse form after a pledge from Cardan to keep it secret. This was in 1539. In 1542 Cardan and his pupil Lodovico Ferrari (1522–65), on the occasion of a visit by della Nave, determined that dal Ferro's method was the same as Tartaglia's. Despite his pledge, Cardan published his version of the method in his *Ars Magna*. In Chapter 11 he says that, "Scipio Ferro of Bologna well-nigh thirty years ago discovered this rule and handed it on to Antonio Maria Fior of Venice whose contest with Niccolò Tartaglia of Brescia gave Niccolò occasion to discover it. He gave it to me in response to my entreaties, though withholding the demonstration. Armed with this assistance, I sought out its demonstration in [various] forms. This was very difficult. My version of it follows."

Tartaglia protested the breach of promise, and in *Quesiti ed invenzioni diverse* (1546) presented his own case. However, neither in this book nor in his *General trattato de' numeri e misure* (1556), which is a good presentation of the arithmetical and geometrical knowledge of the times, did he give more on the cubic equation itself. The dispute as to who first solved the cubic led to an open conflict between Tartaglia and Ferrari that exhausted itself in wild quarreling in which Cardan took no part. Tartaglia himself was not above reproach; he published a translation of some of Archimedes' work which he actually took over from William of Moerbecke (d. *c.* 1281), and he claimed to have discovered the law of motion of an object on an inclined plane— which really came from Jordanus Nemorarius.

The method Cardan published he first illustrates with the equation $x^3 + 6x = 20$. However, to see the generality of the method we shall consider

$$(4) \qquad\qquad x^3 + mx = n$$

with m and n positive. Cardan introduces two quantities t and u and lets

$$(5) \qquad\qquad t - u = n$$

and

$$(6) \qquad\qquad (tu) = \left(\frac{m}{3}\right)^3.$$

He then asserts that

$$(7) \qquad\qquad x = \sqrt[3]{t} - \sqrt[3]{u}.$$

By elimination, using (5) and (6), and by solving the resulting quadratic equation, he obtains

$$(8) \qquad t = \sqrt{\left(\frac{n}{2}\right)^2 + \left(\frac{m}{3}\right)^3} + \frac{n}{2}, \qquad u = \sqrt{\left(\frac{n}{2}\right)^2 + \left(\frac{m}{3}\right)^3} - \frac{n}{2}.$$

Here we have taken the positive radical as Cardan does. Having obtained t and u Cardan takes the positive cube root of each and by (7) obtains one value of x. Presumably this is the same root that Tartaglia got.

The above is Cardan's method. However, he had to prove that (7) gives a correct value for x. His proof is geometrical; for Cardan, t and u were volumes of cubes whose sides were $\sqrt[3]{t}$ and $\sqrt[3]{u}$, and the product $\sqrt[3]{t} \cdot \sqrt[3]{u}$ was a rectangle formed by the two sides whose area was $m/3$. Also, where we said $t - u = n$, Cardan says the difference in the volumes is n. Then he says that the solution x is the difference in the edges of the two cubes, i.e. $x = \sqrt[3]{t} - \sqrt[3]{u}$. To prove that this value of x is correct, he states and proves a

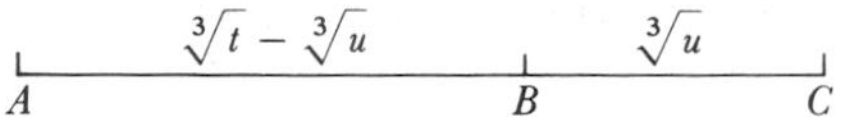

Figure 13.2

$$\underset{A}{\llcorner}\quad\overset{\sqrt[3]{t}-\sqrt[3]{u}}{\underset{B}{\rule{0pt}{0pt}}}\quad\overset{\sqrt[3]{u}}{\underset{C}{\rule{0pt}{0pt}}}\quad\lrcorner$$

geometrical lemma to the effect that, if from a line segment AC (Fig. 13.2) a segment BC is cut off, then the cube on AB will equal the cube on AC minus the cube on BC minus three times the right parallelepiped whose edges are AC, AB, and BC. This geometrical lemma is of course no more than that

(9) $$(\sqrt[3]{t} - \sqrt[3]{u})^3 = t - u - 3(\sqrt[3]{t} - \sqrt[3]{u})\sqrt[3]{t}\sqrt[3]{u}.$$

Granted this lemma (which, by using the binomial theorem, we see must be correct, but which Cardan establishes by citing theorems of Euclid), Cardan has but to observe that if he lets $x = \sqrt[3]{t} - \sqrt[3]{u}$, $t - u = n$, and $\sqrt[3]{t}\sqrt[3]{u} = \dfrac{m}{3}$, then the lemma says that $x^3 = n - mx$. Hence if he chooses t and u to satisfy the conditions (5) and (6), the value of x given by (7) in terms of t and u will satisfy the cubic. He then gives a purely verbal arithmetical rule for the method, which tells us to form $\sqrt[3]{t} - \sqrt[3]{u}$ where t and u are given by (8) in terms of m and n.

Cardan also solves (as did Tartaglia) particular equations of the types

$$x^3 = mx + n, \qquad x^3 + mx + n = 0, \qquad x^3 + n = mx.$$

He has to treat each of these cases separately and all three separately from equation (4) because, first, up to this time the Europeans wrote equations so that only terms with positive numbers appeared in them, and second, because he had to give a separate geometrical justification for the rule in each case.

Cardan also shows how to solve equations such as $x^3 + 6x^2 = 100$. He knew how to eliminate the x^2 term; that is, since the coefficient was 6, he replaced x by $y - 2$ and obtained $y^3 = 12y + 84$. He also recognized that one can treat an equation such as $x^6 + 6x^4 = 100$ as a cubic by letting $x^2 = y$. Throughout the book he gives positive and negative roots, despite the fact that he calls negative numbers fictitious. However he ignored complex roots. In fact, in Chapter 37 he calls problems leading to neither true nor false (positive or negative) roots false problems. The book is detailed— even boring, to a modern reader—because Cardan treats separately the many cases not only of the cubic equation but also of the auxiliary quadratic equations he must solve to find t and u. In each case, he writes the equation so that the coefficients of the terms are positive.

There is a difficulty with Cardan's solution of the cubic, which he observed but did not resolve. When the roots of the cubic are all real and distinct, it can be shown that t and u will be complex because the radicand in (8) is negative; yet we need $\sqrt[3]{t}$ and $\sqrt[3]{u}$ to obtain x. This means that real numbers can be expressed in terms of the cube roots of complex numbers. However, these three real roots cannot be obtained by algebraic means, that is, by radicals. This case was called irreducible by Tartaglia. One would think that the fact that real numbers can be expressed as combinations of complex numbers would have caused Cardan to take complex numbers seriously, but it did not.

Vieta, in *De Aequationum Recognitione et Emendatione*, written in 1591 and published in 1615,[8] was able to solve the irreducible case of cubics by using a trigonometric identity, and so avoided the use of Cardan's formula. This method is also used today. He started with the identity

$$(10) \qquad\qquad \cos 3A \equiv 4 \cos^3 A - 3 \cos A.$$

By letting $z = \cos A$ the identity becomes

$$(11) \qquad\qquad z^3 - \frac{3}{4} z - \frac{1}{4} \cos 3A \equiv 0.$$

Suppose the given cubic is (Vieta worked with $x^3 - 3a^2x = a^2b$ and $a > b/2$)

$$(12) \qquad\qquad y^3 + py + q = 0.$$

By introducing $y = nz$, where n is at our disposal, we can make the coefficients of (12) the same as those of (11). Substituting $y = nz$ in (12) gives

$$(13) \qquad\qquad z^3 + \frac{p}{n^2} z + \frac{q}{n^3} = 0.$$

Now we require of n that $p/n^2 = -3/4$ so that

$$(14) \qquad\qquad n = \sqrt{-4p/3}.$$

With this value of n chosen, we select a value A so that

$$(15) \qquad\qquad \frac{q}{n^3} = -\frac{1}{4} \cos 3A$$

or so that

$$(16) \qquad\qquad \cos 3A = -\frac{4q}{n^3} = \frac{-q/2}{\sqrt{-p^3/27}}.$$

One can show that if the three roots are real then p is negative so that n is real. Moreover, one can show that $|\cos 3A| < 1$. Hence one can find $3A$ from a table.

Whatever the value of A, $\cos A$ satisfies (11) because (11) is an identity. Now A was selected so that (13) is a particular case of (11). For this value of A, $\cos A$ satisfies (13). However, the value of A is determined by (16), and

8. On the Review and Correction of Equations, *Opera*, 82–162.

this fixes $3A$. But for any given A that satisfies (16), so do $A + 120°$ and $A + 240°$. Since $z = \cos A$, there are, then, three values satisfying (13):

$$\cos A, \qquad \cos (A + 120°), \quad \text{and} \quad \cos (A + 240°).$$

The three values satisfying (12) are n times these values of z, where n is given by (14). Vieta obtained just one root.

Of course the cubic equation has three roots. It was Leonhard Euler who, in 1732, gave the first complete discussion of Cardan's solution of the cubic, in which he emphasized that there are always three roots and pointed out how these are obtained.[9] If ω and ω^2 are the complex roots of $x^3 - 1 = 0$, that is, the roots of $x^2 + x + 1 = 0$, then the three cube roots of t and u in (8) are

$$\sqrt[3]{t}, \; \omega\sqrt[3]{t}, \; \omega^2\sqrt[3]{t} \quad \text{and} \quad \sqrt[3]{u}, \; \omega\sqrt[3]{u}, \; \omega^2\sqrt[3]{u}.$$

We must now choose a member of the first set and a member of the second so that the product is the real number $m/3$. (See equation (6) in Cardan's solution.) Since ω and ω^2 are roots of unity, $\omega \cdot \omega^2 = \omega^3 = 1$; hence the proper choices for x, in view of (7), are

(17) $\qquad x_1 = \sqrt[3]{t} - \sqrt[3]{u}, \qquad x_2 = \omega\sqrt[3]{t} - \omega^2\sqrt[3]{u}, \qquad x_3 = \omega^2\sqrt[3]{t} - \omega\sqrt[3]{u}.$

Success in solving the cubic equation was followed almost immediately by success in solving the quartic. The method is due to Lodovico Ferrari and was published in Cardan's *Ars Magna*; we describe it in modern notation and with literal coefficients to show the generality. The equation is

(18) $\qquad\qquad\qquad x^4 + bx^3 + cx^2 + dx + e = 0.$

By transposition we get

(19) $\qquad\qquad\qquad x^4 + bx^3 = -cx^2 - dx - e.$

Now complete the square on the left side by adding $\left(\dfrac{1}{2}bx\right)^2$. This gives

(20) $\qquad\qquad \left(x^2 + \dfrac{1}{2}bx\right)^2 = \left(\dfrac{1}{4}b^2 - c\right)x^2 - dx - e.$

Now add $\left(x^2 + \dfrac{1}{2}bx\right)y + \dfrac{1}{4}y^2$ to each side. Thus

(21) $\qquad \left(x^2 + \dfrac{1}{2}bx\right)^2 + \left(x^2 + \dfrac{1}{2}bx\right)y + \dfrac{1}{4}y^2$

$$= \left(\dfrac{1}{4}b^2 - c + y\right)x^2 + \left(\dfrac{1}{2}by - d\right)x + \dfrac{1}{4}y^2 - e.$$

9. *Comm. Acad. Sci. Petrop.*, 6, 1732/33, 217–31, pub. 1738 = *Opera*, (1), 6, 1–19.

By making the discriminant of the quadratic expression in x on the right side zero, we can make this side a perfect square of a first degree expression in x. Hence we set

(22) $$\left(\frac{1}{2}by - d\right)^2 - 4\left(\frac{1}{4}b^2 - c + y\right)\left(\frac{1}{4}y^2 - e\right) = 0.$$

This is a *cubic* equation in y. Choose any root of this cubic and substitute it for y in (21). Using the fact that the left side is also a perfect square, and taking the square root, we get a quadratic in x equaling either of two linear functions of x, one the negative of the other. Solving these two quadratics gives 4 roots for x. Choosing another root from (22) would give a different equation in (21) but the same four roots.

In presenting Ferrari's method Cardan, in Chapter 39 of *Ars Magna*, solves a multitude of special cases, each with numerical coefficients. Thus he solves equations of the type

$$x^4 = bx^2 + ax + n, \qquad x^4 = bx^2 + cx^3 + n,$$
$$x^4 = cx^3 + n, \qquad x^4 = ax + n.$$

As in the case of the cubic equation, he gives a geometrical proof of the basic algebraic steps and then gives the rule for solution in words.

Cardan, Tartaglia, and Ferrari showed, by solving numerous instances of cubic and quartic equations, that they had sought and obtained methods that would work for all cases of the respective degrees. The interest in generality is a new feature. Their work preceded Vieta's introduction of literal coefficients, so they could not take advantage of this device. Vieta, who had already made possible a generality in proof by introducing literal coefficients, now sought another kind of generality. He noted that the methods of solving the second, third, and fourth degree equations were quite different. He therefore sought a method that would work for equations of each degree. His first idea was to get rid of the term next to the highest in degree by a substitution. Tartaglia had done this for the cubic but did not try it for all equations.

In the *Isagoge* Vieta does the following. To solve the quadratic equation

$$x^2 + 2bx = c$$

he lets

$$x + b = y.$$

Then

$$y^2 = x^2 + 2bx + b^2.$$

In view of the original equation,

$$y = \sqrt{c + b^2}.$$

Then

$$x = y - b = \sqrt{c + b^2} - b.$$

In the case of the third degree equation

$$x^3 + bx^2 + cx + d = 0,$$

Vieta starts by letting $x = y - b/3$. This substitution gives the reduced cubic

(23) $$y^3 + py + q = 0.$$

Next he introduces a further transformation, indeed the one we learn today. He lets

(24) $$y = z - \frac{p}{3z}$$

and obtains

$$z^3 - \frac{p^3}{27z^3} + q = 0.$$

He then solves the quadratic in z^3 and obtains

$$z^3 = -\frac{q}{2} \pm \sqrt{R} \quad \text{where} \quad R = \left(\frac{p}{3}\right)^3 + \left(\frac{q}{2}\right)^2.$$

Here, as in Cardan's method, there are two values for z^3. Though Vieta used only the positive cube root of z^3, one can use all six (complex) roots. The use of (24) would show that only three distinct values of y result from the six values of z.

To solve the general fourth degree equation

$$x^4 + bx^3 + cx^2 + dx + e = 0,$$

Vieta lets $x = y - b/4$ and reduces the equation to

$$x^4 + px^2 + qx + r = 0.$$

He then transposes the last three terms and adds $2x^2y^2 + y^4$ to both sides. This makes the left side a perfect square and, as in Ferrari's method, by properly choosing y he makes the right side a perfect square of the form $(Ax + B)^2$. To choose y properly he applies the discriminant condition for a quadratic equation and is led to a sixth degree equation in y, which fortunately is a cubic in y^2. This step and the rest of the work is precisely the same as in Ferrari's method.

Another general method explored by Vieta was to factor the polynomial into first degree factors just as we might factor $x^2 + 5x + 6$ into $(x + 2)(x + 3)$. He was unsuccessful partly because he rejected all but positive roots and partly because he did not have enough theory, such as the factor theorem, on which to base a general method. Thomas Harriot had the same idea and failed for the same reasons.

The search for general algebraic methods turned next to solving equations of degree higher than four. James Gregory, who had given his own methods of solving cubic and quartic equations, tried to use them to solve the quintic equation. He and Ehrenfried Walter von Tschirnhausen (1651–1708) tried to use transformations to reduce higher-degree equations to two terms, a power of x and a constant. These attempts to solve equations beyond the quartic failed. In later work on integration Gregory surmised that one could not solve algebraically the general nth-degree equation for $n > 4$.

5. The Theory of Equations

The work on methods of solving equations produced a number of related theorems and observations that are studied nowadays in the elementary theory of equations. The number of roots an equation can have was considered. Cardan had introduced complex roots and it seemed to him for a while that an equation might have any number of roots. But he soon recognized that a third degree equation has 3 roots, a fourth degree 4, and so on. In *L'Invention nouvelle*, Albert Girard inferred and stated that an nth-degree polynomial equation has n roots if one counts the impossible roots, that is, the complex roots, and if one takes into account the repeated roots. But Girard gave no proof. Descartes, in the third book of *La Géométrie*, said that an equation can have as many distinct roots as the number of dimensions (degree) of the unknown. He said "can have" because he considered negative roots as false roots. Later, by including imaginary and negative roots for the purpose of counting roots, he concluded that there are as many as the degree.

The next significant question was how to predict the number of positive, negative, and complex roots. Cardan observed that the complex roots of an equation (with real coefficients) occur in pairs. Newton proved this in his *Arithmetica Universalis*. Descartes in *La Géométrie* stated without proof the rule of signs, known as Descartes's rule, which states that the maximum number of positive roots of $f(x) = 0$, where f is a polynomial, is the number of alterations in sign of the coefficients and that the maximum number of negative roots is the number of times two $+$ signs or two $-$ signs occur in succession. As now given, the latter part of the rule states that the maximum number of negative roots is the number of alterations in $f(-x) = 0$. The rule was proved by several eighteenth-century mathematicians. The proof usually presented today is due to Abbé Jean-Paul de Gua de Malves (1712–85), who showed also that the absence of $2m$ successive terms indicates $2m + 2$ or $2m$ complex roots, according as the two terms between which the deficiency occurs have like or unlike signs.

In his *Arithmetica Universalis* Newton described but did not prove another method for determining the maximum number of positive and negative real roots and hence the least possible number of complex roots. This method is more complicated to apply but gives better results than Descartes's rule of signs. It was finally proved as a special case of a more general theorem by Sylvester.[10] Somewhat earlier Gauss showed that if the number of positive roots falls short of the number of variations of sign, it falls short by an even number.

Another class of results concerns the relationships between the roots and coefficients of an equation. Cardan discovered that the sum of the roots is the negative of the coefficient of x^{n-1}, that the sum of the products two at a time is the coefficient of x^{n-2}, and so forth. Both Cardan and Vieta (in *De Aequationum Recognitione et Emendatione*) used the first relationship between roots and coefficients of low degree equations to eliminate the x^{n-1} term in polynomial equations in manners we have described earlier. Newton stated the relationship between roots and coefficients in his *Arithmetica Universalis*; so did James Gregory in a letter to John Collins (1625–83), the secretary of the Royal Society. However, no proofs were given by any of these men.

Vieta and Descartes constructed equations whose roots were more or less than the roots of a given equation. The process is merely to replace x by $y + m$. Both men also used the transformation $y = mx$ to obtain an equation whose roots are m times the roots of a given equation. For Descartes the former process had the significance we mentioned earlier, namely, that false (negative) roots can be made true (positive) roots, and conversely.

Descartes also proved that if a cubic equation with rational coefficients has a rational root, then the polynomial can be expressed as the product of factors with rational coefficients.

Another major result is now known as the factor theorem. In the third book of *La Géométrie* Descartes asserted that $f(x)$ is divisible by $x - a$, a positive, if and only if a is a root of $f(x) = 0$, and by $x + a$ if a is a false root. With this fact and others he had asserted, Descartes established the modern method of finding the rational roots of a polynomial equation. After making the highest coefficient 1, he made all the coefficients integral by multiplying the roots of the given equation by the necessary factor. This is done by using the rule he had given of replacing x by y/m in the equation. The rational roots of the original equation must now be integral factors of the constant term in the new equation. If by trial one finds that a, say, is a root, then by the factor theorem, $y - a$ is a factor of the new polynomial in y. Descartes points out that by eliminating this factor one reduces the degree of the equation and can then work with the reduced equation.

Newton in *Arithmetica Universalis* and others earlier gave theorems on the upper bound for the roots of equations. One of these theorems involves

10. *Proc. London Math. Soc.*, 1, 1865, 1–16 = *Math. Papers*, 2, 498–513.

the calculus and will be stated in Chapter 17 (sec. 7). Newton discovered the relation between the roots and discriminant of an equation, namely that, for example, $ax^2 + bx + c = 0$ has equal, real, or nonreal roots according as $b^2 - 4ac$ equals 0, is greater than 0 or is less than 0.

In *La Géométrie* Descartes introduced the principle of undetermined co-efficients. The principle can be illustrated thus: To factor $x^2 - 1$ into two linear factors, one supposes that

$$x^2 - 1 = (x + b)(x + d).$$

By multiplying out the right-hand side and equating coefficients of like powers of x, one finds that

$$b + d = 0$$
$$bd = -1.$$

One can now solve for b and d. Descartes stressed the usefulness of this method.

One other method, the method of mathematical induction, entered algebra explicitly in the late sixteenth century. Of course the method is implicit even in Euclid's proof of the infinitude of the number of primes. As he proves the theorem, he shows that if there are n primes, there must be $n + 1$ primes; and since there is a first prime, the number of primes must be infinite. The method was recognized explicitly by Maurolycus in his *Arithmetica* of 1575 and was used by him to prove, for example, that $1 + 3 + 5 + \cdots + (2n - 1) = n^2$. Pascal in one of his letters ackowledged Maurolycus's introduction of the method and used it himself in his *Traité du triangle arithmétique* (1665), wherein he presents what we now call the Pascal triangle (sec. 6).

6. *The Binomial Theorem and Allied Topics*

The binomial theorem for positive integral exponents, that is, the expansion of $(a + b)^n$ for n positive integral, was known by the Arabs of the thirteenth century. About 1544 Stifel introduced the term "binomial coefficient" and showed how to calculate $(1 + a)^n$ from $(1 + a)^{n-1}$. The arrangement of numbers

$$
\begin{array}{ccccccc}
 & & & 1 & & & \\
 & & 1 & & 1 & & \\
 & & 1 & 2 & 1 & & \\
 & 1 & 3 & & 3 & 1 & \\
 1 & 4 & & 6 & & 4 & 1 \\
1 & 5 & 10 & & 10 & 5 & 1
\end{array}
$$

in which each number is the sum of the two immediately above it, already known to Tartaglia, Stifel, and Stevin, was used by Pascal (1654) to obtain the coefficients of the binomial expansion. Thus the numbers in the fourth row are the coefficients in the expansion of $(a + b)^3$. Despite the fact that this arrangement was known to many predecessors, it has been called Pascal's triangle.

Newton in 1665 showed that we may compute $(1 + a)^n$ directly without reference to $(1 + a)^{n-1}$. Then he became convinced that the expansion held for fractional and negative n (it is an infinite series in this case) and so stated, but never proved, this generalization. He did verify that the series of $(1 + x)^{1/2}$ times itself gave $1 + x$, but neither he nor James Gregory (who arrived at the theorem independently) thought a proof necessary. In two letters, of June 6 and October 4, 1676, to Henry Oldenburg (c. 1615–77), secretary of the Royal Society, Newton stated the more general result, which he knew before 1669, namely, the expansion of $(P + PQ)^{m/n}$. He thought of it as a useful method of extracting roots, because if Q is less than 1 (the P being factored out), the successive terms, being powers of Q, are smaller and smaller in value.

Independently of the work on the binomial theorem, the formulas for the number of permutations and the number of combinations of n things taken r at a time appeared in the works of a number of mathematicians, for example, Bhāskara and the Frenchman Levi ben Gerson (1321). Pascal observed that the formula for combinations, often denoted by $_nC_r$ or $\binom{n}{r}$, also gives the binomial coefficients. That is, for n fixed and r running from 0 to n, the formula yields the successive coefficients. James Bernoulli, in his *Ars Conjectandi* (1713), extended the theory of combinations and then proved the binomial theorem for the case of n positive integral by using the formula for combinations.

The work on permutations and combinations is connected with another development, the theory of probability, which was to assume major importance in the late nineteenth century but which barely warrants mention in the sixteenth and seventeenth centuries. The problem of the probability of throwing a particular number on a throw of two dice had been raised even in medieval times. Another problem, how to divide the stake between two players when the stake is to go to the player who first wins n points, but the play is interrupted after the first player has made p points and the second q points, appears in Pacioli's *Summa* and in books by Cardan, Tartaglia, and others. This problem acquired some importance when, after it was proposed to Pascal by Antoine Gombaud, Chevalier de Méré (1610–85), Pascal and Fermat corresponded about it. The problem and their solutions are unimportant, but their work on it does mark the beginning of the theory of probability. Both applied the theory of combinations.

The first significant book on probability was Bernoulli's *Ars Conjectandi*. The most important new result, still called Bernoulli's theorem, states that if p is the probability of a single event happening and q the probability of its failing to happen, then the probability of the event happening at least m times in n trials is the sum of the terms in the expansion of $(p + q)^n$ from p^n to the term involving $p^m q^{n-m}$.

7. The Theory of Numbers

While practical interests stimulated the improvements in calculations, symbolism, and the theory of equations, interest in purely mathematical problems led to renewed activity in the theory of numbers. This subject had, of course, been initiated by the classical Greeks; and Diophantus had added the topic of indeterminate equations. The Hindus and Arabs had, at least, kept the subject from falling into oblivion. Though almost all the Renaissance algebraists we have mentioned made conjectures and observations, the European who first made extensive and impressive contributions to the theory of numbers and gave the subject enormous impetus was Pierre de Fermat (1601–65).

Born to a family of tradespeople, he was trained as a lawyer in the French city of Toulouse and made his living in this profession. For a time he was a councillor of the parliament of Toulouse. Though mathematics was but a hobby for Fermat and he could devote only spare time to it, he contributed first-class results to the theory of numbers and the calculus, was one of the two creators of coordinate geometry, and, together with Pascal, as we have seen, initiated work on probability. Like all mathematicians of his century, he worked on problems of science and made a lasting contribution to optics: Fermat's Principle of Least Time (Chap. 24, sec. 3). Most of Fermat's results are known through letters he wrote to friends. He published only a few papers, but some of his books and papers were published after his death.

Fermat believed that the theory of numbers had been neglected. He complained on one occasion that hardly anyone propounded or understood arithmetical questions and asked, "Is it due to the fact that up to now arithmetic has been treated geometrically rather than arithmetically?" Even Diophantus, he observed, was somewhat tied to geometry. Arithmetic, he believed, had a special domain of its own, the theory of integral numbers.

Fermat's work in the theory of numbers determined the direction of the work in this area until Gauss made his contributions. Fermat's point of departure was Diophantus. Many translations of the latter's *Arithmetica* had been made by Renaissance mathematicians. In 1621 Bachet de Méziriac published the Greek text and a Latin translation. This was the edition that Fermat owned; he noted most of his results in the margins of the book, though a few were communicated in letters to friends. The copy with Fermat's marginal notes was published in 1670 by his son.

Fermat stated many theorems on the theory of numbers but only in one case did he give a proof, and this was a sketch. The best mathematicians of the eighteenth century worked hard to prove his results, (Chap. 25, sec. 4). These all turned out to be correct except for one error (which we shall note later) and one still-unproved famous "theorem," for which the indications are all favorable. There is no doubt that he had great intuition, but it is unlikely that he had proofs for all of his affirmations.

A document discovered in 1879 among the manuscripts of Huygens gives a famous method, called the method of infinite descent, which was introduced and used by Fermat. To understand the method let us consider the theorem asserted by Fermat in a letter to Marin Mersenne (1588–1648) of December 25, 1640, which states that a prime of the form $4n + 1$ can be expressed in one and only one way as the sum of two squares. Thus $17 = 16 + 1$ and $29 = 25 + 4$. The method proceeds by showing that if there is one prime of the form $4n + 1$ that does *not* possess the required property, then there will be a smaller prime of the form $4n + 1$ not possessing it. Then, since n is arbitrary, there must be a still smaller one. By descending through the positive integral values of n one must reach $n = 1$ and thus the prime $4 \cdot 1 + 1$ or 5. Then 5 cannot possess the required property. But, since 5 is expressible as a sum of two squares and in only one way, so is every prime of the form $4n + 1$. This sketch of his method Fermat sent to his friend Pierre de Carcavi (d. 1684) in 1659. Fermat said that he used the method to prove the theorem just described, but his proof was never found. He also said that he proved other theorems by this method.

The method of infinite descent differs from mathematical induction. First of all, the method does not require that one exhibit even one case in which the proposed theorem is satisfied because one can conclude the argument by the fact that the case $n = 1$ merely leads to a contradiction of some other known fact. Moreover, after the appropriate hypothesis is made for one value of n, the method shows that there is a smaller, but not necessarily the next, value of n for which the hypothesis is true. Finally the method disproves certain assertions and is in fact more useful for this purpose.

Fermat also stated that no prime of the form $4n + 3$ can be expressed as a sum of two squares. In a note in his copy of Diophantus and in the letter to Mersenne, Fermat generalized on the well-known 3, 4, 5 right-triangle relationship by asserting the following theorems: A prime of the form $4n + 1$ is the hypotenuse of one and only one right triangle with integral arms. The square of $(4n + 1)$ is the hypotenuse of two and only two such right triangles; its cube, of three; its biquadrate, of 4; and so on, ad infinitum. As an example, consider the case of $n = 1$. Then $4n + 1 = 5$ and 3, 4, 5 are the sides of the one and only right triangle with 5 as hypotenuse. However, 5^2 is the hypotenuse of the two, and only two, right triangles 15, 20, 25 and 7, 24, 25. Also 5^3 is the hypotenuse of the three, and only three, right triangles 75, 100, 125; 35, 120, 125; and 44, 117, 125.

In the letter to Mersenne, Fermat declared that the same prime number $4n + 1$ and its square are each the sum of two squares in one way only; its cube and biquadrate, each in two ways; its fifth and sixth powers, each in three ways, and so on ad infinitum. Thus for $n = 1$, $5 = 4 + 1$ and $5^2 = 9 + 16$; $5^3 = 4 + 121 = 25 + 100$; and so forth. The letter continues: If a prime number that is a sum of two squares be multiplied into another prime that is also the sum of two squares, the product will be the sum of two squares in two ways. If the first prime be multiplied into the square of the second one, the product will be the sum of two squares in three ways; if multiplied into the cube of the second one, then the product will be the sum of two squares in four ways; and so on ad infinitum.

Fermat stated many theorems on the representation of prime numbers in the form $x^2 + 2y^2$, $x^2 + 3y^2$, $x^2 + 5y^2$, $x^2 - 2y^2$, and other such forms, which are extensions of the representation as a sum of squares. Thus every prime of the form $6n + 1$ can be represented as $x^2 + 3y^2$; every prime of the form $8n + 1$ and $8n + 3$ can be represented as $x^2 + 2y^2$. An odd prime number (every prime but 2) can be expressed as the difference of two squares in one and only one way.

Two theorems asserted by Fermat have since been referred to as the minor and major theorems, the latter also known as the last theorem. The minor one, communicated by Fermat in a letter of October 18, 1640, to his friend Bernard Frénicle de Bessy (1605–75), states that if p is any prime and if a is any integer, then $a^p - a$ is divisible by p.

The major Fermat "theorem," which he believed he had proved, states that for $n > 2$ no integral solutions of $x^n + y^n = z^n$ are possible. This theorem was stated by Fermat in a marginal note in his copy of Diophantus alongside Diophantus' problem: To divide a given square number into (a sum of) two squares. Fermat added, "On the other hand it is impossible to separate a cube into two cubes, or a biquadrate into two biquadrates, or generally any power except a square into two powers with the same exponent. I have discovered a truly marvelous proof of this, which however the margin is not large enough to contain." Unfortunately, Fermat's proof, if he had one, was never found and hundreds of the best mathematicians have not been able to prove it. Fermat stated in a letter to Carcavi that he had used the method of infinite descent to prove the case of $n = 4$ but did not give full details. Frénicle, using Fermat's few indications, did give a proof for that case in 1676, in his posthumously published *Traité des triangles rectangles en nombres* (Treatise on Numerical Properties of Right Triangles).[11]

If we may anticipate, Euler proved the theorem for $n = 3$ (Chap. 25, sec. 4). Since the theorem is true for $n = 3$, it is true for any multiple of 3;

11. *Mém. de l'Acad. des Sci., Paris*, 5, 1729, 83–166.

for if it were not true for $n = 6$, say, then there would be integers x, y, and z such that

$$x^6 + y^6 = z^6.$$

But then

$$(x^2)^3 + (y^2)^3 = (z^2)^3,$$

and the theorem would be false for $n = 3$. Hence we know that Fermat's theorem is true for an infinite number of values of n, but we still do not know that it is true for all values of n. It is actually necessary to prove the theorem only for $n = 4$ and for n an odd prime. For suppose first that n is not divisible by an odd prime; it must then be a power of 2, and since it is larger than 2 it must be 4 or divisible by 4. Let $n = 4m$. Then the equation $x^n + y^n = z^n$ becomes

$$(x^m)^4 + (y^m)^4 = (z^m)^4.$$

If the theorem were not true for n, it would therefore not be true for $n = 4$. Hence if it is true for $n = 4$, it is true for all n not divisible by an odd prime. If $n = pm$ where p is an odd prime, then if the theorem were not true for n it would not be true for the exponent p. Hence if true for $n = p$ it is true for any n divisible by an odd prime.

Fermat did make some mistakes. He believed that he had found a solution to the long-standing problem of producing a formula that would yield primes for values of the variable n. Now it is not hard to show that $2^m + 1$ cannot be a prime unless m is a power of 2. In many letters dating from 1640 on[12] Fermat asserted the converse—namely, that $2^{2n} + 1$ represents a series of primes—though he admitted that he could not prove this assertion. Later he doubted its correctness. Thus far only the five primes 3, 5, 17, 257, and 65,537 yielded by the formula are known. (See Chap. 25, sec. 4.)

Fermat stated and sketched[13] the proof by infinite descent of the theorem: The area of a right-angled triangle the sides of which are rational numbers cannot be a square number. This sketch is the only detailed one ever given by him, and it follows as a corollary that the solution of $x^4 + y^4 = z^4$ in integers is impossible.

On polygonal numbers Fermat stated in his copy of Diophantus the important theorem that every positive integer is itself triangular or the sum of 2 or 3 triangular numbers; every positive integer is itself square or a sum of 2, 3, or 4 squares; every positive integer is either pentagonal or a sum of 2, 3, 4, or 5 pentagonal numbers; and so on for higher polygonal numbers. Much work was required to prove these results, which are correct

12. Œuvres, 2, 206.
13. Œuvres, 1, 340; 3, 271.

only if 1 is included as a polygonal number. Fermat asserts that he proved them by the method of infinite descent.

Perfect numbers, as we know, were studied by the Greeks, and Euclid gave the basic result that $2^{n-1}(2^n - 1)$ is perfect if $2^n - 1$ is prime. For $n = 2$, 3, 5, and 7, the values of $2^n - 1$ are prime so that 6, 28, 496, and 8128 are perfect numbers (as noted by Nichomachus). A manuscript of 1456 correctly gave 33,550,336 as the fifth perfect number; it corresponds to $n = 13$. In his *Epitome* (1536) Hudalrich Regius also gave this fifth perfect number. Pietro Antonio Cataldi (1552–1626) noted in 1607 that $2^n - 1$ is composite if n is composite, and verified that $2^n - 1$ is prime for $n = 13$, 17, and 19. Marin Mersenne in 1644 gave other values. Fermat worked on the subject of perfect numbers also. He considered when $2^n - 1$ is prime and in a letter to Mersenne of June 1640 stated these theorems: (a) If n is not a prime, $2^n - 1$ is not a prime. (b) If n is a prime, $2^n - 1$ is divisible only by primes of the form $2kn + 1$ if divisible at all. About twenty perfect numbers are now known. Whether any odd ones exist is an open question.

By rediscovering a rule that was first stated by Tâbit ibn Qorra, Fermat in 1636 gave a second pair of amicable numbers, 17,296 and 18,416 (the first, 220 and 284, was given by Pythagoras), and Descartes in a letter to Mersenne gave a third pair, 9,363,548 and 9,437,506.

Fermat rediscovered the problem of solving $x^2 - Ay^2 = 1$, wherein A is integral and not a square. The problem has a long history among the Greeks and Hindus. In a letter of February 1657 to Frénicle, Fermat stated the theorem that $x^2 - Ay^2 = 1$ has an unlimited number of solutions when A is positive and not a perfect square.[14] Euler erroneously called the equation Pell's equation; it is now so known. In the same letter[15] Fermat challenged all mathematicians to find an infinity of integral solutions. Lord Brouncker gave solutions, though he did not prove that there was an infinity of them. Wallis did solve the full problem and gave his solutions in letters of 1657 and 1658[16] and in Chapter 98 of his *Algebra*. Fermat also asserted that he could show when $x^2 - Ay^2 = B$, for given A and B, is solvable and could solve it. We do not know how Fermat solved either equation though he said in a letter of 1658 that he used the method of infinite descent for the former.

8. *The Relationship of Algebra to Geometry*

Algebra, as we can see, expanded enormously during the sixteenth and seventeenth centuries. Because it had been tied to geometry, prior to 1500 equations of higher degree than the third were considered unreal. When the study of higher-degree equations was forced upon mathematicians (as for

14. *Œuvres*, 2, 333–35.
15. *Œuvres*, 2, 333–35; 3, 312–13.
16. Fermat, *Œuvres*, 3, 457–80, 490–503.

example by the use of trigonometric identities to aid in the calculation of tables) or suggested as the natural extension of third degree equations, the idea struck many mathematicians as absurd. Thus Stifel in his edition of Rudolff's *Coss* (Algebra) says, "Going beyond the cube just as if there were more than three dimensions . . . is against nature."

Nevertheless, algebra did rise above the limitations imposed by geometric thinking. But the relationship of algebra to geometry remained complicated. The major problem was how to justify algebraic reasoning; and the answer, during the sixteenth century and a good deal of the seventeenth, was to fall back upon the equivalent geometrical meaning of the algebra. Pacioli, Cardan, Tartaglia, Ferrari, and others gave geometrical proofs of algebraic rules. Vieta too was largely tied to geometry. Thus he writes $A^3 + 3B^2A = Z^3$, where A is the unknown and B and Z are constants, in order that each term be of the third degree and so represent a volume. However, as we shall see, Vieta's position on algebra was transitional. Barrow and Pascal actually objected to algebra, and later to analytic methods in coordinate geometry and the calculus, because the algebra lacked justification.

The dependence of algebra on geometry began to be reversed somewhat when Vieta, and later Descartes, used algebra to help solve geometric construction problems. The motivation for much of the algebra that appears in Vieta's *In Artem Analyticam Isagoge*, is solving geometric problems and systematizing geometrical constructions. Typical of the application of algebra to geometry by Vieta is the following problem from his *Zeteticorum Libri Quinque*: Given the area of a rectangle and the ratio of its sides, to find the sides of the rectangle. He takes the area as B *planum* and the ratio of the larger side to the smaller as S to R. Let A be the larger side. Then RA/S is the smaller side. Hence B *planum* equals $(R/S)(A$ squared$)$. Multiplying by S gives the final equation, $BS = RA^2$. Vieta then shows how from this equation A can be constructed by ruler and compass starting from the known quantities B and R/S. The idea here is that if one finds that a desired length x satisfies the equation $ax^2 + bx + c = 0$, one knows that

$$x = \frac{-b + \sqrt{b^2 - 4ac}}{2a}$$

and one can construct x by performing on a, b, and c the geometrical constructions called for by the algebraic expression on the right.

Algebra for Vieta meant a special procedure for discovery; it was analysis in the sense of Plato, who opposed it to synthesis. Theon of Alexandria, who introduced the term "analysis," defined it as the process that begins with the assumption of what is sought and by deduction arrives at a known truth. This is why Vieta called his algebra the analytic art. It performed the process of analysis, particularly for geometric problems. In fact this was the starting point of Descartes's thinking on coordinate geometry and his work on the theory of equations was motivated by the desire to further their use in solving geometric constructions.

The interdependence of algebra and geometry can be seen also in the work of Marino Ghetaldi (1566–1627), a pupil of Vieta. He made a systematic study of the algebraic solution of determinate geometric problems in one book of his *Apollonius Redivivus* (Apollonius Modernized, 1607). Conversely, he gave geometric proofs of algebraic rules. He also constructed geometrically the roots of algebraic equations. A full work on this subject is his *De resolutione et compositione mathematica* (1630), published posthumously.

We also find in the sixteenth and seventeenth centuries the recognition that algebra had to be developed to replace the geometrical methods introduced by the Greeks. Vieta saw the possibility of using algebra to deal with equality and proportion of magnitudes, no matter whether these magnitudes arose in geometrical, physical, or commercial problems. Hence he did not hesitate to consider higher-degree equations and algebraic methodology; he envisioned a deductive science of magnitudes employing symbolism. While algebra was to Vieta largely a royal road to geometry, his vision was great enough to see that algebra had a life and meaning of its own. Bombelli gave algebraic proofs acceptable for his time, without the use of geometry. Stevin asserted that what could be done in geometry could be done in arithmetic and algebra. Harriot's book *Artis Analyticae Praxis* (1631) extended, systematized, and brought out some of the implications of Vieta's work. The book is much like a modern text on algebra; it is more analytical than any algebra preceding it and presents a great advance in symbolism. It was widely used.

Descartes too began to see great potentialities in algebra. He says he begins where Vieta left off. He does not regard algebra as a science in the sense of giving knowledge of the physical world. Indeed such knowledge, he says, consists of geometry and mechanics; he sees in algebra a powerful method wherewith to carry on reasoning, particularly about abstract and unknown quantities. In his view algebra mechanizes mathematics so that thinking and processes become simple and do not require a great effort of the mind. Mathematical creation might become almost automatic.

Algebra, for Descartes, precedes the other branches of mathematics. It is an extension of logic useful for handling quantity, and in this sense more fundamental even than geometry; that is, it is logically prior to geometry. He therefore sought an independent and systematic algebra instead of an unplanned and unfounded collection of symbols and procedures tied to geometry. There is a sketch of a treatise on algebra, known as *Le Calcul* (1638), written either by Descartes himself or under his direction, that treats algebra as a distinct science. His algebra is devoid of meaning. It is a technique of calculation, or a method, and is part of his general search for method.

Descartes's view of algebra as an extension of logic in treating quantity suggested to him that a broader science of algebra might be created, which would embrace other concepts than quantity and be used to approach all problems. Even the logical principles and methods might be expressed symbolically, and the whole system employed to mechanize all reasoning. Descartes called this idea a "universal mathematics." The idea is vague in his works and was not pursued far by him. Nevertheless, he was the first to assign to algebra a fundamental place in the system of knowledge.

This view of algebra, first envisioned fully by Leibniz and ultimately developed into symbolic logic (Chap. 51, sec. 4), was also entertained by Isaac Barrow, though in more limited scope. Barrow, Newton's friend, teacher, and predecessor in the Lucasian chair of mathematics at Cambridge, did not regard algebra as part of mathematics proper but rather as a formalization of logic. To him only geometry was mathematical, and arithmetic and algebra dealt with geometrical magnitudes expressed in symbols.

No matter what philosophy of algebra may have been entertained by Descartes and Barrow, and no matter what potentiality they may have seen in it as a universal science of reasoning, the practical effect of the expanding use of arithmetic and algebraic techniques was to set up algebra as a branch of mathematics independent of geometry. For the time, it was a significant step that Descartes used a^2 to represent a length as well as an area, whereas Vieta had insisted that a second power must represent an area. Descartes called attention to his use of a^2 as a number, noting explicitly that he had departed from his predecessors. He says x^2 is a quantity such that $x^2 : x = x : 1$. Likewise, he says in *La Géométrie* that a product of lines can be a line; he was thinking of the quantities involved and not geometrically, as had the Greeks. It was clear to him that algebraic calculation was independent of geometry.

John Wallis, influenced by Vieta, Descartes, Fermat, and Harriot, went far beyond these men in freeing arithmetic and algebra from geometric representation. In his *Algebra* (1685) he derived all the results of Book V of Euclid algebraically. He too abandoned the limitation to homogeneous algebraic equations in x and y, a concept that had been maintained because such equations were derived from geometrical problems. He saw in algebra brevity and perspicuity.

Though Newton loved geometry, in his *Arithmetica Universalis* we find for the first time an affirmation of the basic importance of arithmetic and algebra as opposed to geometry; Descartes and Barrow had still favored geometry as the fundamental branch of mathematics. Newton needed and used the algebraic language for the development of the calculus, which could be handled best algebraically. And as far as the supremacy of algebra over geometry was concerned, the needs of the differential and integral calculus were to be decisive.

By 1700, then, algebra had reached the point where it could stand on its own feet. The only difficulty was that there was no ground on which to place them. From Egyptian and Babylonian times, intuition and trial and error had supplied some working rules; the reinterpretation of Greek geometrical algebra had supplied others; and independent algebraic work in the sixteenth and seventeenth centuries, partly guided by geometric interpretation, led to many new results. But logical foundations of algebra analogous to those Euclid had provided for geometry were nonexistent. The general lack of concern, apart from the objections of Pascal and Barrow, is surprising, in view of the fact that the Europeans were now fully aware of what rigorous deductive mathematics called for.

How did the mathematicians know what was correct? The properties of the positive integers and fractions are so readily derived from experience with collections of objects that they seem self-evident. Even Euclid failed to supply a logical basis for those books in the *Elements* that dealt with the theory of numbers. As new types of numbers were added to the number system, the rules of operation already accepted for the positive integers and fractions were applied to the new elements, with geometrical thinking as a handy guide. Letters, when introduced, were just representations of numbers and so could be treated as such. The more complicated algebraic techniques seemed justified either by geometrical arguments like those Cardan used or by sheer induction on specific cases. But none of these procedures was logically satisfactory. Geometry, even where called upon, did not supply the logic for negative, irrational, and complex numbers, nor did it justify arguing, for example, that if a polynomial is negative for $x = a$ and positive for $x = b$, that it must be zero between a and b.

Nevertheless, the mathematicians proceeded blithely and confidently to employ the new algebra. Wallis, in fact, affirmed that the procedures of algebra were not less legitimate than those of geometry. Without realizing it the mathematicians were about to enter a new era in which induction, intuition, trial and error, and physical arguments were to be the bases for proof. The problem of building a logical foundation for the number system and algebra was a difficult one, far more difficult than any seventeenth-century mathematician could possibly have appreciated. And it is fortunate that the mathematicians were so credulous and even naive, rather than logically scrupulous. For free creation must precede formalization and logical foundations, and the greatest period of mathematical creativity was already under way.

Bibliography

Ball, W. W. R.: *A Short Account of the History of Mathematics*, Dover (reprint), 1960, Chap. 12.

Boyer, Carl B.: *History of Analytic Geometry*, Scripta Mathematica, 1956, Chap. 4.

—————: *A History of Mathematics*, John Wiley and Sons, 1968, Chaps. 15–16.

Cajori, Florian: *Oughtred, A Great Seventeenth Century Teacher of Mathematics*, Open Court, 1916.

—————: *A History of Mathematics*, 2nd ed., Macmillan, 1919, pp. 130–59.

—————: *A History of Mathematical Notations*, Open Court, 1928, Vol. 1.

Cantor, Moritz: *Vorlesungen über Geschichte der Mathematik*, 2nd ed., B. G. Teubner, 1900, Johnson Reprint Corp., 1965, Vol. 2, pp. 369–806.

Cardan, G.: *Opera Omnia*, 10 vols., 1663, Johnson Reprint Corp., 1964.

Cardano, Girolamo: *The Great Art*, trans. T. R. Witmer, Massachusetts Institute of Technology Press, 1968.

Coolidge, Julian L.: *The Mathematics of Great Amateurs*, Dover (reprint), 1963, Chaps. 6–7.

David, F. N.: *Games, Gods and Gambling: The Origins and History of Probability*, Hafner, 1962.

Descartes, René: *The Geometry*, Dover (reprint), 1954, Book 3.

Dickson, Leonard E.: *History of the Theory of Numbers*, Carnegie Institution, 1919–23, Chelsea (reprint), 1951, Vol. 2, Chap. 26.

Fermat, Pierre de: *Œuvres*, 4 vols. and Supplement, Gauthier-Villars, 1891–1912, 1922.

Heath, Sir Thomas L.: *Diophantus of Alexandria*, 2nd ed., Cambridge University Press, 1910; Dover (reprint), 1964, Supplement, Secs. 1–5.

Hobson, E. W.: *John Napier and the Invention of Logarithms*, Cambridge University Press, 1914.

Klein, Jacob: *Greek Mathematical Thought and the Origin of Algebra*, Massachusetts Institute of Technology Press, 1968. Contains a translation of Vieta's *Isagoge*.

Knott, C. G.: *Napier Tercentenary Memorial Volume*, Longmans Green, 1915.

Montucla, J. F.: *Histoire des mathématiques*, Albert Blanchard (reprint), 1960, Vol. 1, Part 3, Book 3; Vol. 2, Part 4, Book 1.

Mordell, J. L.: *Three Lectures on Fermat's Last Theorem*, Cambridge University Press, 1921.

Morley, Henry: *Life of Cardan*, 2 vols., Chapman and Hall, 1854.

Newton, Sir Isaac: *Mathematical Works*, Vol. 2, ed. D. T. Whiteside, Johnson Reprint Corp., 1967. This volume contains a translation of *Arithmetica Universalis*.

Ore, Oystein: *Cardano: The Gambling Scholar*, Princeton University Press, 1953.

—————: "Pascal and the Invention of Probability Theory," *Amer. Math. Monthly*, 67, 1960, 409–19.

Pascal, Blaise: *Œuvres complètes*, Hachette, 1909.

Sarton, George: *Six Wings: Men of Science in the Renaissance*, Indiana University Press, 1957, "Second Wing."

Schneider, I.: "Der Mathematiker Abraham de Moivre (1667–1754)," Archive for History of Exact Sciences, 5, 1968, 177–317.

Scott, J. F.: *A History of Mathematics*, Taylor and Francis, 1958, Chaps. 6 and 9.

—————: *The Scientific Work of René Descartes*, Taylor and Francis, 1952, Chap. 9.

—————: *The Mathematical Work of John Wallis*, Oxford University Press, 1938.

Smith, David Eugene: *History of Mathematics*, Dover (reprint), 1958, Vol. 1, Chaps. 8–9; Vol. 2, Chaps. 4 and 6.

————: *A Source Book in Mathematics*, 2 vols., Dover (reprint), 1959.

Struik, D. J.: *A Source Book in Mathematics, 1200–1800*, Harvard University Press, 1969, Chaps. 1–2.

Todhunter, I.: *A History of the Mathematical Theory of Probability*, Chelsea (reprint), 1949, Chaps. 1–7.

Turnbull, H. W.: *James Gregory Tercentenary Memorial Volume*, Bell and Sons, 1939.

————: *The Mathematical Discoveries of Newton*, Blackie and Son, 1945.

Vieta, F.: *Opera mathematica*, (1646), Georg Olms (reprint), 1970.

14
The Beginnings of Projective Geometry

> I freely confess that I never had a taste for study or research either in physics or geometry except in so far as they could serve as a means of arriving at some sort of knowledge of the proximate causes . . . for the good and convenience of life, in maintaining health, in the practice of some art . . . having observed that a good part of the arts is based on geometry, among others the cutting of stone in architecture, that of sundials, that of perspective in particular.
>
> GIRARD DESARGUES

1. *The Rebirth of Geometry*

The resurgence of significant creative activity in geometry lagged behind that in algebra. Apart from the creation of the mathematical system of perspective and the incidental geometrical work of the Renaissance artists (Chap. 12, sec. 2), very little of consequence was done in geometry from the time of Pappus to about 1600. Some interest was created by the appearance of numerous printed editions of Apollonius' *Conic Sections*, especially the notable Latin translation by Federigo Commandino (1509–75) of Books I to IV in 1566. Books V to VII were made available by other translators, and a number of men, including Vieta, Willebrord Snell (1580–1626), and Ghetaldi, undertook to reconstruct the lost eighth book.

What was needed and did arise to direct the minds of mathematicians into new channels was new problems. One problem had already been raised by Alberti: What geometrical properties do two sections of the same projection of an actual figure have in common? A large number of problems came from science and practical needs. Kepler's use of the conic sections in his work of 1609 gave enormous impetus to the reexamination of these curves and to the search for properties useful in astronomy. Optics, an interest of mathematicians since Greek times, received greatly increased attention after the invention of the telescope and microscope in the beginning of the seventeenth century. The design of lenses for these instruments became a major problem; this meant an interest in the shapes of the surfaces or, since these surfaces are figures of revolution, the shapes of the generating curves. The geographical explorations had created a need for maps and for studying the paths of ships as represented on the sphere and on the map. The introduction of the notion of a moving earth called for new principles of mechanics to account for the paths of moving objects; this, too, meant a study of curves. Among moving objects, projectiles became more important because cannons could now fire balls over hundreds of yards; prediction of path and range was vital. The practical problem of finding areas and volumes began to attract more attention. Kepler's *Nova Stereometria Doliorum Vinariorum* (The New Science of Measuring Volumes of Wine Casks, 1615) initiated a new burst of activity in this area.

Another kind of problem came to the fore as a consequence of assimilation of the Greek works. Mathematicians began to realize that the Greek methods of proof lacked generality. A special method had to be devised for nearly every theorem. This point had been made by Agrippa von Nettesheim (1486–1532), as far back as 1527, and by Maurolycus, who had translated Greek works and had written books of his own on the conic sections and other mathematical subjects.

Much of the response to the new problems resulted in minor variations on old themes. The approach to the conic sections was altered. The curves were defined at the outset as loci in the plane instead of sections of a cone as in Apollonius. Guidobaldo del Monte, for example, in 1579 defined the ellipse as the locus of points the sum of whose distances from the foci is a constant. Not only the conics but older Greek curves such as the conchoid of Nicomedes, the cissoid of Diocles, the spiral of Archimedes, and the quadratrix of Hippias were restudied. New curves were introduced, notably the cycloid (see Chap. 17, sec. 2). While all of this work was helpful in disseminating the Greek contributions, none of it offered new theorems or new methods of proof. The first innovation of consequence came in answer to the problems raised by the painters.

2. *The Problems Raised by the Work on Perspective*

The basic idea in the system of focused perspective created by the painters is the principle of projection and section (Chap. 12, sec. 1). A real scene is viewed by the eye regarded as a point. The lines of light from various points of the scene to the eye are said to constitute a projection. According to the system, the painting itself must contain a section of that projection, the section being mathematically what a plane passing through the projection would contain.

Now suppose the eye at O (Fig. 14.1) looks at the horizontal rectangle

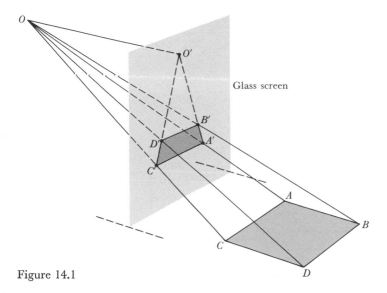

Figure 14.1

ABCD. The lines from *O* to the points on the four sides of this rectangle constitute a projection of which *OA*, *OB*, *OC*, and *OD* are typical lines. If a plane is now interposed between the eye and the rectangle, the lines of the projection will cut through the plane and mark out on it the quadrangle *A'B'C'D'*. Since the section, namely *A'B'C'D'*, creates the same impression on the eye as does the original rectangle, it is reasonable to ask, as Alberti did, what geometrical properties do the section and the original rectangle have in common? It is intuitively apparent that the original figure and the section will be neither congruent nor similar; nor will they contain the same area. In fact the section need not be a rectangle.

There is an extension of this problem: Suppose two different sections of this same projection are made by two different planes that cut the projection at any angle. What properties would the two sections have in common?

The problem may be further extended. Suppose a rectangle *ABCD* is viewed from two different locations *O'* and *O"* (Fig. 14.2). Then there are two projections, one determined by *O'* and the rectangle and the second determined by *O"* and the rectangle. If a section is made of each projection, then, in view of the fact that each section should have some geometrical properties in common with the rectangle, the two sections should have some common geometrical properties.

Some of the seventeenth-century geometers undertook to answer these questions. They viewed the methods and results they obtained as part of Euclidean geometry. However, these methods and results, while indeed contributing much to that subject, proved to be the beginning of a new

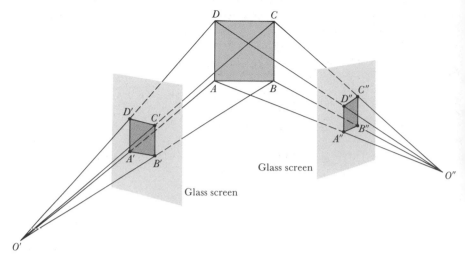

Figure 14.2

branch of geometry, which in the nineteenth century became known as projective geometry. We shall refer to the work by that name, but the distinction between Euclidean and projective geometry was not made in the seventeenth century.

3. *The Work of Desargues*

The man who first took up directly the problems just sketched was the self-educated Girard Desargues (1591–1661), who became an army officer and later an engineer and architect. Desargues knew the work of Apollonius and thought that he could introduce new methods for proving theorems about conics. He did and, indeed, was fully aware of the power of these methods. Desargues was also concerned to improve the education and technique of artists, engineers, and stonecutters; he had little use for theory for its own sake. He began by organizing numerous useful theorems and at first distributed his results through letters and handbills. He also gave lectures gratis in Paris. Afterwards he wrote several books, one of which taught children how to sing. Another applied geometry to masonry and stonecutting.

His major text, which was preceded in 1636 by a pamphlet on perspective, was *Brouillon project d'une atteinte aux événemens des rencontres du cône avec un plan* (Proposed Draft of an Attempt to Treat the Results of a Cone Intersecting a Plane, 1639).[1] This book deals with what we would now call

1. *Œuvres*, 1, 103–230.

projective methods in geometry. Desargues printed about fifty copies and circulated them among his friends. Not long afterwards all copies were lost. A manuscript copy made by La Hire was found accidentally in 1845 by Michel Chasles and reproduced by N. G. Poudra, who edited Desargues's work in 1864. However, an original edition of the 1639 work was discovered about 1950 by Pierre Moisy in the Bibliothèque nationale (Paris) and has been reproduced. This newly discovered copy also contains an appendix and errata by Desargues that are important. Desargues's main theorem on triangles and other theorems of his were published in 1648 in an appendix to a book on perspective by his friend Abraham Bosse (1602–76). In this book, *Manière universelle de M. Desargues, pour pratiquer la perspective,* Bosse tried to popularize Desargues's practical methods.

Desargues used curious terminology, some of which had appeared in Alberti's work. A straight line he called a "palm." When points were marked out on the line he called it a "trunk." However, a line with three pairs of points in involution (see below) was a "tree." Desargues's intent in introducing this new terminology was to gain clarity by avoiding the ambiguities in more customary terms. But the language and the strange ideas made his book difficult reading. His contemporaries, except for his friends Mersenne, Descartes, Pascal, and Fermat, called him crazy. Even Descartes, when he heard that Desargues was introducing a new method to treat conics, wrote to Mersenne that he could not see how anyone could do anything new on conics except with the aid of algebra. However, after Descartes learned the details of what Desargues was doing, he respected it highly. Fermat regarded Desargues as the real founder of the theory of conic sections and found his book, which Fermat apparently owned, rich in ideas. But the general lack of appreciation disgusted Desargues and he retired to his estate.

Before we note some of the theorems Desargues established, we must introduce one new convention on parallel lines. Alberti had pointed out that two lines that are parallel in some actual scene being painted (unless they are parallel to the glass screen or canvas) must be drawn on the canvas so as to meet in a point. Thus the lines $A'B'$ and $C'D'$ in Figure 14.1 above, which correspond to the parallel lines AB and CD, must, according to the principle of projection and section, meet at some point O'. As a matter of fact, O and AB determine a plane and so do O and CD. These two planes cut the glass screen in $A'B'$ and $C'D'$ but, since they meet at O, these two planes must have a line in common; this line cuts the glass screen at some point O', which is also the intersection of $A'B'$ and $C'D'$. The point O' does not correspond to any ordinary point on AB or CD. In fact, the line OO' is horizontal, and so parallel to AB and CD. The point O' is called a vanishing point because it does not have a corresponding point on the lines AB or CD, whereas any other point on the lines $A'B'$ or $C'D'$ corresponds to some definite point on AB or CD, respectively.

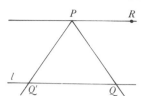

Figure 14.3

To complete the correspondence between points on the lines $A'B'$ and AB as well as between those on $C'D'$ and CD, Desargues introduced a new point on AB and on CD. Called the point at infinity, it is additional to the usual points on the two lines, and is to be regarded as the common point of the two lines. Moreover any other line parallel to AB or CD is to have this same point on it and meet AB and CD at this point. Any set of parallel lines having a different direction from that of AB or CD is likewise to have a common point at infinity. Since each set of parallel lines has a point in common and there is an infinite number of different sets of such lines, Desargues's convention introduces an infinity of new points in the Euclidean plane. He made the further assumption that all these new points lie on one line, which corresponds to the horizon line or vanishing line on the section. Thus a new line is added to the already existing lines in the Euclidean plane. A set of parallel planes is assumed to have the line at infinity on each in common; that is, all parallel planes meet on one line.

The addition of a new point on each line does not contradict any axiom or theorem of Euclidean geometry, though it does call for a change in wording. Non-parallel lines continue to meet in ordinary points while parallel lines meet in the "point at infinity" on each of the lines. The agreement as to the points at infinity is actually a convenience in Euclidean geometry, in that it avoids special cases. For example, one could now say that any two lines meet in exactly one point. We shall soon see more fully the advantages of this convention.

Kepler, too, decided (1604) to add the point at infinity to parallel lines but for a different reason. To each line through P (Fig. 14.3) and cutting l there is one point, Q, on l. However, no point on l corresponds to PR, the parallel to l through P. By adding a point at infinity common to PR and l, Kepler could affirm that every line through P cuts l. Moreover, after Q has moved to "infinity" on the right and PQ has become PR, the point of intersection of PR and l can be thought of as being at infinity to the left of P; and as PR continues to rotate around P, the point of intersection Q' of PR and l will move in from the left. Thus continuity of the intersection of PR and l is maintained. In other words, Kepler (and Desargues) regarded the two "ends" of the line as meeting at "infinity" so that the line has the structure of a circle. In fact, Kepler actually thought of a line as a circle with its center at infinity.

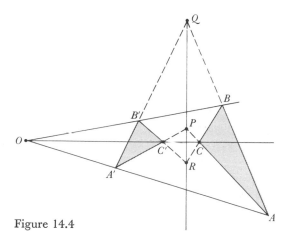

Figure 14.4

Having introduced his points and lines at infinity, Desargues was able to state a basic theorem, still called Desargues's theorem. Consider the point O (Fig. 14.4) and triangle ABC. The lines from O to the various points on the sides of the triangle constitute, as we know, a projection. A section of this projection will then contain a triangle $A'B'C'$, where A' corresponds to A, B' to B, and C' to C. The two triangles ABC and $A'B'C'$ are said to be perspective from the point O. Desargues's theorem then states: The pairs of corresponding sides AB and $A'B'$, BC and $B'C'$, and AC and $A'C'$ (or their prolongations) of two triangles perspective from a point meet in three points which lie on one straight line. Conversely, if the three pairs of corresponding sides of the two triangles meet in three points which lie on one straight line, then the lines joining corresponding vertices meet in one point. With specific reference to the figure, the theorem proper says that because AA', BB', and CC' meet in a point O, the sides AC and $A'C'$ meet in a point P; AB and $A'B'$ meet in a point Q; and BC and $B'C'$ meet in a point R; and P, Q, and R lie on a straight line.

Though the theorem is true whether the triangles ABC and $A'B'C'$ lie in the same or different planes, its proof is simple only in the latter case. Desargues proved the theorem and its converse for both the two- and three-dimensional cases.

In the appendix to Bosse's 1648 work, there appears another fundamental result due to Desargues, the invariance of cross ratio under projection. The cross ratio of the line segments formed by four points A, B, C, and D on one line (Fig. 14.5) is by definition $\dfrac{BA}{BC}\bigg/\dfrac{DA}{DC}$. Pappus had already introduced this ratio (Chap. 5, sec. 7) and proved that it is the same on the two lines AD and $A'D'$. Menelaus also had a similar theorem about arcs of great circles on

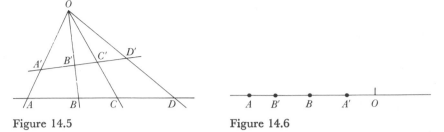

Figure 14.5   Figure 14.6

a sphere (Chap. 5, sec. 6). But neither of these men thought in terms of projection and section. Desargues did, and proved that the cross ratio is the same for every section of the projection.

In his major work (1639), he treated the concept of involution, which Pappus had also introduced but which Desargues named. Two pairs of points A, B and A', B' are said to be an involution if there is a special point O, called the center of the involution, on the line containing the four such that $OA \cdot OB = OA' \cdot OB'$ (Fig. 14.6). Likewise, three pairs of points, A, B, A', B', and A'', B'' are said to be an involution if $OA \cdot OB = OA' \cdot OB' = OA'' \cdot OB''$. The points A and B, A' and B', and A'' and B'' are said to be conjugate points. If there is a point E such that $OA \cdot OB = \overline{OE}^2$, then E is called a double point. In this case there is a second double point F, and O is the midpoint of EF. The conjugate of O is the point at infinity. Desargues used Menelaus' theorem on a line cutting the sides of a triangle to prove that if the four points A, B, A', and B' are an involution (Fig. 14.7), and if they are projected from P onto the points A_1, B_1, A_1', and B_1' of another line, then the second set of four points is also an involution.

In the subject of involutions, Desargues proved a major theorem. To get at it, let us first consider the concept of a complete quadrilateral, a notion already partly treated by Pappus. Let B, C, D, and E be any four points in a plane (Fig. 14.8), no three collinear. Then they determine six lines that are the sides of the complete quadrilateral. Opposite sides are two sides that do not have one of the four points in common. Thus BC and DE are opposite,

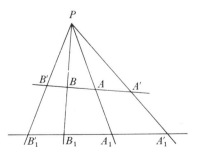

Figure 14.7

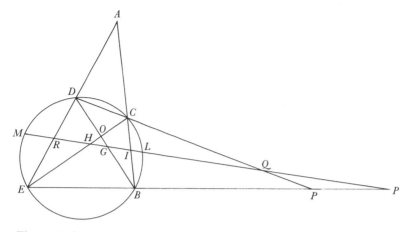

Figure 14.8

as are *CD* and *BE* and *BD* and *CE*. The intersections *O*, *F*, and *A* of the
three pairs of opposite sides are the diagonal points of the quadrilateral. Now
take the four vertices *B*, *C*, *D*, *E* on a circle. Suppose a line *PM* meets the
pairs of sides in *P*, *Q*; *I*, *K*; and *G*, *H*; and the circle in the pair *L*, *M*. These
four pairs of points are four pairs of an involution.

Further, suppose the entire figure is projected from some point outside
the plane of the figure and a section is made of this projection. The circle
will give rise to a conic in the section, and each line in the original figure will
give rise to some line in the section. In particular, the quadrilateral in the
circle will give rise to a quadrilateral in the conic. Since an involution
projects into an involution, there follows the important and general result:
If a quadrilateral is inscribed in a conic, any straight line not passing through
a vertex intersects the conic and the pairs of opposite sides of the complete
quadrilateral in four pairs of points of an involution.

Desargues next introduces the notion of a harmonic set of points. The
points *A*, *B*, *E*, and *F* form a harmonic set if *A* and *B* are a pair of conjugate
points with respect to the double points *E* and *F* of an involution. (The
current definition that the cross ratio of a harmonic set is -1 is a later
approach.)[2] Since an involution projects into an involution, so does a
harmonic set project into a harmonic set. Then Desargues shows that if one
member of a harmonic set of points is the point at infinity (on the line of the
four), the other point of that pair bisects the line segment joining the other
two points of the set. Further if *A*, *B*, *A'*, and *B'* are a harmonic set (Fig. 14.9),
and if the projection from *O* is formed, then if *OA'* is perpendicular to *OB'*,
OA' bisects angle *AOB*, and *OB'* bisects the supplementary angle.

With the notion of harmonic set available, Desargues proceeds to pole

2. It is due to Möbius: *Barycentrische Calcul* (1827), p. 269.

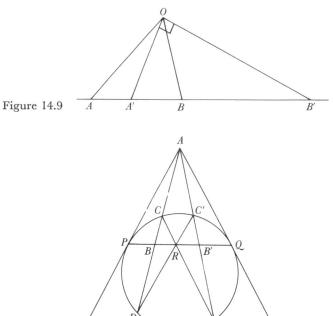

Figure 14.9

Figure 14.10

and polar theory, a subject already introduced by Apollonius. Desargues starts with a circle and the point A outside (Fig. 14.10). Then on any line from A cutting the circle in C and D, say, there is a fourth harmonic point B. For all such lines from A the fourth harmonic points lie on one line, the polar of the point A. Thus the line BB' is the polar of A. Moreover, suppose we introduce any complete quadrilateral for which A is one diagonal point and whose four vertices are on the circle. Thus we could choose C, D, D', and C' as the vertices of such a complete quadrilateral. Then the polar of A will go through the other two diagonal points of this complete quadrilateral. (In the figure, R is one of the two diagonal points.) The same assertions hold when the point A is inside the circle. If the point A is outside the circle, the polar of A joins the points of contact of the tangents from A to the circle (the points P and Q in the figure).

Having proved the above assertions for the circle, Desargues again uses projection from a point outside the plane of the figure and a section of the projection to prove that the assertions hold for any conic.

Desargues treats a diameter of a conic as the polar of a point at infinity. We shall see in a moment that this definition agrees with Apollonius'. Consider a family of parallel lines that cut the conic (Fig. 14.11). These lines have a point at infinity in common. If $A'B'$ is one such line, the har-

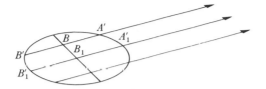

Figure 14.11

monic conjugate with respect to A' and B' of the point at infinity on $A'B'$ is the midpoint B of the chord $A'B'$. Likewise B_1, the harmonic conjugate of the point at infinity with respect to A'_1 and B'_1, is the midpoint of the chord $A'_1B'_1$. These midpoints of a family of parallel chords lie on one straight line, which is a diameter in Apollonius' definition also. Desargues then proves a number of facts about diameters, conjugate diameters, and asymptotes to hyperbolas.

As we can now see, Desargues not only introduced new concepts, notably the elements at infinity, and many new theorems, but above all he introduced projection and section as a new method of proof and unified the approach to the several types of conics through projection and section, whereas Apollonius had treated each type of conic separately. Desargues was one of the most original mathematicians in a century rich in genius.

4. *The Work of Pascal and La Hire*

The second major contributor to projective geometry was Blaise Pascal (1623–62). Born in Clermont, France, he was a sickly child and in poor health during his short life. His father, Etienne Pascal, intended to keep his son from mathematics until he was fifteen or sixteen because he believed that a subject should not be introduced to a child until he was old enough to absorb it. But at the age of twelve Blaise insisted on knowing what geometry was and, on being told, set to work on it himself.

The family had moved to Paris when Pascal was eight. Even as a child he went with his father to the weekly meetings of the "Académie Mersenne," which later became the Académie Libre and, in 1666, the Académie des Sciences. Among the members were Father Mersenne, Desargues, Roberval (professor of mathematics at the Collège de France), Claude Mydorge (1585–1647), and Fermat.

Pascal devoted considerable time and energy to projective geometry. He was one of the founders of the calculus and in this subject influenced Leibniz. As we have seen, he also took a hand in the start of the work on probability. At the age of nineteen he invented the first calculating machine to help his father in the latter's work as a tax assessor. He also contributed to physics some experimental work on an original device for the creation of vacuums, the fact that the weight of the air decreases as the altitude increases, and a clarification of the concept of pressure in liquids. The originality of the work in physics has been questioned; in fact, some historians of science have described it as popularization, or plagiarism, depending upon whether or not they wished to be charitable.

Pascal was great in many other fields. He became a master of French prose; his *Pensées* and *Lettres provinciales* are literary classics. He also became famous as a polemicist in theology. From childhood on, he attempted to reconcile religious faith with the rationalism of mathematics and science, and throughout his life the two interests competed for his energy and time. Pascal, like Descartes, believed that truths of science must either appeal clearly and distinctly to the senses or the reason or be logical consequences of such truths. He saw no room for mystery-mongering in matters of science and mathematics. "Nothing that has to do with faith can be the concern of reason." In matters of science, in which only our natural thinking is involved, authority is useless; reason alone has grounds for such knowledge. However, the mysteries of faith are hidden from sense and reason and must be accepted on the authority of the Bible. He condemns those who use authority in science or reason in theology. However, the level of faith was above the level of reason.

Religion dominated his thoughts after the age of twenty-four though he continued to do mathematical and scientific work. He believed that the pursuit of science for mere enjoyment was wrong. To make enjoyment the chief end of research was to corrupt the research, for then one acquired "a greed or lust for learning, a profligate appetite for knowledge. . . . Such a study of science sprang from a prior concern for self as the center of things rather than a concern for seeking out, amid all surrounding natural phenomena, the presence of God and His glory."

In mathematical work he was largely intuitive; he anticipated great results, made superb guesses, and saw shortcuts. Later in life, he favored intuition as a source of all truths. Several of his declarations bearing on this have become famous. "The heart has its own reasons, which reason does not know." "Reason is the slow and tortuous method by which those who do not know the truth discover it." "Humble thyself, impotent reason."

If one may judge from a letter Pascal wrote to Fermat on August 10, 1660, toward the end of his life, Pascal seems to have turned somewhat against mathematics. He wrote: "To speak freely of mathematics, I find it the highest exercise of the spirit; but at the same time I know that it is so useless that I make little distinction between a man who is only a mathematician and a common artisan. Also, I call it the most beautiful occupation in the world; but it is only an occupation; and I have often said that it is good to make the attempt [to study mathematics], but not to use our forces: so that I would not take two steps for mathematics, and I am confident that

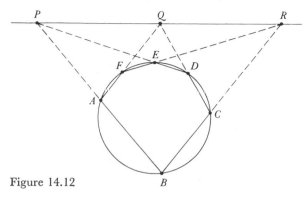

Figure 14.12

you are strongly of my opinion." Pascal was a man of manifold but contradictory qualities.

Desargues urged Pascal to work on the method of projection and section and suggested in particular the goal of reducing the many properties of the conic sections to a small number of basic propositions. Pascal took up these recommendations. In 1639, at the age of sixteen, he wrote a work on conics that used projective methods, that is projection and section. This work is now lost, but Leibniz did see it in Paris in 1676 and described it to Pascal's nephew. An eight-page *Essay on Conics* (1640), known to a few of Pascal's contemporaries, was also lost until 1779, when it was recovered.[3] Descartes, who saw the 1640 essay, regarded it as so brilliant that he could not believe it was written by so young a man.

Pascal's most famous result in projective geometry, which appeared in both of the works just mentioned, is a theorem now named after him. The theorem in modern language asserts the following: If a hexagon is inscribed in a conic, the three points of intersection of the pairs of opposite sides lie on one line. Thus (Fig. 14.12) P, Q, and R lie on one straight line. If the opposite sides of the hexagon are parallel, P, Q, and R will lie on the line at infinity.

We have only indications of how Pascal proved this theorem. He says that since it is true of the circle, it must by projection and section be true of all conics. And it is clear that if one forms a projection of the above figure from a point outside the plane and then a section of this projection, the section will contain a conic and a hexagon inscribed in it. Moreover, the opposite sides of this hexagon will meet in three points of a straight line, the points and straight line that correspond to P, Q, R, and the line PQR of the original figure. Incidentally Pappus' theorem (Chap. 5, sec. 7), which refers to three points on each of two lines, is a special case of the above

3. *Œuvres*, 1, 1908, 243–60.

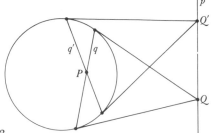

Figure 14.13

theorem. When the conic degenerates into two straight lines, as when a hyperbola degenerates into its asymptotes, then the situation described by Pappus results.

The converse of Pascal's theorem, namely, if a hexagon is such that the points of intersection of its three pairs of opposite sides lie on one straight line, then the vertices of the hexagon lie on a conic, is also correct but was not considered by Pascal. There are other results in Pascal's 1640 work but they do not warrant attention here.

The method of projection and section was taken up by Philippe de La Hire (1640–1718), who was a painter in his youth and then turned to mathematics and astronomy. Like Pascal, La Hire was influenced by Desargues and did a considerable amount of work on the conic sections. Some of it, in publications of 1673 and 1679, employed the synthetic manner of the Greeks but with new approaches, such as the focus-distance definition of the ellipse and hyperbola, and some of it used the analytic geometry of Descartes and Fermat. His greatest work, however, is the *Sectiones Conicae* (1685) and this is devoted to projective geometry.

Like Desargues and Pascal, La Hire first proved properties of the circle, chiefly involving harmonic sets, and then carried these properties over to the other conic sections by projection and section. Thus he could carry the properties of the circle over to any type of conic section in one method of proof. Though there were a few omissions, such as Desargues's involution theorem and Pascal's theorem, in this 1685 work of La Hire we find practically all the now familiar properties of conic sections synthetically proved and systematically established. In fact, La Hire proves almost all of Apollonius' 364 theorems on the conics. He also has the harmonic properties of quadrilaterals. In all, La Hire proved about 300 theorems. He tried to show that projective methods were superior to those of Apollonius and to the new analytic methods of Descartes and Fermat (Chap. 15) which had already been created.

On the whole, La Hire's results do not go beyond Desargues's and Pascal's. However, in pole and polar theory he has one major new result.

He proves that if a point traces a straight line, then the polar of the point will rotate around the pole of that straight line. Thus if Q (Fig. 14.13) moves along the line p, then the polar of Q rotates around the pole P of the line p.

5. The Emergence of New Principles

Over and above the specific theorems created by men such as Desargues, Pascal, and La Hire, several new ideas and outlooks were beginning to appear. The first is the idea of *continuous change of a mathematical entity* from one state to another; in the present instance, the entity is a geometrical figure. It was Kepler, in his *Astronomiae pars Optica*[4] of 1604, who first seemed to grasp the fact that parabola, ellipse, hyperbola, circle, and the degenerate conic consisting of a pair of lines are continuously derivable from each other. To effect this process of continuous change from one figure to another, say from ellipse to parabola and then to hyperbola, Kepler imagined that one focus is held fixed and the other is allowed to move along the straight line joining them. By letting the moving focus go off to infinity (while letting the eccentricity approach 1), the ellipse becomes the parabola; then, by having the moving focus reappear on the line but on the other side of the stationary focus, the parabola becomes a hyperbola. When the two foci of an ellipse move together, the ellipse becomes a circle; and when the two foci of a hyperbola move together, the hyperbola degenerates into two straight lines. So that one focus could go off to infinity in one direction and then reappear on the other side, Kepler assumed, as already noted, that the straight line extended in either direction meets itself at infinity at a single point, so that the line has the property of a circle. Though it is not intuitively satisfying to view the line in this manner, the idea is logically sound, and indeed became fundamental in nineteenth-century projective geometry. Kepler also pointed out that the various sections of the cone can be obtained by continuously varying the inclination of the plane that makes the section.

The notion of continuous change in a figure was also employed by Pascal. He allowed two consecutive vertices of his hexagon to approach each other so that the figure became a pentagon. He then argued from properties of the hexagon to properties of the pentagon by considering what happened to these properties under the continuous change. In the same manner he passed from pentagons to quadrilaterals.

The second idea to emerge clearly from the work of the projective geometers is that of *transformation and invariance*. To project a figure from some point and then take a section of that projection is to transform the figure to a new one. The properties of the original figure that are of interest are those that remain invariant under the transformation. Other geometers of the

4. *Ad Vitellionem Paralipomena, quibus Astronomiae pars Optica Traditur* (Supplement to Vitello, Giving the Optical Part of Astronomy).

seventeenth century, for example, Gregory of St. Vincent (1584–1667) and Newton,[5] introduced transformations other than projection and section.

The projective geometers, too, undertook the search for general methods that had been initiated by the algebraists, particularly Vieta. In Greek times the methods of proof had been limited in power. Each theorem required another plan of attack. Neither Euclid nor Apollonius seemed to have been concerned with general methods. But Desargues emphasized projection and section because he saw in it a general procedure for proving theorems about all conics once they were proved for the circle. And in the notions of involution and harmonic sets, he saw more general concepts than those of Euclidean geometry. Indeed a harmonic set of four points, when one of the four is at infinity, reduces to three points, one of which is the midpoint of the line segment joining the other two. Hence, the notion of harmonic set and theorems about harmonic sets are more general than the notion of a point bisecting a line segment. Desargues and Pascal sought to derive the largest possible number of results from a single theorem. Bosse says that Desargues derived sixty theorems of Apollonius from his involution theorem and that Pascal praised him for it. Pascal, by seeking the relationships of different figures, such as the hexagon and pentagon, also sought some common approach to these figures. In fact, he was supposed to have deduced some 400 corollaries from his theorem on the hexagon, by examining the consequences of the theorem for related figures. However, there are no extant works of his in which this is done. The interest in method is clear in the 1685 work of La Hire, for its major point was to show that the method of projection and section was superior to Apollonius' methods and even to Descartes's algebraic methods. This search for generality in results and methods became a powerful force in subsequent mathematics.

Though the geometers were emphasizing generality of method, they were unconsciously unearthing another kind of generality. Many of the theorems, such as Desargues's triangle theorem, deal with the intersection of points and lines rather than with the sizes of line segments, angles, and areas, as is true of Euclidean geometry. The fact that lines intersect is logically prior to any consideration of size because the very fact of intersection determines the formation of a figure. A new and fundamental branch of geometry, in which location and intersection properties, rather than size or metric properties, mattered, was being born. However, the seventeenth-century workers in projective geometry used Euclidean geometry as a base, in particular the concepts of distance and size of angle. Moreover, these geometers, far from thinking in terms of a new geometry, were in fact trying to improve the methods of Euclidean geometry. The realization that a new branch of geometry was implicit in their work did not come about until the nineteenth century.

5. *Principia*, 3rd ed., Book 1, Lemma 22 and Prop. 25.

Although at the outset the motivation for the work in projective geometry was a desire to help the painters, its goal became diverted to and merged with the rising interest in the conic sections. But pure mathematics was not congenial to the temper of the seventeenth century, whose mathematicians were far more concerned with understanding and mastering nature—in short, with scientific problems. Algebraic methods of working with mathematical problems proved in general to be more effective, and in particular to yield the quantitative knowledge that science and technology needed. The qualitative results the projective geometers produced by their synthetic methods were not nearly so helpful. Hence projective geometry was abandoned in favor of algebra, analytic geometry, and the calculus, which themselves blossomed into still other subjects of central importance in modern mathematics. The results of Desargues, Pascal, and La Hire were forgotten and had to be rediscovered later, chiefly in the nineteenth century, by which time intervening creations and new points of view enabled mathematicians to fructify the large ideas still dormant in projective geometry.

Bibliography

Chasles, Michel: *Aperçu historique des méthodes en géométrie*, 3rd ed., Gauthier-Villars et Fils, 1889, pp. 68–95, 118–37. (Same as first edition of 1837.)

Coolidge, Julian L.: *A History of Geometrical Methods*, Dover (reprint), 1963, pp. 88–92.

————: *A History of the Conic Sections and Quadric Surfaces*, Dover (reprint), 1968, Chap. 3.

————: "The Rise and Fall of Projective Geometry," *Amer. Math. Monthly*, 41, 1934, 217–28.

Desargues, Girard: *Œuvres*, 2 vols., Leiber, 1864.

Ivins, W. M., Jr.: "A Note on Girard Desargues," *Scripta Math.*, 9, 1943, 33–48.

————: "A Note on Desargues's Theorem," *Scripta Math.*, 13, 1947, 203–10.

Mortimer, Ernest: *Blaise Pascal: The Life and Work of a Realist*, Harper and Bros., 1959.

Pascal, Blaise: *Œuvres*, Hachette, 1914–21.

Smith, David Eugene: *A Source Book in Mathematics*, Dover (reprint), 1959, Vol. 2, pp. 307–14, 326–30.

Struik, D. J.: *A Source Book in Mathematics, 1200–1800*, Harvard University Press, 1969, pp. 157–68.

Taton, René: *L'Œuvre mathématique de G. Desargues*, Presses Universitaires de France, 1951.

15
Coordinate Geometry

> . . . I have resolved to quit only abstract geometry, that is to say, the consideration of questions which *serve only to exercise the mind*, and this, in order to study another kind of geometry, which has for its object the explanation of the phenomena of nature.
>
> RENÉ DESCARTES

1. *The Motivation for Coordinate Geometry*

Fermat and Descartes, the two men primarily responsible for the next major creation in mathematics, were, like Desargues and his followers, concerned with general methods for studying curves. But Fermat and Descartes were very much involved in scientific work, keenly aware of the need for quantitative methods, and impressed with the power of algebra to supply that method. And so Fermat and Descartes turned to the application of algebra to the study of geometry. The subject they created is called coordinate, or analytic, geometry; its central idea is the association of algebraic equations with curves and surfaces. This creation ranks as one of the richest and most fruitful veins of thought ever struck in mathematics.

That the needs of science and an interest in methodology motivated both Fermat and Descartes is beyond doubt. Fermat's contributions to the calculus such as the construction of tangents to curves and the calculation of maxima and minima, were, as we shall see more clearly in connection with the history of the calculus, designed to answer scientific problems; he was also a first-rate contributor to optics. His interest in methodology is attested to by an explicit statement in his brief book, *Ad Locos Planos et Solidos Isagoge* (Introduction to Plane and Solid Loci[1]), written in 1629 but published by 1637.[2] He says there that he sought a universal approach to problems involving curves. As for Descartes, he was one of the greatest seventeenth-century scientists, and he made methodology a prime objective in all of his work.

---

1. Fermat uses these terms in the sense explained by Pappus. See Chap. 8, sec. 2.
2. *Œuvres*, 1, 91–103.

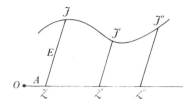

Figure 15.1

2. The Coordinate Geometry of Fermat

In his work on the theory of numbers, Fermat started with Diophantus. His work on curves began with his study of the Greek geometers, notably Apollonius, whose lost book, *On Plane Loci*, he, among others, had reconstructed. Having contributed to algebra, he was prepared to apply it to the study of curves, which he did in *Ad Locos*. He says that he proposed to open up a general study of loci, which the Greeks had failed to do. Just how Fermat's ideas on coordinate geometry evolved is not known. He was familiar with Vieta's use of algebra to solve geometric problems, but it is more likely that he translated Apollonius' results directly into algebraic form.

He considers any curve and a typical point J on it (Fig. 15.1). The position of J is fixed by a length A, measured from a point O on a base line to a point Z, and the length E from Z to J. Thus Fermat uses what we call oblique coordinates, though no y-axis appears explicitly and no negative coordinates are used. His A and E are our x and y.

Fermat had stated earlier his general principle: "Whenever in a final equation two unknown quantities are found we have a locus, the extremity of one of these describing a line straight or curved." Thus the extremities $J, J', J'', \ldots$ of E in its various positions describe the "line." His unknown quantities, A and E, are really variables or, one can say, the equation in A and E is indeterminate. Here Fermat makes use of Vieta's idea of having a letter stand for a class of numbers. Fermat then gives various algebraic equations in A and E and states what curves they describe. Thus he writes "*D in A aequetur B in E*" (in our notation, $Dx = By$) and states that this represents a straight line. He also gives (in our notation) the more general equation $d(a - x) = by$ and affirms that this too represents a straight line. The equation "*B quad. − A quad. aequetur E quad.*" (in our notation, $B^2 - x^2 = y^2$) represents a circle. Similarly (in our notation), $a^2 - x^2 = ky^2$ represents an ellipse; $a^2 + x^2 = ky^2$ and $xy = a$ represent hyperbolas; and $x^2 = ay$ represents a parabola. Since Fermat did not use negative coordinates, his equations could not represent the full curve that he said they described. He did appreciate that one can translate and rotate axes, because he gives more complicated second-degree equations and states the simpler forms to which they can be reduced. In fact, he affirms that an equation of the first degree in A and E has a straight-line locus and all second degree equations in A and E have conics as their loci. In his *Methodus ad Disquirendam Maximam et Minimam* (Method of Finding Maxima and Minima, 1637),[3] he introduced the curves of $y = x^n$ and $y = x^{-n}$.

3. René Descartes

Descartes was the first great modern philosopher, a founder of modern biology, a first-rate physicist, and only incidentally a mathematician. However, when a man of his power of intellect devotes even part of his time to a subject, his work cannot but be significant.

He was born in La Haye in Touraine on March 31, 1596. His father, a moderately wealthy lawyer, sent him at the age of eight to the Jesuit school of La Flèche in Anjou. Because he was of delicate health, he was allowed to spend the mornings in bed, during which time he worked. He followed this custom throughout his life. At sixteen he left La Flèche and at twenty he was graduated from the University of Poitiers as a lawyer and went to Paris. There he met Mydorge and Father Marin Mersenne and spent a year with them in the study of mathematics. However, Descartes became restless and entered the army of Prince Maurice of Orange in 1617. During the next nine years he alternated between service in several armies and carousing in Paris, but throughout this period continued to study mathematics. His ability to solve a problem that had been posted on a billboard in Breda in the Netherlands as a challenge convinced him that he had mathematical ability and he began to think seriously in this subject. He returned to Paris and, having become excited by the power of the telescope, secluded himself to study the theory and construction of optical instruments. In 1628 he moved to Holland to secure a quieter and freer intellectual atmosphere. There he lived for twenty years and wrote his famous works. In 1649 he was invited to instruct Queen Christina of Sweden. Tempted by the honor and the glamor of royalty, he accepted. He died there of pneumonia in 1650.

His first work, *Regulae ad Directionem Ingenii* (Rules for the Direction of the Mind),[4] was written in 1628 but published posthumously. His next major work was *Le Monde* (System of the World, 1634), which contains a cosmological theory of vortices to explain how the planets are kept in motion and in their paths around the sun. However, he did not publish it for fear of persecution by the Church. In 1637 he published his *Discours de la méthode pour bien conduire sa raison, et chercher la vérité dans les sciences*.[5] This book, a

3. *Œuvres*, 1, 133–79; 3, 121–56.
4. Published in Dutch in 1692; *Œuvres*, 10, 359–469.
5. *Œuvres*, 6, 1–78.

classic of literature and philosophy, contains three famous appendices, *La Géométrie*, *La Dioptrique*, and *Les Météores*. *La Géométrie*, which is the only book Descartes wrote on mathematics, contains his ideas on coordinate geometry and algebra, though he did communicate many other ideas on mathematics in numerous letters. The *Discours* brought him great fame immediately. As time passed, both he and his public became more impressed with his work. In 1644 he published *Principia Philosophiae*, which is devoted to physical science and especially to the laws of motion and the theory of vortices. It contains material from his *System*, which he believed he had now made more acceptable to the Church. In 1650 he published *Musicae Compendium*.

Descartes's scientific ideas came to dominate the seventeenth century. His teachings and writings became popular even among non-scientists because he presented them so clearly and attractively. Only the Church rejected him. Actually Descartes was devout, and happy to have (as he believed) established the existence of God. But he had taught that the Bible was not the source of scientific knowledge, that reason alone sufficed to establish the existence of God, and that man should accept only what he could understand. The Church reacted to these teachings by putting his books on the *Index of Prohibited Books* shortly after his death and by preventing a funeral oration on the occasion of his interment in Paris.

Descartes approached mathematics through three avenues, as a philosopher, as a student of nature, and as a man concerned with the uses of science. It is difficult and perhaps artificial to try to separate these three lines of thought. He lived when the Protestant-Catholic controversy was at its height and when science was beginning to reveal laws of nature that challenged major religious doctrines. Hence Descartes began to doubt all the knowledge he had acquired at school. As early as the conclusion of his course of study at La Flèche, he decided that his education had advanced only his perplexity. He found himself so beset with doubts that he was convinced he had progressed no further than to recognize his ignorance. And yet, because he had been in one of the most celebrated schools in Europe, and because he believed he had not been an inferior student, he felt justified in doubting whether there was any sure body of knowledge anywhere. He then pondered the question: How do we know anything?

He soon decided that logic in itself was barren: "As for Logic, its syllogisms and the majority of its other precepts are of avail rather in the communication of what we already know, or ... even in speaking without judgment of things of which we are ignorant, than in the investigation of the unknown." Logic, then, did not supply the fundamental truths.

But where were these to be found? He rejected the current philosophy, largely Scholastic, which, though appealing, seemed to have no clear-cut foundations and employed reasoning that was not always irreproachable.

Philosophy, he decided, afforded merely "the means of discoursing with an appearance of truth on all matters." Theology pointed out the path to heaven and he aspired to go there as much as any man, but was the path correct?

The method of establishing truths in all fields came to him, he says, in a dream, on November 10, 1619, when he was on one of his military campaigns; it was the method of mathematics. Mathematics appealed to him because the proofs based on its axioms were unimpeachable and because authority counted for naught. Mathematics provided the method of achieving certainties and effectively demonstrating them. Moreover, he saw clearly that the method of mathematics transcended its subject matter. He says, "It is a more powerful instrument of knowledge than any other that has been bequeathed to us by human agency, as being the source of all others." In the same vein he continues:

> ... All the sciences which have for their end investigations concerning order and measure are related to mathematics, it being of small importance whether this measure be sought in numbers, forms, stars, sounds, or any other object; that accordingly, there ought to exist a general science which should explain all that can be known about order and measure, considered independently of any application to a particular subject, and that, indeed, this science has its own proper name, consecrated by long usage, to wit, mathematics. And a proof that it far surpasses in facility and importance the sciences which depend upon it is that it embraces at once all the objects to which these are devoted and a great many others besides. ...

And so he concluded that "The long chains of simple and easy reasonings by means of which geometers are accustomed to reach the conclusions of their most difficult demonstrations had led me to imagine that all things to the knowledge of which man is competent are mutually connected in the same way."

From his study of mathematical method he isolated in his *Rules for the Direction of the Mind* the following principles for securing exact knowledge in any field. He would accept nothing as true that was not so clear and distinct in his own mind as to exclude all doubt; he would divide difficulties into smaller ones; he would proceed from the simple to the complex; and, lastly, he would enumerate and review the steps of his reasoning so thoroughly that nothing could be omitted.

With these essentials of method, which he distilled from the practice of mathematicians, Descartes hoped to solve problems in philosophy, physics, anatomy, astronomy, mathematics, and other fields. Although he did not succeed in this ambitious program, he did make remarkable contributions to philosophy, science, and mathematics. The mind's immediate apprehension of basic, clear, and distinct truths, this intuitive power, and the deduction of consequences are the essence of his philosophy of knowledge. Purported knowledge otherwise obtained should be rejected as suspect of error and dangerous. The three appendices to his *Discours* were intended to show that his method is effective; he believed that he had shown this.

Descartes inaugurated modern philosophy. We cannot pursue his system except to note a few points relevant to mathematics. In philosophy he sought as axioms truths so clear to him that he could accept them readily. He finally decided on four: (a) *cogito, ergo sum* (I think, therefore I am); (b) each phenomenon must have a cause; (c) an effect cannot be greater than its cause; (d) the mind has innate in it the ideas of perfection, space, time, and motion. The idea of perfection, of a perfect being, could not be derived from or created by the imperfect mind of man in view of axiom (c). It could be obtained only from a perfect being. Hence God exists. Since God would not deceive us, we can be sure that the axioms of mathematics, which are clear to our intuitions, and the deductions we make from them by purely mental processes, really apply to the physical world and so are truths. It follows, then, that God must have established nature according to mathematical laws.

As for mathematics itself, he believed that he had distinct and clear mathematical ideas, such as that of a triangle. These ideas did exist and were eternal and immutable. They did not depend on his thinking them or not. Thus mathematics had an external, objective existence.

Descartes's second major interest, shared by most thinkers of his age, was the understanding of nature. He devoted many years to scientific problems and even experimented extensively in mechanics, hydrostatics, optics, and biology. His theory of vortices was the dominant cosmological theory of the seventeenth century. He is the founder of the philosophy of mechanism—that all natural phenomena, including the functioning of the human body, reduce to motions obeying the laws of mechanics—though Descartes exempted the soul. Optics, and the design of lenses in particular, was of special interest to him; part of *La Géométrie* is devoted to optics, as is *La Dioptrique*. Descartes shares with Willebrord Snell the honor of discovering the correct law of refraction of light. As in philosophy, his work in science was basic and revolutionary.

Also important in Descartes's scientific work is his emphasis on putting the fruits of science to use (Chap. 11, sec. 5). In this attitude he breaks clearly and openly with the Greeks. To master nature for the good of man, he pursued many scientific problems. And, being impressed with the power of mathematics, he naturally sought to use that subject; for him it was not contemplative discipline but a constructive and useful science. Unlike Fermat, he cared little for its beauty and harmony; he did not value pure mathematics. He says that mathematical method applied only to mathematics is without value because it is not a study of nature. Those who cultivate mathematics for its own sake are idle searchers given to a vain play of the spirit.

4. *Descartes's Work in Coordinate Geometry*

Having decided that method was important and that mathematics could be effectively employed in scientific work, Descartes turned to the application of method to geometry. Here his general interest in method and his particular knowledge of algebra joined forces. He was disturbed by the fact that every proof in Euclidean geometry called for some new, often ingenious, approach. He explicitly criticized the geometry of the ancients as being too abstract, and so much tied to figures "that it can exercise the understanding only on condition of greatly fatiguing the imagination." The algebra that he found prevalent he also criticized because it was so completely subject to rules and formulas "that there results an art full of confusion and obscurity calculated to hamper instead of a science fitted to improve the mind." Descartes proposed, therefore, to take all that was best in geometry and algebra and correct the defects of one with the help of the other.

Actually it was the use of algebra in geometry that he undertook to exploit. He saw fully the power of algebra and its superiority over the Greek geometrical methods in providing a broad methodology. He also stressed the generality of algebra and its value in mechanizing the reasoning processes and minimizing the work in solving problems. He saw its potential as a universal science of method. The product of his application of algebra to geometry was *La Géométrie*.

Though in this book Descartes used the improvements in algebraic notation already noted in Chapter 13, the essay is not easy reading. Much of the obscurity was deliberate; Descartes boasted that few mathematicians in Europe would understand his work. He indicated the constructions and demonstrations, leaving it to others to fill in the details. In one of his letters he compares his writing to that of an architect who lays the plans and prescribes what should be done but leaves the manual work to the carpenters and bricklayers. He says also, "I have omitted nothing inadvertently but I have foreseen that certain persons who boast that they know everything would not miss the opportunity of saying that I have written nothing that they did not already know, were I to make myself sufficiently intelligible for them to understand me." He gave other reasons in *La Géométrie*, such as not wishing to deprive his readers of the pleasure of working things out for themselves. Many explanatory commentaries were written to make Descartes's book clear.

His ideas must be inferred from a number of examples worked out in the book. He says that he omits the demonstration of most of his general statements because if one takes the trouble to examine systematically these

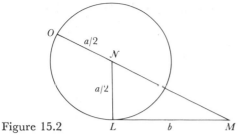

Figure 15.2

examples, the demonstrations of the general results will become apparent, and it is of more value to learn them in that way.

He begins in *La Géométrie* with the use of algebra to solve geometrical construction problems in the manner of Vieta; only gradually does the idea of the equation of a curve emerge. He points out first that geometrical constructions call for adding, subtracting, multiplying, and dividing lines and taking the square root of particular lines. Since all of these operations also exist in algebra, they can be expressed in algebraic terms.

In tackling a given problem, Descartes says we must suppose the solution of the problem already known and represent with letters all the lines, known and unknown, that seem necessary for the required construction. Then, making no distinction between known and unknown lines, we must "unravel" the difficulty by showing in what way the lines are related to each other, aiming at expressing one and the same quantity in two ways. This gives an equation. We must find as many equations as there are unknown lines. If several equations remain, we must combine them until there remains a single unknown line expressed in terms of known lines. Descartes then shows how to construct the unknown line by utilizing the fact that it satisfies the algebraic equation.

Thus, suppose a geometric problem leads to finding an unknown length x, and after algebraic formulation x is found to satisfy the equation $x^2 = ax + b^2$ where a and b are known lengths. Then we know by algebra that

(1)
$$x = \frac{a}{2} + \sqrt{\frac{a^2}{4} + b^2}.$$

(Descartes ignored the second root, which is negative.) Descartes now gives a construction for x. He constructs the right triangle NLM (Fig. 15.2) with $LM = b$ and $NL = a/2$, and prolongs MN to O so that $NO = NL = a/2$. Then the solution x is the length OM. The proof that OM is the correct length is not given by Descartes but it is immediately apparent for

$$OM = ON + MN = \frac{a}{2} + \sqrt{\frac{a^2}{4} + b^2}.$$

Thus the expression (1) for x, which was obtained by solving an algebraic equation, indicates the proper construction for x.

In the first half of Book I, Descartes solves only classical geometric construction problems with the aid of algebra. This is an application of algebra to geometry, but not analytic geometry in our present sense. The problems thus far are what one might call determinate construction problems because they lead to a unique length. He considers next indeterminate construction problems, that is, problems in which there are many possible lengths that serve as answers. The endpoints of the many lengths fill out a curve; and here Descartes says, "It is also required to discover and trace the curve containing all such points." This curve is described by the final indeterminate equation expressing the unknown lengths y in terms of the arbitrary lengths x. Moreover, Descartes stresses that for each x, y satisfies a determinate equation and so can be constructed. If the equation is of the first or second degree, y can be constructed by the methods of Book I, using only lines and circles. For higher-degree equations, he says he will show in Book III how y can be constructed.

Descartes uses the problem of Pappus (Chap. 5, sec. 7) to illustrate what happens when a problem leads to one equation in two unknowns. This problem, which had not been solved in full generality, is as follows: Given the position of three lines in a plane, find the position of all points (the locus) from which we can construct lines, one to each of the given lines and making a known angle with each of these given lines (the angle may be different from line to line), such that the rectangle contained by two of the constructed lines has a given ratio to the square on the third constructed line; if there are four given lines, then the constructed lines, making given angles with the given lines, must be such that the rectangle contained by two must have a given ratio to the rectangle contained by the other two; if there are five given lines, then the five constructed lines, each making a given angle with one of the given lines, must be such that the product of three of them has a given ratio to the product of the remaining two. The condition on the locus when there are more than five given lines is an obvious extension of the above.

Pappus had declared that when three or four lines are given, the locus is a conic section. In Book II Descartes treats the Pappus problem for the case of four lines. The given lines (Fig. 15.3) are AG, GH, EF, and AD. Consider a point C and the four lines from C to each of the four given lines and making a specified angle with each of the four given lines. The angle can be different from one line to another. Let us denote the four lines by CP, CQ, CR, and CS. It is required to find the locus of C satisfying the condition $CP \cdot CR = CS \cdot CQ$.

Descartes denotes AP by x and PC by y. By simple geometric considerations, he obtains the values of CR, CQ, and CS in terms of known

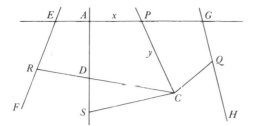

Figure 15.3

quantities. He uses these values to form $CP \cdot CR = CS \cdot CQ$ and obtains a second degree equation in x and y of the form

$$(2) \qquad y^2 = Ay + Bxy + Cx + Dx^2$$

where A, B, C, and D are simple algebraic expressions in terms of the known quantities. Now Descartes points out that if we select any value of x we have a quadratic equation for y that can be solved for y; and then y can be constructed by straightedge and compass as he has shown in Book I. Hence if one takes an infinite number of values for x, one obtains an infinite number of values for y and hence an infinite number of points C. The locus of all these points C is a curve whose equation is (2).

What Descartes has done is to set up one line (AG in the above figure) as a base line with an origin at the point A. The x-values are then lengths along this line, and the y-values are lengths that start at this base line and make a fixed angle with it. This coordinate system is what we now call an oblique system. Descartes's x and y stand for positive numbers only; yet his equations cover portions of curves in other than what we would call the first quadrant. He simply assumes that the locus lies primarily in the first quadrant and makes passing reference to what might happen elsewhere. That there is a length for each positive real number is assumed unconsciously.

Having arrived at the idea of the equation of a curve, Descartes now develops it. It is easily demonstrated, he asserts, that the degree of a curve is independent of the choice of the reference axis; he advises choosing this axis so that the resulting equation is as simple as possible. In another great stride, he considers two different curves, expresses their equations with respect to the same reference axis, and finds the points of intersection by solving the equations simultaneously.

Also in Book II, Descartes considers critically the Greek distinctions among plane, solid, and linear curves. The Greeks had said plane curves were those constructible by straightedge and compass; the solid curves were the conic sections; and the linear curves were all the others, such as the conchoid, spiral, quadratrix, and cissoid. The linear curves were also called mechanical by the Greeks because some special mechanism was required to construct them. But, Descartes says, even the straight line and circle require some instrument. Nor can the accuracy of the mechanical construction matter, because in mathematics only the reasoning counts. Possibly, he continues, the ancients objected to linear curves because they were insecurely defined. On these grounds, Descartes rejects the idea that only the curves constructible with straightedge and compass[6] are legitimate and even proposes some new curves generated by mechanical constructions. He concludes with the highly significant statement that geometric curves are those that can be expressed by a unique algebraic equation (of finite degree) in x and y. Thus Descartes accepts the conchoid and cissoid. All other curves, such as the spiral and the quadratrix, he calls mechanical.

Descartes's insistence that an acceptable curve is one that has an algebraic equation is the beginning of the elimination of constructibility as a criterion of existence. Leibniz went farther than Descartes. Using the words "algebraic" and "transcendental" for Descartes's terms "geometrical" and "mechanical," he protested the requirement that a curve must have an algebraic equation.[7] Actually Descartes and his contemporaries ignored the requirement and worked just as enthusiastically with the cycloid, the logarithmic curve, the logarithmic spiral ($\log \rho = a\theta$), and other non-algebraic curves.

In broadening the concept of admissible curves, Descartes made a major step. He not only admitted curves formerly rejected but opened up the whole field of curves, because, given any algebraic equation in x and y, one can find its curve and so obtain totally new curves. In *Arithmetica Universalis* Newton says (1707), "But the Moderns advancing yet much further [than the plane, solid and linear loci of the Greeks] have received into Geometry all Lines that can be expressed by Equations."

Descartes next considers the classes of geometric curves. Curves of the first and second degree in x and y are in the first and simplest class. Descartes says, in this connection, that the equations of the conic sections are of the second degree, but does not prove this. Curves whose equations are of the third and fourth degree constitute the second class. Curves whose equations are of the fifth and sixth degree are of the third class and so on. His reason for grouping third and fourth, as well as fifth and sixth degree curves, is that he believed the higher one in each class could be reduced to the lower, as the solution of quartic equations could be effected by the solution of cubics. This belief was of course incorrect.

The third book of *La Géométrie* returns to the theme of Book I. Its objective is the solution of geometric construction problems, which, when formulated algebraically, lead to determinate equations of third and higher degree and which, in accordance with the algebra, call for the conic sections

6. Compare the discussion in Chap. 8, sec. 2.
7. *Acta Erud.*, 1684, pp. 470, 587; 1686, p. 292 = *Math. Schriften*, 5, 127, 223, 226.

and higher-degree curves. Thus Descartes considers the construction problem of finding the two mean proportionals between two given quantities a and q. The special case when $q = 2a$ was attempted many times by the classical Greeks and was important because it is a way to solve the problem of doubling the cube. Descartes proceeds as follows: Let z be one of these mean proportionals; then z^2/a must be the second, for we must have

$$\frac{a}{z} = \frac{z}{z^2/a} = \frac{z^2/a}{z^3/a^2}.$$

Then, if we take z^3/a^2 to be q, we have the equation z must satisfy. Hence, given q and a, we must find z such that

$$(3) \qquad\qquad z^3 = a^2 q,$$

or, we must solve a cubic equation. Descartes now shows that such quantities z and z^2/a can be obtained by a geometrical construction that utilizes a parabola and a circle.

As the construction is described by Descartes, seemingly no coordinate geometry is involved. However, the parabola is not constructible with straightedge and compass, except point by point, and so one must use the equation to plot the curve accurately.

Descartes does *not* obtain z by writing the equations in x and y of circle and parabola and finding the coordinates of the point of intersection by solving equations simultaneously. In other words, he is not solving equations graphically in our sense. Rather he uses purely geometric constructions (except for supposing that a parabola can be drawn), the knowledge of the fact that z satisfies an equation, and the geometric properties of the circle and parabola (which can be more readily seen through their equations). Descartes does here just what he did in Book I, except that he is now solving geometric construction problems in which the unknown length satisfies a third or higher-degree equation instead of a first or second degree equation. His solution of the purely algebraic aspect of the problem and the subsequent construction is practically the same one the Arabs gave, except that he was able to use the equations of the conic sections to deduce facts about the curves and to draw them.

Descartes not only wished to show how some solid problems could be solved with the aid of algebra and the conic sections but was interested in classifying problems so that one would know what they involved and how to go about solving them. His classification is based on the degree of the algebraic equation to which one is led when the construction problem is formulated algebraically. If that degree is one or two, then the construction can be performed with straight line and circle. If the degree is three or four,

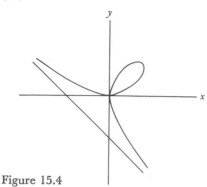

Figure 15.4

the conic sections must be employed. He does affirm, incidentally, that all cubic problems can be reduced to trisecting the angle and doubling the cube and that no cubic problems can be solved without the use of a curve more complex than the circle. If the degree of the equation is higher than four, curves more complicated than the conic sections may be required to perform the construction.

Descartes also emphasized the degree of the equation of a curve as the measure of its simplicity. One should use the simplest curve, that is, the lowest degree possible, to solve a construction problem. The emphasis on the degree of a curve became so strong that a complicated curve such as the folium of Descartes (Fig. 15.4), whose equation is $x^3 + y^3 - 3axy = 0$, was considered simpler than $y = x^4$.

What is far more significant than Descartes's insight into construction problems and their classification is the importance he assigned to algebra. This key makes it possible to recognize the typical problems of geometry and to bring together problems that in geometrical form would not appear to be related at all. Algebra brings to geometry the most natural principles of classification and the most natural hierarchy of method. Not only can questions of solvability and geometrical constructibility be decided elegantly, quickly, and fully from the parallel algebra, but without it they cannot be decided at all. Thus, system and structure were transferred from geometry to algebra.

Part of Book II of *La Géométrie* as well as *La Dioptrique* Descartes devoted to optics, using coordinate geometry as an aid. He was very much concerned with the design of lenses for the telescope, microscope, and other optical instruments because he appreciated the importance of these instruments for astronomy and biology. His *Dioptrique* takes up the phenomenon of refraction. Kepler and Alhazen before him had noted that the belief that the angle of refraction is proportional to the angle of incidence, the proportionality constant being dependent on the medium doing the refracting, was

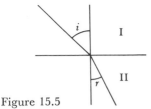

Figure 15.5

incorrect for large angles, but they did not discover the true law. Before 1626 Willebrord Snell discovered but did not publish the correct relationship,

$$\frac{\sin i}{\sin r} = \frac{v_1}{v_2},$$

where v_1 is the velocity of light in the first medium (Fig. 15.5) and v_2 the velocity in the medium into which the light passes. Descartes gave this same law in 1637 in the *Dioptrique*. There is some question as to whether he discovered it independently. His argument was wrong, and Fermat immediately attacked both the law and the proof. A controversy arose between them which lasted ten years. Fermat was not satisfied that the law was correct, until he derived it from his Principle of Least Time (Chap. 24, sec. 3).

In *La Dioptrique*, after describing the operation of the eye, Descartes considers the problem of designing properly focusing lenses for telescopes, microscopes, and spectacles. It was well known even in antiquity that a spherical lens will not cause parallel rays or rays diverging from a source S to focus on one point. Hence the question was open as to what shape would so focus the incoming rays. Kepler had suggested that some conic section would serve. Descartes sought to design a lens that would focus the rays perfectly.

He proceeded to solve the general problem of what surface should separate two media such that light rays starting from one point in the first medium would strike the surface, refract into the second medium, and there converge to one point. He discovered that the curve generating the desired surface of revolution is an oval, now known as the oval of Descartes. This curve and its refracting properties are discussed in *La Dioptrique*, and the discussion is supplemented in Book II of *La Géométrie*.

The modern definition is that the curve is the locus of points M satisfying the condition

$$FM \pm nF'M = 2a$$

where F and F' are fixed points, $2a$ is any real number larger than FF', and n is any real number. If $n = 1$ the curve becomes an ellipse. In the general case, the equation of the oval is of the fourth degree in x and y, and the curve

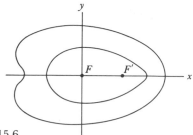

Figure 15.6

consists of two closed, distinct portions (Fig. 15.6) without common point and one inside the other. The inner curve is convex like an ellipse and the outer one can be convex or may have points of inflection, as in the figure.

As we can now see, Descartes's approach to coordinate geometry differs profoundly from Fermat's. Descartes criticized and proposed to break with the Greek tradition, whereas Fermat believed in continuity with Greek thought and regarded his work in coordinate geometry only as a reformulation of the work of Apollonius. The real discovery—the power of algebraic methods—is Descartes's; and he realized he was supplanting the ancient methods. Though the idea of equations for curves is clearer with Fermat than with Descartes, Fermat's work is primarily a technical achievement that completes the work of Apollonius and uses Vieta's idea of letters to represent classes of numbers. Descartes's methodology is universally applicable and potentially applies to the transcendental curves, too.

Despite these significant differences in approach to coordinate geometry and in goals, Descartes and Fermat became embroiled in controversy as to priority of discovery. Fermat's work was not published until 1679; however, his discovery of the basic ideas of coordinate geometry in 1629 predates Descartes's publication of *La Géométrie* in 1637. Descartes was by this time fully aware of many of Fermat's discoveries, but he denied having learned his ideas from Fermat. Descartes's ideas on coordinate geometry, according to the Dutch mathematician Isaac Beeckman (1588–1637), went back to 1619; and furthermore, there is no question about the originality of many of his basic ideas in coordinate geometry.

When *La Géométrie* was published, Fermat criticized it because it omitted ideas such as maxima and minima, tangents to curves, and the construction of solid loci, which, he had decided, merited the attention of all geometers. Descartes in turn said Fermat had done little, in fact no more than could be easily arrived at without industry or previous knowledge, whereas he himself had used a full knowledge of the nature of equations, which he had expounded in the third book of *La Géométrie*. Descartes referred sarcastically to Fermat as *vostre Conseiller De Maximis et Minimis* and said Fermat was indebted to him. Roberval, Pascal, and others sided with Fermat, and Mydorge and Desargues sided with Descartes. Fermat's friends wrote bitter letters against Descartes. Later the attitudes of the two men toward each other softened, and in a work of 1660, Fermat, while calling attention to an error in *La Géométrie*, declared that he admired that genius so much that even when he made mistakes Descartes's work was worth more than that of others who did correct things. Descartes had not been so generous.

The emphasis placed by posterity on *La Géométrie* was not what Descartes had intended. While the salient idea for the future of mathematics was the association of equation and curve, for Descartes this idea was just a means to an end—the solution of geometric construction problems. Fermat's emphasis on the equations of loci is, from the modern standpoint, more to the point. The geometric construction problems that Descartes stressed in Books I and III have dwindled in importance, largely because construction is no longer used, as it was by the Greeks, to establish existence.

One portion of Book III has also found a permanent place in mathematics. Since Descartes solved geometric construction problems by first formulating them algebraically, solving the algebraic equations, and then constructing what the solutions called for, he gathered together work of his own and of others on the theory of equations that might expedite their solution. Because algebraic equations continued to arise in hundreds of different contexts having nothing to do with geometrical construction problems, this theory of equations has become a basic part of elementary algebra.

5. *Seventeenth-Century Extensions of Coordinate Geometry*

The main idea of coordinate geometry—the use of algebraic equations to represent and study curves—was not eagerly seized upon by mathematicians for many reasons. Fermat's book, the *Ad Locos*, though circulated among friends, was not published until 1679. Descartes's emphasis on the solution of geometric construction problems obscured the main idea of equation and curve. In fact, many of his contemporaries thought of coordinate geometry primarily as a tool for solving the construction problems. Even Leibniz spoke of Descartes's work as a regression to the ancients. Descartes himself did realize that he had contributed far more than a new method of solving construction problems. In the introduction to *La Géométrie* he says, "Moreover, what I have given in the second book on the nature and properties of curved lines, and the method of examining them, is, it seems to me, as far beyond the treatment of ordinary geometry as the rhetoric of Cicero is beyond the a, b, c of children." Nevertheless, the uses he made of the equations of the curves, such as solving the Pappus problem, finding normals to curves, and obtaining properties of the ovals, were far overshadowed by the attention given to the construction problems. Another reason for the slow spread of analytic geometry was Descartes's insistence on making his presentation difficult to follow.

In addition many mathematicians objected to confounding algebra and geometry, or arithmetic and geometry. This objection had been voiced even in the sixteenth century, when algebra was on the rise. For example, Tartaglia insisted on the distinction between the Greek operations with geometrical objects and operations with numbers. He reproached a translator of Euclid for using interchangeably *multiplicare* and *ducere*. The first belongs to numbers, he says, and the second to magnitude. Vieta, too, considered the sciences of number and of geometric magnitudes as parallel but distinct. Even Newton, in his *Arithmetica Universalis*, objected to confounding algebra and geometry, though he contributed to coordinate geometry and used it in the calculus. He says,[8]

> Equations are expressions of arithmetical computation and properly have no place in geometry except insofar as truly geometrical quantities (that is, lines, surfaces, solids and proportions) are thereby shown equal, some to others. Multiplications, divisions and computations of that kind have been recently introduced into geometry, unadvisedly and against the first principles of this science. . . . Therefore these two sciences ought not to be confounded, and recent generations by confounding them have lost that simplicity in which all geometrical elegance consists.

A reasonable interpretation of Newton's position is that he wanted to keep algebra out of elementary geometry but did find it useful to treat the conics and higher-degree curves.

Still another reason for the slowness with which coordinate geometry was accepted was the objection to the lack of rigor in algebra. We have already mentioned Barrow's unwillingness to accept irrational numbers as more than symbols for continuous geometrical magnitudes (Chap. 13, sec. 2). Arithmetic and algebra found their logical justification in geometry; hence algebra could not replace geometry or exist as its equal. The philosopher Thomas Hobbes (1588–1679), though only a minor figure in mathematics, nevertheless spoke for many mathematicians when he objected to the "whole herd of them who apply their algebra to geometry." Hobbes said that these algebraists mistook the symbols for geometry and characterized John Wallis's book on the conics as scurvy and as a "scab of symbols."

Despite the hindrances to appreciation of what Descartes and Fermat had contributed, a number of men gradually took up and expanded coordinate geometry. The first task was to explain Descartes's idea. A Latin translation of *La Géométrie* by Frans van Schooten (1615–60), first published in 1649 and republished several times, not only made the book available in

8. *Arithmetica Universalis*, 1707, p. 282.

the language all scholars could read but contained a commentary which expanded Descartes's compact presentation. In the edition of 1659–61, van Schooten actually gave the algebraic form of a transformation of coordinates from one base line (x-axis) to another. He was so impressed with the power of Descartes's method that he claimed the Greek geometers had used it to derive their results. Having the algebraic work, the Greeks, according to van Schooten, saw how to obtain the results synthetically—he showed how this could be done—and then published their synthetic methods, which are less perspicuous than the algebraic, to amaze the world. Van Schooten may have been misled by the word "analysis," which to the Greeks meant analyzing a problem, and the term "analytic geometry," which specifically described Descartes's use of algebra as a method.

John Wallis, in *De Sectionibus Conicis* (1655), first derived the equations of the conics by translating Apollonius' geometric conditions into algebraic form (much as we did in Chap. 4, sec. 12) in order to elucidate Apollonius' results. He then defined the conics as curves corresponding to second degree equations in x and y and proved that these curves were indeed the conic sections as known geometrically. He was probably the first to use equations to prove properties of the conics. His book helped immensely to spread the idea of coordinate geometry and to popularize treatment of the conics as curves in the plane instead of as sections of a cone, though the latter approach persisted. Moreover, Wallis emphasized the validity of the algebraic reasoning whereas Descartes, at least in his *Géométrie*, really rested on the geometry, regarding algebra as just a tool. Wallis was also the first to consciously introduce negative abscissas and ordinates. Newton, who did this later, may have gotten the idea from Wallis. We can contrast van Schooten's remark on method with one by Wallis, who said that Archimedes and nearly all the ancients so hid from posterity their method of discovery and analysis that the moderns found it easier to invent a new analysis than to seek out the old.

Newton's *The Method of Fluxions and Infinite Series*, written about 1671 but first published in an English translation by John Colson (d. 1760) under the above title in 1736, contains many uses of coordinate geometry, such as sketching curves from equations. One of the original ideas it offers is the use of new coordinate systems. The seventeenth- and even many of the eighteenth-century men generally used one axis, with the y-values drawn at an oblique or right angle to that axis. Among the new coordinate systems introduced by Newton is the location of points by reference to a fixed point and a fixed line through that point. The scheme is essentially our polar coordinate system. The book contains many variations on the polar coordinate idea. Newton also introduced bipolar coordinates. In this scheme a point is located by its distance from two fixed points (Fig. 15.7). Because this work of Newton did not become known until 1736, credit for the discovery of polar coordinates

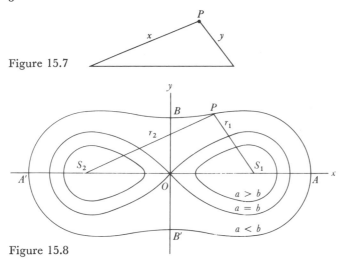

Figure 15.7

Figure 15.8

is usually given to James (Jakob) Bernoulli who published a paper on what was essentially this scheme in the *Acta Eruditorum* of 1691.

Many new curves and their equations were introduced. In 1694 Bernoulli introduced the lemniscate,[9] which played a major role in eighteenth-century analysis. This curve is a special case of a class of curves called the Cassinian ovals (general lemniscates) introduced by Jean-Dominique Cassini (1625–1712), though they did not appear in print until his son Jacques (1677–1756) published the *Eléments d'astronomie* in 1749. The Cassinian ovals (Fig. 15.8) are defined by the condition that the product $r_1 r_2$ of the distances of any point on the curve from two fixed points S_1 and S_2 equals b^2 where b is a constant. Let the distance $S_1 S_2$ be $2a$. Then if $b > a$ we get the non-self-intersecting oval. If $b = a$ we get the lemniscate introduced by James Bernoulli. And if $b < a$ we get the two separate ovals. The rectangular coordinate equation of the Cassinian ovals is of the fourth degree. Descartes himself introduced the logarithmic spiral,[10] which in polar coordinates has the equation $\rho = a^\theta$, and discovered many of its properties. Still other curves, among them the catenary and cycloid, will be noted in other connections.

The beginning of an extension of coordinate geometry to three dimensions was made in the seventeenth century. In Book II of his *Géométrie* Descartes remarks that his ideas can easily be made to apply to all those curves that can be conceived of as generated by the regular movements of a point in three-dimensional space. To represent such curves algebraically his plan is to drop perpendiculars from each point of the curve upon two planes

9. *Acta Erud.*, Sept. 1694 = *Opera*, 2, 608–12.
10. Letter to Mersenne of Sept. 12, 1638 = *Œuvres*, 2, 360.

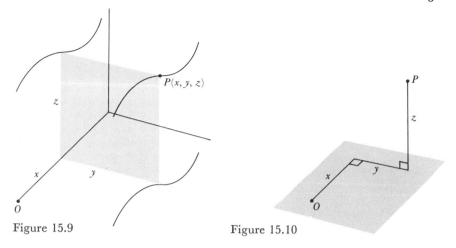

Figure 15.9   Figure 15.10

intersecting at right angles (Fig. 15.9). The ends of these perpendiculars will each describe a curve in the respective plane. These plane curves can then be treated by the method already given. Earlier in Book II Descartes observes that one equation in three unknowns for the determination of the typical point C of a locus represents a plane, a sphere, or a more complex surface. Clearly he appreciated thas his method could be extended to curves and surfaces in three-dimensional space, but he did not himself go further with the extension.

Fermat, in a letter of 1643, gave a brief sketch of his ideas on analytic geometry of three dimensions. He speaks of cyclindrical surfaces, elliptic paraboloids, hyperboloids of two sheets, and ellipsoids. He then says that, to crown the introduction of plane curves, one should study curves on surfaces. "This theory is susceptible of being treated by a general method which if I have leisure I will explain." In a work of half a page, *Novus Secundarum*,[11] he says that an equation in three unknowns gives a surface.

La Hire, in his *Nouveaux élémens des sections coniques* (1679), was a little more specific about three-dimensional coordinate geometry. To represent a surface, he first represented a point P in space by the three coordinates indicated in Figure 15.10 and actually wrote the equation of a surface. However, the development of three-dimensional coordinate geometry is the work of the eighteenth century and will be discussed later.

6. *The Importance of Coordinate Geometry*

In light of the fact that algebra had made considerable progress before Fermat and Descartes entered the mathematical scene, coordinate geometry

11. *Œuvres*, 1, 186–87; 3, 161–62.

was not a great technical achievement. For Fermat it was an algebraic rephrasing of Apollonius. With Descartes it arose as an almost accidental discovery as he continued the work of Vieta and others in expediting the solution of determinate construction problems by the introduction of algebra. But coordinate geometry changed the face of mathematics.

By arguing that a curve is any locus that has an algebraic equation, Descartes broadened in one swoop the domain of mathematics. When one considers the variety of curves that have come to be accepted and used in mathematics and compares this assemblage with what the Greeks had accepted, one sees how important it was that the Greek barriers be stormed.

Through coordinate geometry Descartes sought to introduce method in geometry. He achieved far more than he envisioned. It is commonplace today to recognize not only how readily one can prove, with the aid of the algebra, any number of facts about curves, but also that the method of approaching the problems is almost automatic. The methodology is even more powerful. When letters began to be used by Wallis and Newton to stand for positive and negative numbers and later even for complex numbers, it became possible to subsume under one algebraic treatment many cases that pure geometry would have had to treat separately. For example, in synthetic geometry, to prove that the altitudes of a triangle meet in a point, intersections inside and outside the triangle are considered separately. In coordinate geometry they are considered together.

Coordinate geometry made mathematics a double-edged tool. Geometric concepts could be formulated algebraically and geometric goals attained through the algebra. Conversely, by interpreting algebraic statements geometrically one could gain an intuitive grasp of their meanings as well as suggestions for the deduction of new conclusions. These virtues were cited by Lagrange in his *Leçons élémentaires sur les mathématiques*:[12] "As long as algebra and geometry travelled separate paths their advance was slow and their applications limited. But when these two sciences joined company, they drew from each other fresh vitality and thenceforward marched on at a rapid pace towards perfection." Indeed the enormous power mathematics developed from the seventeenth century on must be attributed, to a very large extent, to coordinate geometry.

The most significant virtue of coordinate geometry was that it provided science with just that mathematical facility it had always sorely needed and which, in the seventeenth century, was being openly demanded—quantitative tools. The study of the physical world does seem to call primarily for geometry. Objects are basically geometrical figures, and the paths of moving bodies are curves. Indeed Descartes himself thought that all of physics could be reduced to geometry. But, as we have pointed out, the uses of science in geodesy, navigation, calendar-reckoning, astronomical predictions, projectile motion,

12. *Œuvres*, 7, 183–287, p. 271 in part.

and even the design of lenses, which Descartes himself undertook, call for quantitative knowledge. Coordinate geometry made possible the expression of shapes and paths in algebraic form, from which quantitative knowledge could be derived.

Thus algebra, which Descartes had thought was just a tool, an extension of logic rather than part of mathematics proper, became more vital than geometry. In fact, coordinate geometry paved the way for a complete reversal of the roles of algebra and geometry. Whereas from Greek times until about 1600 geometry dominated mathematics and algebra was subordinate, after 1600 algebra became the basic mathematical subject; in this transposition of roles the calculus was to be the decisive factor. The ascendancy of algebra aggravated the difficulty to which we have already called attention, namely, that there was no logical foundation for arithmetic and algebra; but nothing was done about it until the late nineteenth century.

The fact that algebra was built up on an empirical basis has led to confusion in mathematical terminology. The subject created by Fermat and Descartes is usually referred to as analytic geometry. The word "analytic" is inappropriate; coordinate geometry or algebraic geometry (which now has another meaning) would be more suitable. The word "analysis" had been used since Plato's time to mean the process of analyzing by working backward from what is to be proved until one arrives at something known. In this sense it was opposed to "synthesis," which describes the deductive presentation. About 1590 Vieta rejected the word "algebra" as having no meaning in the European language and proposed the term "analysis" (Chap. 13, sec. 8); the suggestion was not adopted. However, for him and for Descartes, the word "analysis" was still somewhat appropriate to describe the application of algebra to geometry because the algebra served to analyze the geometric construction problem. One assumed the desired geometric length was known, found an equation that this length satisfied, manipulated the equation, and then saw how to construct the required length. Thus Jacques Ozanam (1640–1717) said in his *Dictionary* (1690) that moderns did their analysis by algebra. In the famous eighteenth-century *Encyclopédie*, d'Alembert used "algebra" and "analysis" as synonyms. Gradually, "analysis" came to mean the algebraic method, though the new coordinate geometry, up to about the end of the eighteenth century, was most often formally described as the application of algebra to geometry. By the end of the century the term "analytic geometry" became standard and was frequently used in titles of books.

However, as algebra became the dominant subject, mathematicians came to regard it as having a much greater function than the analysis of a problem in the Greek sense. In the eighteenth century the view that algebra as applied to geometry was more than a tool—that algebra itself was a basic method of introducing and studying curves and surfaces (the supposed view of Fermat as opposed to Descartes)—won out, as a result of the work of Euler, Lagrange, and Monge. Hence the term "analytic geometry" implied proof as well as the use of the algebraic method. Consequently we now speak of analytic geometry as opposed to synthetic geometry, and we no longer mean that one is a method of invention and the other of proof. Both are deductive.

In the meantime the calculus and extensions such as infinite series entered mathematics. Both Newton and Leibniz regarded the calculus as an extension of algebra; it was the algebra of the infinite, or the algebra that dealt with an infinite number of terms, as in the case of infinite series. As late as 1797, Lagrange, in *Théorie des fonctions analytiques*, said that the calculus and its developments were only a generalization of elementary algebra. Since algebra and analysis had been synonyms, the calculus was referred to as analysis. In a famous calculus text of 1748 Euler used the term "infinitesimal analysis" to describe the calculus. This term was used until the late nineteenth century, when the word "analysis" was adopted to describe the calculus and those branches of mathematics built on it. Thus we are left with a confusing situation in which the term "analysis" embraces all the developments based on limits, but "analytic geometry" involves no limit processes.

Bibliography

Boyer, Carl B.: *History of Analytic Geometry*, Scripta Mathematica, 1956.

Cantor, Moritz: *Vorlesungen über Geschichte der Mathematik*, 2nd ed., B. G. Teubner, 1900, Johnson Reprint Corp., 1965, Vol. 2, pp. 806–76.

Chasles, Michel: *Aperçu historique sur l'origine et le développement des méthodes en géométrie*, 3rd ed., Gauthier-Villars et Fils, 1889, Chaps. 2–3 and relevant notes.

Coolidge, Julian L.: *A History of Geometrical Methods*, Dover (reprint), 1963, pp. 117–31.

Descartes, René: *La Géométrie* (French and English), Dover (reprint), 1954.

————: *Œuvres*, 12 vols., Cerf, 1897–1913.

Fermat, Pierre de: *Œuvres*, 4 vols. and Supplement, Gauthier-Villars, 1891–1912; Supplement, 1922.

Montucla, J. F.: *Histoire des mathématiques*, Albert Blanchard (reprint), 1960, Vol. 2, pp. 102–77.

Scott, J. F.: *The Scientific Work of René Descartes*, Taylor and Francis, 1952.

Smith, David E.: *A Source Book in Mathematics*, Dover (reprint), 1959, Vol. 2, pp. 389–402.

Struik, D. J.: *A Source Book in Mathematics, 1200–1800*, Harvard University Press, 1969, pp. 87–93, 143–57.

Wallis, John: *Opera*, 3 vols., (1693–99), Georg Olms (reprint), 1968.

Vrooman, Jack R.: *René Descartes: A Biography*, G. P. Putnam's Sons, 1970.

Vuillemin, Jules: *Mathématiques et métaphysique chez Descartes*, Presses Universitaires de France, 1960.

16

The Mathematization of Science

So that we may say the door is now opened, for the first
time, to a new method fraught with numerous and wonderful
results which in future years will command the attention of
other minds.　　　　　　　　　　　GALILEO GALILEI

1. *Introduction*

By 1600 the European scientists were unquestionably impressed with the
importance of mathematics for the study of nature. The strongest evidence
of this conviction was the willingness of Copernicus and Kepler to overturn
the accepted laws of astronomy and mechanics and religious doctrines for
the sake of a theory which in their time had only mathematical advantages.
However, the astonishing successes of modern science and the enormous
impetus to creative work that mathematics derived from that source probably
would not have come about if science had continued in the footsteps of the
past. But in the seventeenth century two men, Descartes and Galileo,
revolutionized the very nature of scientific activity. They selected the concepts science should employ, redefined the goals of scientific activity, and
altered the methodology of science. Their reformulation not only imparted
unexpected and unprecedented power to science but bound it indissolubly
to mathematics. In fact, their plan practically reduced theoretical science to
mathematics. To understand the spirit that animated mathematics from
the seventeenth through the nineteenth centuries, we must first examine
the ideas of Descartes and Galileo.

2. *Descartes's Concept of Science*

Descartes proclaimed explicitly that the essence of science was mathematics.
He says that he "neither admits nor hopes for any principles in Physics
other than those which are in Geometry or in abstract mathematics, because thus all phenomena of nature are explained and some demonstrations
of them can be given." The objective world is space solidified, or geometry
incarnate. Its properties should therefore be deducible from the first principle
of geometry.

325

Descartes elaborated on why the world must be accessible and reducible to mathematics. He insisted that the most fundamental and reliable properties of matter are shape, extension, and motion in space and time. Since shape is just extension, Descartes asserted, "Give me extension and motion and I shall construct the universe." Motion itself resulted from the action of forces on molecules. Descartes was convinced that these forces obeyed invariable mathematical laws; and, since extension and motion were mathematically expressible, all phenomena were mathematically describable.

Descartes's mechanistic philosophy extended even to the functioning of the human body. He believed that laws of mechanics would explain life in man and animals, and in his work in physiology he used heat, hydraulics, tubes, valves, and the mechanical actions of levers to explain the actions of the body. However, God and the soul were exempt from mechanism.

If Descartes regarded the external world as consisting only of matter in motion, how did he account for tastes, smells, colors, and the qualities of sounds? Here he adopted the old Greek doctrine of primary and secondary qualities which, as stated by Democritus, maintained that "sweet and bitter, cold and warm, as well as the colors, all these things exist but in opinion and not in reality; what really exist are unchangeable particles, atoms, and their motions in empty space." The primary qualities, matter and motion, exist in the physical world; the secondary qualities are only effects the primary qualities induce in the sense organs of human beings by the impact of external atoms on these organs.

Thus to Descartes there are two worlds; one, a huge, harmoniously designed mathematical machine existing in space and time, and the other, the world of thinking minds. The effect of elements in the first world on the second produces the nonmathematical or secondary qualities of matter. Descartes affirmed further that the laws of nature are invariable, since they are but part of a predetermined mathematical pattern, and that God could not alter invariable nature. Here Descartes denied the prevailing belief that God continually intervened in the functioning of the universe.

Though Descartes's philosophical and scientific doctrines subverted Aristotelianism and Scholasticism, he was a Scholastic in one fundamental respect: he drew from his own mind propositions about the nature of being and reality. He believed that there are a priori truths and that the intellect by its own power may arrive at a perfect knowledge of all things; he stated laws of motion, for example, on the basis of a priori reasoning. (Actually in his biological work he did experiment, and he drew vital conclusions from the experiments.) However, apart from his reliance upon a priori principles, he did promulgate a general and systematic philosophy that shattered the hold of Scholasticism and opened up fresh channels of thought. His attempt to sweep away all preconceptions and prejudices was a clear declaration of revolt from the past. By reducing natural phenomena to purely physical happenings he did much to rid science of mysticism and occult forces. Descartes's writings were highly influential; his deductive and systematic philosophy pervaded the seventeenth century and impressed Newton, especially, with the importance of motion. Daintily bound expositions of his philosophy even adorned ladies' dressing tables.

3. Galileo's Approach to Science

Though Galileo Galilei's philosophy of science agreed in large part with Descartes's, it was Galileo who formulated the more radical, more effective, and more concrete procedures for modern science and who by his own work demonstrated their effectiveness.

Galileo (1564–1642), born in Pisa to a cloth merchant, entered the University of Pisa to study medicine. The courses there were still at about the level of the medieval curriculum; Galileo learned his mathematics privately from a practical engineer, and at the age of seventeen switched from medicine to mathematics. After about eight years of study he applied for a teaching position at the University of Bologna but was refused as not sufficiently distinguished. He did secure a professorship of mathematics at Pisa. While there, he began to attack Aristotelian science; and he did not hesitate to express his views even though his criticisms alienated his colleagues. He had also begun to write important mathematical papers that aroused jealousy in the less competent. Galileo was made to feel uncomfortable and left in 1592 to accept the position of professor of mathematics at the University of Padua. There he wrote a short book, Le mecaniche (1604). After eighteen years at Padua he was invited to Florence by the Grand Duke Cosimo II de' Medici, who appointed him Chief Mathematician of his court, gave him a home and handsome salary, and protected him from the Jesuits, who dominated the papacy and had already threatened Galileo because he championed the Copernican theory. To express his gratitude, Galileo named the satellites of Jupiter, which he discovered in the first year of his service under Cosimo, the Medicean stars. In Florence Galileo had the leisure to pursue his studies and to write.

His advocacy of the Copernican theory irked the Roman Inquisition, and in 1616 he was called to Rome. His teachings on the heliocentric theory were condemned by the Inquisition; he had to promise not to publish any more on this subject. In 1630 Pope Urban VIII did give him permission to publish if he would make his book mathematical and not doctrinal. Thereupon, in 1632, he published his classic Dialogo dei massimi sistemi (Dialogue on the Great World Systems). The Roman Inquisition summoned him again in 1633 and under the threat of torture impelled him to recant his advocacy of the heliocentric theory. He was again forbidden to publish and required to live practically under house arrest. But he undertook to write up his years of thought and work on the phenomena of motion and on the strength of materials. The manuscript, entitled *Discorsi e dimostrazioni matematiche intorno à due nuove scienze* (Discourses and Mathematical Demonstrations Concerning Two New Sciences, also referred to as Dialogues Concerning Two New Sciences), was secretly transported to Holland and published there in 1638. This is the classic in which Galileo presented his new scientific method. He defended his actions with the words that he had never "declined in piety and reverence for the Church and my own conscience."

Galileo was an extraordinary man in many fields. He was a keen astronomical observer. He is often called the father of modern invention; though he did not invent the telescope or "perplexive glasses," as Ben Jonson called them, he was immediately able to construct one when he heard of the idea. He was an independent inventor of the microscope, and he designed the first pendulum clock. He also designed and made a compass with scales that automatically yielded the results of numerical computations so the user could read the scales and avoid having to do the calculations. This device was so much in demand that he produced many for sale.

Galileo was the first important modern student of sound. He suggested a wave theory of sound and began work on pitch, harmonics, and the vibrations of strings. This work was continued by Mersenne and Newton and became a major inspiration for mathematical work in the eighteenth century.

Galileo's major writings, though concerned with scientific subjects, are still regarded as literary masterpieces. His *Sidereus Nuncius* (Sidereal Messenger) of 1610, in which he announced his astronomical observations and declared himself in support of Copernican theory, was an immediate success, and he was elected to the prestigious Academy of the Lynx-like in Rome. His two greatest classics, the *Dialogue on the Great World Systems* and *Dialogues Concerning Two New Sciences*, are clear, direct, witty, yet profound. In both, Galileo has one character present the current views, against which another argues cleverly and tenaciously to show the fallacies and weaknesses of these views and the strengths of the new ones.

In his philosophy of science Galileo broke sharply from the speculative and mystical in favor of a mechanical and mathematical view of nature. He also believed that scientific problems should not become enmeshed in and beclouded by theological arguments. Indeed, one of his achievements in science, though somewhat apart from the method we are about to examine, is that he recognized clearly the domain of science and severed it sharply from religious doctrines.

Galileo, like Descartes, was certain that nature is mathematically designed. His statement of 1610 is famous:

> Philosophy [nature] is written in that great book which ever lies before our eyes—I mean the universe—but we cannot understand it if we do

not first learn the language and grasp the symbols in which it is written. The book is written in the mathematical language, and the symbols are triangles, circles and other geometrical figures, without whose help it is impossible to comprehend a single word of it; without which one wanders in vain through a dark labyrinth.[1]

Nature is simple and orderly and its behavior is regular and necessary. It acts in accordance with perfect and immutable mathematical laws. Divine reason is the source of the rational in nature; God put into the world that rigorous mathematical necessity that men reach only laboriously. Mathematical knowledge is therefore not only absolute truth, but as sacrosanct as any line of Scripture. In fact it is superior, for there is much disagreement about the Scriptures, but there can be none about mathematical truths.

Another doctrine, the atomism of the Greek Democritus, is clearer in Galileo than in Descartes. Atomism presupposed empty space (which Descartes did not accept) and individual, indestructible atoms. Change consisted in the combination and separation of atoms. All qualitative varieties in bodies were due to quantitative variety in number, size, shape, and spatial arrangement of the atoms. The atom's chief properties were impenetrability and indestructibility; these properties served to explain chemical and physical phenomena. Galileo's espousal of atomism placed it in the forefront of scientific doctrines.

Atomism led Galileo to the doctrine of primary and secondary qualities. He says, "If ears, tongues, and noses were removed, I am of the opinion that shape, quantity [size] and motion would remain, but there would be an end of smells, tastes, and sounds, which abstracted from the living creature, I take to be mere words." Thus in one swoop Galileo, like Descartes, stripped away a thousand phenomena and qualities to concentrate on matter and motion, properties that are mathematically describable. It is perhaps not too surprising that in the century in which problems of motion were the most prominent and serious, scientists should find motion to be a fundamental physical phenomenon.

The concentration on matter and motion was only the first step in Galileo's new approach to nature. His next thought, also voiced by Descartes, was that any branch of science should be patterned on the model of mathematics. This implies two essential steps. Mathematics starts with axioms—clear, self-evident truths—and from these proceeds by deductive reasoning to establish new truths. Any branch of science, then, should start with axioms or principles and then proceed deductively. Moreover, one should extract from the axioms as many consequences as possible. This thought, of course, goes back to Aristotle, who also aimed at deductive structure in science with the mathematical model in mind.

However, Galileo departed radically from the Greeks, the medieval

1. *Opere*, 4, 171.

scientists, and even Descartes in his method of obtaining first principles. The pre-Galileans and Descartes had believed that the mind supplied the basic principles; it had but to think about any class of phenomena and it would immediately recognize fundamental truths. This power of the mind was clearly evidenced in mathematics. Axioms such as "equals added to equals give equals" and "two points determine a line" suggested themselves immediately in thinking about number or geometrical figures, and were indubitable truths. So too had the Greeks found some physical principles equally appealing. That all objects in the universe should have a natural place was no more than fitting. The state of rest seemed clearly more natural than the state of motion. It seemed indubitable, too, that force must be applied to put and keep bodies in motion. To believe that the mind supplies fundamental principles did not deny that observations might play a role in obtaining these principles. But the observations merely evoked the correct principles, just as the sight of a familiar face might call to mind facts about that person.

The Greek and medieval scientists were so convinced that there were a priori fundamental principles that when occasional observations did not fit they invented special explanations to preserve the principles but still account for the anomalies. These men, as Galileo put it, first decided how the world should function and then fitted what they saw into their preconceived principles.

Galileo decided that in physics, as opposed to mathematics, first principles must come from experience and experimentation. The way to obtain correct and basic principles is to pay attention to what nature says rather than what the mind prefers. Nature, he argued, did not first make men's brains and then arrange the world to be acceptable to human intellects. To the medieval thinkers who kept repeating Aristotle and debating what he meant, Galileo addressed the criticism that knowledge comes from observation and not from books, and that it was useless to debate about Aristotle. He says, "When we have the decrees of nature, authority goes for nothing. . . ." Of course some Renaissance thinkers and Galileo's contemporary Francis Bacon had also arrived at the conclusion that experimentation was necessary; in this particular aspect of his new method, Galileo was not ahead of all others. Yet the modernist Descartes did not grant the wisdom of Galileo's reliance upon experimentation. The facts of the senses, Descartes said, can only lead to delusion, but reason penetrates such delusions. From the innate general principles supplied by the mind, we can deduce particular phenomena of nature and understand them. Galileo did appreciate that one may glean an incorrect principle from experimentation and that as a consequence the deductions from it could be incorrect. Hence he proposed the use of experiments to check the conclusions of his reasonings as well as to acquire basic principles.

Galileo was actually a transitional figure as far as experimentation is concerned. He, and Isaac Newton fifty years later, believed that a few key or critical experiments would yield correct fundamental principles. Moreover, many of Galileo's so-called experiments were really thought-experiments; that is, he relied upon common experience to imagine what would happen if an experiment were performed. He then drew a conclusion as confidently as if he had actually performed the experiment. When in the *Dialogue on the Great World Systems* he describes the motion of a ball dropped from the mast of a moving ship, he is asked by Simplicio, one of the characters, whether he had made an experiment. Galileo replies, "No, and I do not need it, as without any experience I can confirm that it is so, because it cannot be otherwise." He says in fact that he experimented rarely, and then primarily to refute those who did not follow the mathematics. Though Newton performed some famous and ingenious experiments, he too says that he used experiments to make his *results* physically intelligible and to convince the common people.

The truth of the matter is that Galileo had some preconceptions about nature, which made him confident that a few experiments would suffice. He believed, for example, that nature was simple. Hence when he considered freely falling bodies, which fall with increasing velocity, he supposed that the increase in velocity is the same for each second of fall. This was the simplest "truth." He believed also that nature is mathematically designed, and hence any mathematical law that seemed to fit even on the basis of rather limited experimentation appeared to him to be correct.

For Galileo, as well as for Huygens and Newton, the deductive, mathematical part of the scientific enterprise played a greater part than the experimental. Galileo was no less proud of the abundance of theorems that flow from a single principle than of the discovery of the principle itself. The men who fashioned modern science—Descartes, Galileo, Huygens, and Newton (we can also include Copernicus and Kepler)—approached the study of nature as mathematicians, in their general method and in their concrete investigations. They were primarily speculative thinkers who expected to apprehend broad, deep (but also simple), clear, and immutable mathematical principles either through intuition or through crucial observations and experiments, and then to deduce new laws from these fundamental truths, entirely in the manner in which mathematics proper had constructed its geometry. The bulk of the activity was to be the deductive portion; whole systems of thought were to be so derived.

What the great thinkers of the seventeenth century envisaged as the proper procedure for science did indeed prove to be the profitable course. The rational search for laws of nature produced, by Newton's time, extremely valuable results on the basis of the slimmest observational and experimental knowledge. The great scientific advances of the sixteenth and seventeenth centuries were in astronomy, where observation offered little that was new, and in mechanics, where the experimental results were hardly startling and certainly not decisive, whereas the mathematical theory attained comprehensiveness and perfection. And for the next two centuries scientists produced deep and sweeping laws of nature on the basis of very few, almost trivial, observations and experiments.

The expectation of Galileo, Huygens, and Newton that just a few experiments would suffice can be readily understood. Because these men were convinced that nature is mathematically designed, they saw no reason why they could not proceed in scientific matters much as mathematicians had proceeded in their domain. As John Herman Randall says in *Making of the Modern Mind*, "Science was born of a faith in the mathematical interpretation of nature...."

Galileo did, however, obtain a few principles from experience; and in this work also his approach was a radical departure from that of his predecessors. He decided that one must penetrate to what is fundamental in phenomena and start there. In *Two New Sciences* he says that it is not possible to treat the infinite variety of weights, shapes, and velocities. He had observed that the speeds with which dissimilar objects fall differ less in air than in water. Hence the thinner the medium, the less difference in speed of fall among bodies. "Having observed this I came to the conclusion that in a medium totally devoid of resistance all bodies would fall with the same speed." What Galileo was doing here was to strip away the incidental or minor effects in an effort to get at the major one.

Of course, actual bodies do fall in resisting media. What could Galileo say about such motions? His answer was "... hence, in order to handle this matter in a scientific way, it is necessary to cut loose from these difficulties [air resistance, friction, etc.] and having discovered and demonstrated the theorems in the case of no resistance, to use them and apply them with such limitations as experience will teach."

Having stripped away air resistance and friction, Galileo sought basic laws for motion in a vacuum. Thus he not only contradicted Aristotle and even Descartes by thinking of bodies moving in empty space, but did just what the mathematician does in studying real figures. The mathematician strips away molecular structure, color, and thickness of lines to get at some basic properties and concentrates on these. So did Galileo penetrate to basic physical factors. The mathematical method of abstraction is indeed a step away from reality but, paradoxically, it leads back to reality with greater power than if all the factors actually present are taken into account at once.

Thus far Galileo had formulated a number of methodological principles, many of which were suggested by the approach mathematics had employed in geometry. His next principle was to use mathematics itself, but in a special way. Unlike the Aristotelians and the late medieval scientists, who had fastened upon qualities they regarded as fundamental and studied the acquisition and loss of qualities or debated the meaning of the qualities, Galileo proposed to seek *quantitative* axioms. This change is most important; we shall see the full significance of it later, but an elementary example may be useful now. The Aristotelians said that a ball falls because it has weight, and that it falls to the earth because every object seeks its natural place and the natural place of heavy bodies is the center of the earth. These principles are qualitative. Even Kepler's first law of motion, that the path of each planet is an ellipse, is a qualitative statement. By contrast, let us consider the statement that the speed (in feet per second) with which a ball falls is 32 times the number of seconds it has been falling, or in symbols, $v = 32t$. This is a quantitative statement about how a ball falls. Galileo intended to seek such quantitative statements as his axioms, and he expected to deduce new ones by mathematical means. These deductions would also give quantitative knowledge. Moreover, as we have seen, mathematics was to be his essential medium.

The decision to seek quantitative knowledge expressed in formulas carried with it another radical decision, though first contact with it hardly reveals its full significance. The Aristotelians believed that one of the tasks of science was to explain why things happened; explanation meant unearthing the causes of a phenomenon. The statement that a body falls because it has weight gives the effective cause of the fall and the statement that it seeks its natural place gives the final cause. But the quantitative statement $v = 32t$, for whatever it may be worth, gives no explanation of why a ball falls; it tells only how the speed changes with the time. In other words, formulas do not explain; they describe. The knowledge of nature Galileo sought was descriptive. He says in *Two New Sciences*, "The cause of the acceleration of the motion of falling bodies is not a necessary part of the investigation." More generally, he points out that he will investigate and demonstrate some of the properties of motion without regard to what the causes might be. Positive scientific inquiries were to be separated from questions of ultimate causation, and speculation as to physical causes was to be abandoned.

First reactions to this principle of Galileo are likely to be negative. Description of phenomena in terms of formulas hardly seems to be more than a first step. It would seem that the true function of science had really been grasped by the Aristotelians, namely, to explain why phenomena happened. Even Descartes protested Galileo's decision to seek descriptive formulas. He said, "Everything that Galileo says about bodies falling in empty space is built without foundation: he ought first to have determined the nature of weight." Further, said Descartes, Galileo should reflect on ultimate reasons. But we shall see clearly after a few chapters that Galileo's decision to aim for description was the deepest and most fruitful idea that anyone has had about scientific methodology.

Whereas the Aristotelians had talked in terms of qualities such as fluidity, rigidity, essences, natural places, natural and violent motion, and potentiality, Galileo chose an entirely new set of concepts, which, moreover, were measurable, so that their measures could be related by formulas. Some of them are: distance, time, speed, acceleration, force, mass, and weight. These concepts are too familiar to surprise us. But in Galileo's time they were radical choices, at least as fundamental concepts; and these are the ones that proved most instrumental in the task of understanding and mastering nature.

We have described the essential features of Galileo's program. Some of the ideas in it had been espoused by others; some were entirely original with him. But what establishes Galileo's greatness is that he saw so clearly what was wrong or deficient in the current scientific efforts, shed completely the older ways, and formulated the new procedures so clearly. Moreover, in applying them to problems of motion he not only exemplified the method but succeeded in obtaining brilliant results—in other words, he showed that it worked. The unity of his work, the clarity of his thoughts and expressions, and the force of his argumentation influenced almost all of his contemporaries and successors. More than any other man, Galileo is the founder of the methodology of modern science. He was fully conscious of what he had accomplished (see the chapter legend); so were others. The philosopher Hobbes said of Galileo, "He has been the first to open to us the door to the whole realm of physics."

We cannot pursue the history of the methodology of science. However, since mathematics became so important in this methodology and profited so much from its adoption, we should note how completely Galileo's program was accepted by giants such as Newton. He asserts that experiments are needed to furnish basic laws. Newton is also clear that the function of science, after having obtained some basic principles, is to deduce new facts from these principles. In the preface to his *Principia*, he says:

> Since the ancients (as we are told by Pappus) esteemed the science of mechanics of greatest importance in the investigation of natural things, and the moderns, rejecting substantial forms and occult qualities, have endeavored to subject the phenomena of nature to the laws of mathematics, I have in this treatise cultivated mathematics as far as it relates to philosophy [science] ... and therefore I offer this work as the mathematical principles of philosophy, for the whole burden in philosophy seems to consist in this—from the phenomena of motions to investigate the forces of nature, and then from these forces to demonstrate the other phenomena....

Of course, mathematical principles, to Newton as to Galileo, were quantitative principles. He says in the *Principia* that his purpose is to discover and set forth the exact manner in which "all things had been ordered in measure, number and weight." Newton had good reason to emphasize quantitative mathematical laws, as opposed to physical explanation, because the central physical concept in his celestial mechanics was the force of gravitation, whose action could not be explained at all in physical terms. In lieu of explanation Newton had a quantitative formulation of how gravity acted that was significant and usable. And this is why he says, at the beginning of the *Principia*, "For I here design only to give a mathematical notion of these forces, without considering their physical causes and seats." Toward the end of the book he repeats this thought:

> But our purpose is only to trace out the quantity and properties of this force from the phenomena, and to apply what we discover in some simple cases as principles, by which, in a mathematical way, we may estimate the effects thereof in more involved cases ... We said, in *a mathematical way* [italics Newton's], to avoid all questions about the nature or quality of this force, which we would not be understood to determine by any hypothesis..."

The abandonment of physical mechanism in favor of mathematical description shocked even great scientists. Huygens regarded the idea of gravitation as "absurd," because its action through empty space precluded any mechanism. He expressed surprise that Newton should have taken the trouble to make such a number of laborious calculations with no foundation but the mathematical principle of gravitation. Leibniz attacked gravitation as an incorporeal and inexplicable power; John Bernoulli (James's brother) denounced it as "revolting to minds accustomed to receiving no principle in physics save those which are incontestable and evident." But this reliance on mathematical description even where physical understanding was completely lacking made possible Newton's amazing contributions, to say nothing of subsequent developments.

Because science became heavily dependent upon—almost subordinate to—mathematics, it was the scientists who extended the domain and techniques of mathematics; and the multiplicity of problems provided by science gave mathematicians numerous and weighty directions for creative work.

4. *The Function Concept*

The first mathematical gain from scientific investigations conducted in accordance with Galileo's program came from the study of motion. This problem engrossed the scientists and mathematicians of the seventeenth century. It is easy to see why. Though Kepler's astronomy was accepted early in the seventeenth century, especially after Galileo's observations supplied additional evidence for a heliocentric theory, Kepler's law of elliptical motion is only approximately correct, though it would be exact if there were just the sun and one planet in the heavens. The ideas that the other planets disturb the elliptical motion of any one planet and that the sun disturbs the elliptical motion of the moon around the earth were already being considered; in fact, the notion of a gravitational force acting between any two bodies was suggested by Kepler, among others. Hence the problem of improving the calculation of the planets' positions was open. Moreover, Kepler had obtained his laws essentially by fitting curves to astronomical data, with no explanation in terms of fundamental laws of motion of why the planets moved in elliptical paths. The basic problem of deriving Kepler's laws from principles of motion posed a clear challenge.

The improvement of astronomical theory also had a practical objective. In their search for raw materials and trade, the Europeans had undertaken large-scale navigation that involved sailing long distances out of sight of land. Mariners therefore needed accurate methods of determining latitude and longitude. The determination of latitude can be made by direct observation of the sun or the stars, but determination of longitude is far more difficult. In the sixteenth century the methods of doing it were so inaccurate that navigators were often in error as much as 500 miles. After about 1514, the direction of the moon relative to the stars was used to determine longitude. These directions, as seen from some standard place at various times, were tabulated. A navigator would determine the direction of the moon, which was not affected much by his being in a different location, and determine his local time by using, for example, the directions of the stars. Directly from the tables or by interpolation he could find the time at the standard location when the moon had the measured direction and so compute the difference in time between his position and the standard one. Each hour of difference means a 15-degree difference in longitude. This method, however, was not accurate. Because the ships of those times were constantly heaving, it was difficult to obtain the moon's direction accurately; but, because the moon does not move much relative to the stars in a few hours, the direction of the moon had to be rather precisely determined. A mistake of one minute of angle means an error of half a degree of longitude; but even a measure accurate to within one minute was far beyond the capabilities of those times. Though other methods of determining longitude were suggested and tried, better knowledge of the moon's path to extend and improve the tables seemed indispensable and many scientists, including Newton, worked on the problem. Even in Newton's time the knowledge of the moon's position was so inaccurate that use of the tables led to errors of as much as 100 miles in determining position at sea.

The governments of Europe were very much concerned, because

shipping losses were considerable. In 1675 King Charles II of England set up the Royal Observatory at Greenwich to obtain better observations on the moon's motion and to serve as a fixed station for longitude. In 1712 the British government established a Commission for the Discovery of Longitude and offered rewards of up to £20,000 for ideas on how to measure longitude.

The problem of explaining terrestrial motions also faced seventeenth-century scientists. Under the heliocentric theory the earth was both rotating and revolving around the sun. Why then should objects stay with the earth? Why should dropped objects fall to earth if it was no longer the center of the universe? Moreover, all motions, projectile motion for example, seemed to take place as though the earth were at rest. These questions engaged the attention of many men, including Cardan, Tartaglia, Galileo, and Newton. The paths of projectiles, their ranges, the heights they could reach, and the effect of muzzle velocity on height and range were basic questions and the princes then, like nations now, spent great sums on the solutions. New principles of motion were needed to account for these terrestrial phenomena; and it occurred to the scientists that, since the universe was believed to be constructed according to one master plan, the same principles that explained terrestrial motions would also account for heavenly motions.

From the study of the various problems of motion there emerged the specific problem of designing more accurate methods of measuring time. Mechanical clocks, which had been in use since 1348, were not very accurate. The Flemish cartographer Gemma Frisius (1508–55) had suggested the use of a clock to determine longitude. A ship could carry a clock set to the time of a place of known longitude; since the determination of local time by the sun's position, for example, was relatively simple, the navigator need merely note the difference in time and translate this at once into the difference in longitude. But no durable, accurate, seaworthy clocks were available even by 1600.

The motion of a pendulum seemed to provide the basic mechanism for measuring time. Galileo had observed that the time for one complete oscillation of a pendulum was constant and ostensibly independent of the amplitude of the swing. He prepared the design of a pendulum clock and had his son construct one; but it was Robert Hooke and Huygens who did the basic work on the pendulum. Though the pendulum clock was unsuitable for a ship (an accuracy of two or three seconds a day was needed for the purpose of longitude-reckoning, and pendulums were too much affected by ship's motion), it proved immensely valuable in scientific work, as well as for timekeeping in homes and business. A clock appropriate for navigation was finally designed by John Harrison (1693–1776) in 1761 and began to be used by the end of the eighteenth century. Because a proper clock was not available earlier, accurate determination of the motion of the moon was still the chief scientific problem in that century.

From the study of motion mathematics derived a fundamental concept that was central to practically all of the work for the next two hundred years —the concept of a function or a relation between variables. One finds this notion almost throughout Galileo's *Two New Sciences*, the book in which he founded modern mechanics. Galileo expresses his functional relationships in words and in the language of proportion. Thus in his work on the strength of materials, he has occasion to state, "The areas of two cylinders of equal volumes, neglecting the bases, bear to each other a ratio which is the square root of the ratio of their lengths." Again, "The volumes of right cylinders having equal curved surfaces are inversely proportional to their altitudes." In his work on motion he states, for example, "The spaces described by a body falling from rest with a uniformly accelerated motion are to each other as the squares of the time intervals employed in traversing these distances." "The times of descent along inclined planes of the same height, but of different slopes, are to each other as the lengths of these planes." The language shows clearly that he is dealing with variables and functions; it was but a short step to write these statements in symbolic form. Since the symbolism of algebra was being extended at this time, Galileo's statement on the spaces described by a falling body soon was written as $s = kt^2$ and his statement on times of descent as $t = kl$.

Most of the functions introduced during the seventeenth century were first studied as curves, before the function concept was fully recognized. This was true, for example, of the elementary transcendental functions such as $\log x$, $\sin x$, and a^x. Thus Evangelista Torricelli (1608–47), a pupil of Galileo, in a letter of 1644 described his research on the curve we would represent by $y = ae^{-cx}$ with $x \geq 0$ (the manuscript in which he wrote up this research was not edited until 1900). The curve was suggested to Torricelli by the current work on logarithms. Descartes encountered the same curve in 1639 but did not speak of its connection with logarithms. The sine curve entered mathematics as the companion curve to the cycloid in Roberval's work on the cycloid (Chap. 17, sec. 2) and appears graphed for two periods in Wallis's *Mechanica* (1670). Of course the tabular values of the trigonometric and logarithmic functions were, by this time, known with great precision.

It is also relevant that old and new curves were introduced by means of motions. In Greek times, a few curves, such as the quadratrix and the Archimedean spiral, were defined in terms of motion, but in that period such curves were outside the pale of legitimate mathematics. The attitude was quite different in the seventeenth century. Mersenne in 1615 defined the cycloid (which had been known earlier) as the locus of a point on a wheel that rolls along the ground. Galileo, who had shown that the path of a projectile shot up into the air at an angle to the ground is a parabola, regarded the curve as the locus of a moving point.

With Roberval, Barrow, and Newton the concept of a curve as the path of a moving point attains explicit recognition and acceptance. Newton says in *Quadrature of Curves* (written in 1676), "I consider mathematical quantities in this place not as consisting of very small parts, but as described by a continued motion. Lines [curves] are described, and thereby generated, not by the apposition of parts but by the continued motion of points.... These geneses really take place in the nature of things, and are daily seen in the motion of bodies."

Gradually the terms and symbolism for the various types of functions represented by these curves were introduced. There were many subtle difficulties that were hardly recognized. For example, the use of functions of the form a^x, with x taking on positive and negative integral and fractional values, became common in the seventeenth century. It was assumed (until the nineteenth century, when irrational numbers were first defined) that the function was also defined for irrational values of x, so that no one questioned an expression of the form $2^{\sqrt{2}}$. The implicit understanding was that such a value was intermediate between that obtained for any two rational exponents above and below $\sqrt{2}$.

Descartes's distinction between geometric and mechanical curves (Chap. 15, sec. 4) gave rise to the distinction between algebraic and transcendental functions. Fortunately his contemporaries ignored his banishment of what he called mechanical curves. Through quadratures, the summation of series, and other operations that entered with the calculus, many types of transcendental functions arose and were studied. The distinction between algebraic and transcendental functions was clearly made by James Gregory in 1667, when he sought to show that the area of a circular sector could not be an algebriac function of the radius and the chord. Leibniz showed that sin x cannot be an algebraic function of x and incidentally proved the result sought by Gregory.[2] The full understanding and use of the transcendental functions came gradually.

The most explicit definition of the function concept in the seventeenth century was given by James Gregory in his *Vera Circuli et Hyperbolae Quadratura* (1667). He defined a function as a quantity obtained from other quantities by a succession of algebraic operations or by any other operation imaginable. By the last phrase he meant, as he explains, that it is necessary to add to the five operations of algebra a sixth operation, which he defines as passage to the limit. (Gregory, as we shall see in Chapter 17, was concerned with quadrature problems.) Gregory's concept of function was lost sight of; but in any case, it would soon have proved too narrow, because the series representation of functions became widely used.

From the very beginning of his work on the calculus, that is from 1665

2. *Math. Schriften*, 5, 97–98.

on, Newton used the term "fluent" to represent any relationship between variables. In a manuscript of 1673 Leibniz used the word "function" to mean any quantity varying from point to point of a curve—for example, the length of the tangent, the normal, the subtangent, and the ordinate. The curve itself was said to be given by an equation. Leibniz also introduced the words "constant," "variable," and "parameter," the latter used in connection with a family of curves.[3] In working with functions John Bernoulli spoke from 1697 on of a quantity formed, in any manner whatever, of variables and of constants;[4] by "any manner" he meant to cover algebraic and transcendental expressions. He adopted Leibniz's phrase "function of x" for this quantity in 1698. In his *Historia* (1714), Leibniz used the word "function" to mean quantities that depend on a variable.

As to notation, John Bernoulli wrote X or ξ for a general function of x, though in 1718 he changed to ϕx. Leibniz approved of this, but proposed also x^1 and x^2 for functions of x, the superscript to be used when several functions were involved. The notation $f(x)$ was introduced by Euler in 1734.[5] The function concept immediately became central in the work on the calculus. We shall see later how the concept was extended.

Bibliography

Bell, A. E.: *Christian Huygens and the Development of Science in the Seventeenth Century*, Edward Arnold, 1947.

Burtt, A. E.: *The Metaphysical Foundations of Modern Physical Science*, 2nd ed., Routledge and Kegan Paul, 1932, Chaps. 1–7.

Butterfield, Herbert: *The Origins of Modern Science*, Macmillan, 1951, Chaps. 4–7.

Cohen, I. Bernard: *The Birth of a New Physics*, Doubleday, 1960.

Coolidge, J. L.: *The Mathematics of Great Amateurs*, Dover (reprint), 1963, pp. 119–27.

Crombie, A. C.: *Augustine to Galileo*, Falcon Press, 1952, Chap. 6.

Dampier-Whetham, W. C. D.: *A History of Science and Its Relations with Philosophy and Religion*, Cambridge University Press, 1929, Chap. 3.

Dijksterhuis, E. J.: *The Mechanization of the World Picture*, Oxford University Press, 1961.

Drabkin, I. E., and Stillman Drake: *Galileo Galilei: On Motion and Mechanics*, University of Wisconsin Press, 1960.

Drake, Stillman: *Discoveries and Opinions of Galileo*, Doubleday, 1957.

Galilei, Galileo: *Opere*, 20 vols., 1890–1909, reprinted by G. Barbera, 1964–66.

———: *Dialogues Concerning Two New Sciences*, Dover (reprint), 1952.

Hall, A. R.: *The Scientific Revolution*, Longmans Green, 1954, Chaps. 1–8.

———: *From Galileo to Newton*, Collins, 1963, Chaps. 1–5.

3. *Math. Schriften*, 5, 266–69.
4. *Mém de l'Acad des Sci.*, Paris, 1718, 100 ff. = *Opera*, 2, 235–69, p. 241 in particular.
5. *Comm. Acad. Sci. Petrop.*, 7, 1734/35, 184–200, pub. 1740 = *Opera*, (1), 22, 57–75.

Huygens, C.: *Œuvres complètes*, 22 vols., M. Nyhoff, 1888–1950.

Newton, I.: *Mathematical Principles of Natural Philosophy*, University of California Press, 1946.

Randall, John H., Jr.: *Making of the Modern Mind*, rev. ed., Houghton Mifflin, 1940, Chap. 10.

Scott, J. F.: *The Scientific Work of René Descartes*, Taylor and Francis, 1952, Chaps. 10–12.

Smith, Preserved: *A History of Modern Culture*, Henry Holt, 1930, Vol. 1, Chaps. 3, 6, and 7.

Strong, Edward W.: *Procedures and Metaphysics*, University of California Press, 1936, Chaps. 5–8.

Whitehead, Alfred North: *Science and the Modern World*, Cambridge University Press, 1926, Chap. 3.

Wolf, Abraham: *A History of Science, Technology and Philosophy in the 16th and 17th Centuries*, 2nd ed., George Allen and Unwin, 1950, Chap. 3.

17
The Creation of the Calculus

Who, by a vigor of mind almost divine, the motions and figures of the planets, the paths of comets, and the tides of the seas first demonstrated.　NEWTON'S EPITAPH

1. *The Motivation for the Calculus*

Following hard on the adoption of the function concept came the calculus, which, next to Euclidean geometry, is the greatest creation in all of mathematics. Though it was to some extent the answer to problems already tackled by the Greeks, the calculus was created primarily to treat the major scientific problems of the seventeenth century.

There were four major types of problems. The first was: Given the formula for the distance a body covers as a function of the time, to find the velocity and acceleration at any instant; and, conversely, given the formula describing the acceleration of a body as a function of the time, to find the velocity and the distance traveled. This problem arose directly in the study of motion and the difficulty it posed was that the velocities and the acceleration of concern to the seventeenth century varied from instant to instant. In calculating an instantaneous velocity, for example, one cannot, as one can in the case of average velocity, divide the distance traveled by the time of travel, because at a given instant both the distance traveled and time are zero, and 0/0 is meaningless. Nevertheless, it was clear on physical grounds that moving objects do have a velocity at each instant of their travel. The inverse problem of finding the distance covered, knowing the formula for velocity, involves the corresponding difficulty; one cannot multiply the velocity at any one instant by the time of travel to obtain the distance traveled because the velocity varies from instant to instant.

The second type of problem was to find the tangent to a curve. Interest in this problem stemmed from more than one source; it was a problem of pure geometry, and it was of great importance for scientific applications. Optics, as we know, was one of the major scientific pursuits of the seventeenth century; the design of lenses was of direct interest to Fermat, Descartes, Huygens, and Newton. To study the passage of light through a lens, one

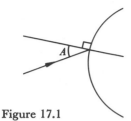

Figure 17.1

must know the angle at which the ray strikes the lens in order to apply the law of refraction. The significant angle is that between the ray and the normal to the curve (Fig. 17.1), the normal being the perpendicular to the tangent. Hence the problem was to find either the normal or the tangent. Another scientific problem involving the tangent to a curve arose in the study of motion. The direction of motion of a moving body at any point of its path is the direction of the tangent to the path.

Actually, even the very meaning of "tangent" was open. For the conic sections the definition of a tangent as a line touching a curve at only one point and lying on one side of the curve sufficed; this definition was used by the Greeks. But it was inadequate for the more complicated curves already in use in the seventeenth century.

The third problem was that of finding the maximum or minimum value of a function. When a cannonball is shot from a cannon, the distance it will travel horizontally—the range—depends on the angle at which the cannon is inclined to the ground. One "practical" problem was to find the angle that would maximize the range. Early in the seventeenth century, Galileo determined that (in a vacuum) the maximum range is obtained for an angle of fire of 45°; he also obtained the maximum heights reached by projectiles fired at various angles to the ground. The study of the motion of the planets also involved maxima and minima problems, such as finding the greatest and least distances of a planet from the sun.

The fourth problem was finding the lengths of curves, for example, the distance covered by a planet in a given period of time; the areas bounded by curves; volumes bounded by surfaces; centers of gravity of bodies; and the gravitational attraction that an *extended* body, a planet for example, exerts on another body. The Greeks had used the method of exhaustion to find some areas and volumes. Despite the fact that they used it for relatively simple areas and volumes, they had to apply much ingenuity, because the method lacked generality. Nor did they often come up with numerical answers. Interest in finding lengths, areas, volumes, and centers of gravity was revived when the work of Archimedes became known in Europe. The method of exhaustion was first modified gradually, and then radically by the invention of the calculus.

2. Early Seventeenth-Century Work on the Calculus

The problems of the calculus were tackled by at least a dozen of the greatest mathematicians of the seventeenth century and by several dozen minor ones. All of their contributions were crowned by the achievements of Newton and Leibniz. Here we shall be able to note only the principal contributions of the precursors of these two masters.

The problem of calculating the instantaneous velocity from a knowledge of the distance traveled as a function of the time, and its converse, were soon seen to be special cases of calculating the instantaneous rate of change of one variable with respect to another and its converse. The first significant treatment of general rate problems is due to Newton; we shall examine it later.

Several methods were advanced to find the tangent to a curve. In his *Traité des indivisibles*, which dates from 1634 (though not published until 1693), Gilles Persone de Roberval (1602–75) generalized a method Archimedes had used to find the tangent at any point on his spiral. Like Archimedes, Roberval thought of a curve as the locus of a point moving under the action of two velocities. Thus a projectile shot from a cannon is acted on by a horizontal velocity, PQ in Figure 17.2, and a vertical velocity, PR. The resultant of these two velocities is the diagonal of the rectangle formed on PQ and PR. Roberval took the line of this diagonal to be the tangent at P. As Torricelli pointed out, Roberval's method used a principle already asserted by Galileo, namely, that the horizontal and vertical velocities acted independently of each other. Torricelli himself applied Roberval's method to obtain tangents to curves whose equations we now write as $y = x^n$.

While the notion of a tangent as a line having the direction of the resultant velocity was more complicated than the Greek definition of a line touching a curve, this newer concept applied to many curves for which the older one failed. It was also valuable because it linked pure geometry and dynamics, which before Galileo's work had been regarded as essentially distinct. On the other hand, this notion of a tangent was objectionable on mathematical grounds, because it based the definition of tangent on physical concepts. Many curves arose in situations having nothing to do with motion and the definition of tangent was accordingly inapplicable. Hence other methods of finding tangents gained favor.

Fermat's method, which he had devised by 1629 and which is found in his 1637 manuscript *Methodus ad Disquirendam Maximam et Minimam* (Method of Finding Maxima and Minima),[1] is in substance the present method. Let PT be the desired tangent at P on a curve (Fig. 17.3). The length TQ is called the subtangent. Fermat's plan is to find the length of TQ, from which one knows the position of T and can then draw TP.

1. *Œuvres*, 1, 133–79; 3, 121–56.

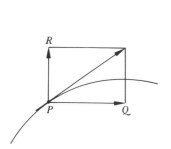

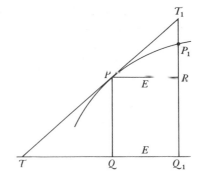

Figure 17.2   Figure 17.3

Let QQ_1 be an increment in TQ of amount E. Since triangle TQP is similar to triangle PRT_1,

$$TQ:PQ = E:T_1R.$$

But, Fermat says, T_1R is almost P_1R; therefore,

$$TQ:PQ = E:(P_1Q_1 - QP).$$

Calling $PQ, f(x)$ in our modern notation, we have

$$TQ:f(x) = E:[f(x + E) - f(x)].$$

Hence

$$TQ = \frac{E \cdot f(x)}{f(x + E) - f(x)}.$$

For the $f(x)$ Fermat treated, it was immediately possible to divide numerator and denominator of the above fraction by E. He then set $E = 0$ (he says, remove the E term) and so obtained TQ.

Fermat applied his method of tangents to many difficult problems. The method has the *form* of the now-standard method of the differential calculus, though it begs entirely the difficult theory of limits.

To Descartes the problem of finding a tangent to a curve was important because it enables one to obtain properties of curves—for example, the angle of intersection of two curves. He says, "This is the most useful, and the most general problem, not only that I know, but even that I have any desire to know in geometry." He gave his method in the second book of *La Géométrie*. It was purely algebraic and did not involve any concept of limit, whereas Fermat's did, if rigorously formulated. However, Descartes's method was useful only for curves whose equations were of the form $y = f(x)$, where $f(x)$ was a simple polynomial. Though Fermat's method was general, Descartes

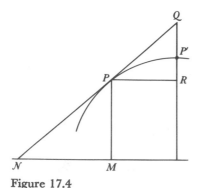

Figure 17.4

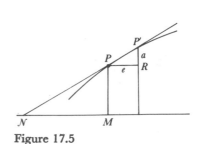

Figure 17.5

thought his own method was better; he criticized Fermat's, which admittedly was not clear as presented then, and tried to interpret it in terms of his own ideas. Fermat in turn claimed his method was superior and saw advantages in his use of the little increments E.

Isaac Barrow (1630–77) also gave a method of finding tangents to curves. Barrow was a professor of mathematics at Cambridge University. Well versed in Greek and Arabic, he was able to translate some of Euclid's works and to improve a number of other translations of the writings of Euclid, Apollonius, Archimedes, and Theodosius. His chief work, the *Lectiones Geometricae* (1669), is one of the great contributions to the calculus. In it he used geometrical methods, "freed," as he put it, "from the loathsome burdens of calculation." In 1669 Barrow resigned his professorship in favor of Newton and turned to theological studies.

Barrow's geometrical method is quite involved and makes use of auxiliary curves. However, one feature is worth noting because it illustrates the thinking of the time; it is the use of what is called the differential, or characteristic, triangle. He starts with the triangle PRQ (Fig. 17.4), which results from the increment PR, and uses the fact that this triangle is similar to triangle PMN to assert that the slope QR/PR of the tangent is equal to PM/MN. However, Barrow says, when the arc PP' is sufficiently small we may safely identify it with the segment PQ of the tangent at P. The triangle PRP' (Fig. 17.5), in which PP' is regarded both as an arc of the curve and as part of the tangent, is the characteristic triangle. It had been used much earlier by Pascal, in connection with finding areas, and by others before him.

In Lecture 10 of the *Lectiones*, Barrow does resort to calculation to find the tangent to a curve. Here the method is essentially the same as Fermat's. He uses the equation of the curve, say $y^2 = px$, and replaces x by $x + e$ and y by $y + a$. Then

$$y^2 + 2ay + a^2 = px + pe.$$

He subtracts $y^2 = px$ and obtains

$$2ay + a^2 = pe.$$

Then he discards higher powers of a and e (where present), which amounts to replacing PRP' of Figure 17.4 by PRP' of Figure 17.5, and concludes that

$$\frac{a}{e} = \frac{p}{2y}.$$

Now he argues that $a/e = PM/NM$, so that

$$\frac{PM}{NM} = \frac{p}{2y}.$$

Since PM is y, he has calculated NM, the subtangent, and knows the position of N.

The work on the third class of problems, finding the maxima and minima of functions, may be said to begin with an observation by Kepler. He was interested in the shape of casks for wine; in his *Stereometria Doliorum* (1615) he showed that, of all right parallelepipeds inscribed in a sphere and having square bases, the cube is the largest. His method was to calculate the volumes for particular choices of dimensions. This in itself was not significant; but he noted that as the maximum volume was approached, the *change* in volume for a fixed change in dimensions grew smaller and smaller.

Fermat in his *Methodus ad Disquirendam* gave his method, which he illustrated with the following example: Given a straight line (segment), it is required to find a point on it such that the rectangle contained by the two segments of the line is a maximum. He calls the whole segment B and lets one part of it be A. Then the rectangle is $AB - A^2$. He now replaces A by $A + E$. The other part is then $B - (A + E)$, and the rectangle becomes $(A + E)(B - A - E)$. He equates the two areas because, he argues, at a maximum the two function values—that is, the two areas—should be equal. Thus

$$AB + EB - A^2 - 2AE - E^2 = AB - A^2.$$

By subtracting common terms from the two sides and dividing by E, he gets

$$B = 2A + E.$$

He then sets $E = 0$ (he says, discard the E term) and gets $B = 2A$. Thus the rectangle is a square.

The method, Fermat says, is quite general; he describes it thus: If A is the independent variable, and if A is increased to $A + E$, then when E becomes indefinitely small and when the function is passing through a maximum or minimum, the two values of the function will be equal. These two values are equated; the equation is divided by E; and E is now made to vanish, so that from the resulting equation the value of A that makes the function a maximum or minimum can be determined. The method is essentially the one he used to find the tangent to a curve. However, the basic fact there is a similarity of two triangles; here it is the equality of two function values. Fermat did not see the need to justify introducing a non-zero E and then, after dividing by E, setting $E = 0$.[2]

The seventeenth-century work on finding areas, volumes, centers of gravity, and lengths of curves begins with Kepler, who is said to have been attracted to the volume problem because he noted the inaccuracy of methods used by wine dealers to find the volumes of kegs. This work (in *Stereometria Doliorum*) is crude by modern standards. For example, the area of a circle is to him the area of an infinite number of triangles, each with a vertex at the center and a base on the circumference. Then from the formula for the area of a regular inscribed polygon, 1/2 the perimeter times the apothem, he obtained the area of the circle. In an analogous manner he regarded the volume of a sphere as the sum of the volumes of small cones with vertices at the center of the sphere and bases on its surface. He then proceeded to show that the volume of the sphere is 1/3 the radius times the surface. The cone he regarded as a sum of very thin circular discs and was able thereby to compute its volume. Stimulated by Archimedes' *Spheroids and Conoids*, he generated new figures by rotation of areas and calculated the volumes. Thus he rotated the segment of a circle cut out by a chord around the chord and found the volume.

The identification of curvilinear areas and volumes with the sum of an infinite number of infinitesimal elements of the same dimension is the essence of Kepler's method. That the circle could be regarded as the sum of an infinite number of triangles was in his mind justified by the principle of continuity (Chap. 14, sec. 5). He saw no difference in kind between the two figures. For the same reason a line and an infinitesimal area were really the same; and he did, in some problems, regard an area as a sum of lines.

In *Two New Sciences* Galileo conceives of areas in a manner similar to Kepler's; in treating the problem of uniformly accelerated motion, he gives an argument to show that the area under the time-velocity curve is the distance. Suppose an object moves with varying velocity $v = 32t$, represented by the straight line in Figure 17.6; then the distance covered in time OA is the area OAB. Galileo arrived at this conclusion by regarding $A'B'$, say, as a typical velocity at some instant and also as the infinitesimal distance covered (as it would be if multiplied by a very small element of time), then arguing that the area OAB, which is made up of lines $A'B'$, must therefore be

2. For the equations that precede his setting $E = 0$, Fermat used the term *adaequalitas*, which Carl B. Boyer in *The Concepts of the Calculus*, p. 156, has aptly translated as "pseudo-equality."

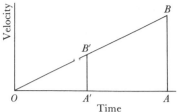

Figure 17.6

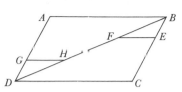

Figure 17.7

the total distance. Since AB is $32t$ and OA is t, the area of OAB is $16t^2$. The reasoning is of course unclear. It was supported in Galileo's mind by philosophical considerations that amount to regarding the area OAB as made up of an infinite number of indivisible units such as $A'B'$. He spent much time on the problem of the structure of continuous magnitudes such as line segments and areas but did not resolve it.

Bonaventura Cavalieri (1598–1647), a pupil of Galileo and professor in a lyceum in Bologna, was influenced by Kepler and Galileo and urged by the latter to look into problems of the calculus. Cavalieri developed the thoughts of Galileo and others on indivisibles into a geometrical method and published a work on the subject, *Geometria Indivisibilibus Continuorum Nova quadam Ratione Promota* (Geometry Advanced by a thus far Unknown Method, Indivisibles of Continua, 1635). He regards an area as made up of an indefinite number of equidistant parallel line segments and a volume as composed of an indefinite number of parallel plane areas; these elements he calls the indivisibles of area and volume, respectively. Cavalieri recognizes that the number of indivisibles making up an area or volume must be indefinitely large but does not try to elaborate on this. Roughly speaking, the indivisibilitists held, as Cavalieri put it in his *Exercitationes Geometricae Sex* (1647), that a line is made up of points as a string is of beads; a plane is made up of lines as a cloth is of threads; and a solid is made up of plane areas as a book is made up of pages. However, they allowed for an infinite number of the constituent elements.

Cavalieri's method or principle is illustrated by the following proposition, which of course can be proved in other ways. To show that the parallelogram $ABCD$ (Fig. 17.7) has twice the area of either triangle ABD or BCD, he argued that when $GD = BE$, then $GH = FE$. Hence triangles ABD and BCD are made up of an equal number of equal lines, such as GH and EF, and therefore must have equal areas.

The same principle is incorporated in the proposition now taught in solid geometry books and known as Cavalieri's Theorem. The principle says that if two solids have equal altitudes and if sections made by planes parallel

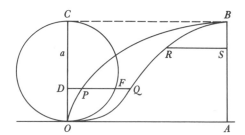

Figure 17.8

to the bases and at equal distances from them always have a given ratio, the volumes of the two solids have this given ratio to each other. Using essentially this principle, Cavalieri proved that the volume of a cone is 1/3 that of the circumscribed cylinder. Likewise he treated the arca under two curves, say $y = f(x)$ and $y = g(x)$ in our notation, and over the same range of x-values; considering the areas as the sums of ordinates, if the ordinates of one are in a constant ratio to those of the other, then, says Cavalieri, the areas are in the same ratio. He showed by his methods in *Centuria di varii problemi* (1639) that, in our notation,

$$\int_0^a x^n \, dx = \frac{a^{n+1}}{n+1}$$

for positive integral values of n up to 9. However, his method was entirely geometrical. He was successful in obtaining correct results because he applied his principle to calculate ratios of areas and volumes where the ratio of the indivisibles making up the respective areas and volumes was constant.

Cavalieri's indivisibles were criticized by contemporaries, and Cavalieri attempted to answer them; but he had no rigorous justification. At times he claimed his method was just a pragmatic device to avoid the method of exhaustion. Despite criticism of the method, it was intensively employed by many mathematicians. Others, such as Fermat, Pascal, and Roberval, used the method and even the language, sum of ordinates, but thought of area as a sum of infinitely small rectangles rather than as a sum of lines.

In 1634 Roberval, who says he studied the "divine Archimedes," used essentially the method of indivisibles to find the area under one arch of the cycloid, a problem Mersenne called to his attention in 1629. Roberval is sometimes credited with independent discovery of the method of indivisibles, but actually he believed in the infinite divisibility of lines, surfaces, and volumes, so that there are no ultimate parts. He called his method the "method of infinities," though he used as the title of his work *Traité des indivisibles*.

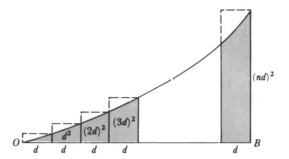

Figure 17.9

Roberval's method of obtaining the area under the cycloid is instructive. Let $OABP$ (Fig. 17.8) be the area under half of an arch of a cycloid. OC is the diameter of the generating circle and P is any point on the arch. Take $PQ = DF$. The locus of Q is called the companion curve to the cycloid. (The curve OQB is, in our notation, $y = a \sin x/a$ where a is the radius of the generating circle, provided the origin is at the midpoint of OQB and the x-axis is parallel to OA.) Roberval affirms that the curve OQB divides the rectangle $OABC$ into two equal parts because, basically, to each line DQ in $OQBC$ there corresponds an equal line RS in $OABQ$. Thus Cavalieri's principle is employed. The rectangle $OABC$ has its base and altitude equal, respectively, to the semicircumference and diameter of the generating circle; hence its area is twice that of the circle. Then $OABQ$ has the same area as the generating circle. Also, the area between OPB and OQB equals the area of the semicircle OFC, since by the very definition of Q, $DF = PQ$, so that these two areas are everywhere of the same width. Hence the area under the half-arch is $1\ 1/2$ times the area of the generating circle. Roberval also found the area under one arch of the sine curve, the volume generated by revolving the arch about its base, other volumes connected with the cycloid, and the centroid of its area.

The most important new method of calculating areas, volumes, and other quantities started with modifications of the Greek method of exhaustion. Let us consider a typical example. Suppose one seeks to calculate the area under the parabola $y = x^2$ from $x = O$ to $x = B$ (Fig. 17.9). Whereas the method of exhaustion used different kinds of rectilinear approximating figures, depending on the curvilinear area in question, some seventeenth-century men adopted a systematic procedure using rectangles as shown. As the width d of these rectangles becomes smaller, the sum of the areas of the rectangles approaches the area under the curve. This sum, if the bases are all d in width, and if one uses the characteristic property of the parabola that the ordinate is the square of the abscissa, is

(1) $$d \cdot d^2 + d(2d)^2 + d(3d)^2 + \cdots + d(nd)^2$$

or
$$d^3(1 + 2^2 + 3^2 + \cdots + n^2).$$

Now the sum of the mth powers of the first n natural numbers had been obtained by Pascal and Fermat for use in just such problems; so the mathematicians could readily replace the last expression by

(2)
$$d^3\left(\frac{2n^3 + 3n^2 + n}{6}\right).$$

But d is the fixed length OB divided by n. Hence (2) becomes

(3)
$$OB^3\left(\frac{1}{3} + \frac{1}{2n} + \frac{1}{6n^2}\right).$$

Now if one argues, as these men did, that the last two terms can be neglected when n is infinite, the correct result is obtained. The limit process had not yet been introduced—or was only crudely perceived—and so the neglect of terms such as the last two was not justified.

We see that the method calls for approximating the curvilinear figure by rectilinear ones, as in the method of exhaustion. However, there is a vital shift in the final step: in place of the indirect proof used in the older method, here the number of rectangles becomes infinite and one takes the limit of (3) as n becomes infinite—though the thinking in terms of limit was at this stage by no means explicit. This new approach, used as early as 1586 by Stevin in his *Statics*, was pursued by many men, including Fermat.[3]

If the curve involved was not the parabola, then one had to replace the characteristic property of the parabola by that of the curve in question and so obtain some other series in place of (1) above. Summing the analogue of (1) to obtain the analogue of (2) did call for ingenuity. Hence the results on areas, volumes, and centers of gravity were limited. Of course the powerful method of evaluating the limit of such sums by reversing differentiation was not yet envisaged.

Using essentially the kind of summation technique we have just illustrated, Fermat knew before 1636 that (in our notation)

$$\int_0^a x^n \, dx = \frac{a^{n+1}}{n+1}$$

for all rational n except -1.[4] This result was also obtained independently by Roberval, Torricelli, and Cavalieri, though in some cases only in geometrical form and for more limited n.

Among those who used summation in geometrical form was Pascal. In 1658 he took up problems of the cycloid.[5] He calculated the area of any

3. *Œuvres*, 1, 255–59; 3, 216–19.
4. *Œuvres*, 1, 255–59; 3, 216–19.
5. *Traité des sinus du quart de cercle*, 1659 = *Œuvres*, 9, 60–76.

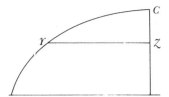

Figure 17.10

segment of the curve cut off by a line parallel to the base, the centroid of the segment, and the volumes of solids generated by such segments when revolved around their bases (YZ in Fig. 17.10) or a vertical line (the axis of symmetry). In this work, as well as in earlier work on areas under the curves of the family $y = x^n$, he summed small rectangles in the manner described in connection with (1) above, though his work and results were stated geometrically. Under the pseudonym of Dettonville, he proposed the problems he had solved as a challenge to other mathematicians, then published his own superior solutions (*Lettres de Dettonville*, 1659).

Before Newton and Leibniz, the man who did most to introduce analytical methods in the calculus was John Wallis (1616–1703). Though he did not begin to learn mathematics until he was about twenty—his university education at Cambridge was devoted to theology—he became professor of geometry at Oxford and the ablest British mathematician of the century, next to Newton. In his *Arithmetica Infinitorum* (1655), he applied analysis and the method of indivisibles to effect many quadratures and obtain broad and useful results.

One of Wallis's notable results, obtained in his efforts to calculate the area of the circle analytically, was a new expression for π. He calculated the area bounded by the axes, the ordinate at x, and the curve for the functions

$$y = (1 - x^2)^0, y = (1 - x^2)^1, y = (1 - x^2)^2, y = (1 - x^2)^3, \cdots$$

and obtained the areas

$$x,\ x - \frac{1}{3}x^3,\ x - \frac{2}{3}x^3 + \frac{1}{5}x^5,\ x - \frac{3}{3}x^3 + \frac{3}{5}x^5 - \frac{1}{7}x^7, \cdots$$

respectively. When $x = 1$, these areas are

(4)
$$1, \frac{2}{3}, \frac{8}{15}, \frac{48}{105}, \cdots.$$

Now the circle is given by $y = (1 - x^2)^{1/2}$. Using induction and interpolation, Wallis calculated its area, and by further complicated reasoning arrived at

$$\frac{\pi}{2} = \frac{2 \cdot 2 \cdot 4 \cdot 4 \cdot 6 \cdot 6 \cdot 8 \cdot 8 \cdots}{1 \cdot 3 \cdot 3 \cdot 5 \cdot 5 \cdot 7 \cdot 7 \cdot 9 \cdots}.$$

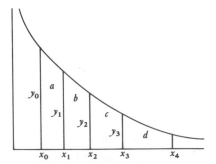

Figure 17.11

Gregory of St. Vincent in his *Opus Geometricum* (1647), gave the basis for the important connection between the rectangular hyperbola and the logarithm function. He showed, using the method of exhaustion, that if for the curve of $y = 1/x$ (Fig. 17.11) the x_i are chosen so that the areas a, b, c, d, ... are equal, then the y_i are in geometric progression. This means that the sum of the areas from x_0 to x_i, which sums form an arithmetical progression, is proportional to the logarithm of the y_i values or, in our notation,

$$\int_{x_0}^{x} \frac{dx}{x} = k \log y.$$

This agrees with our familiar calculus result, because $y = 1/x$. The observation that the areas can be interpreted as logarithms is actually due to Gregory's pupil, the Belgian Jesuit Alfons A. de Sarasa (1618–67), in his *Solutio Problematis a Mersenno Propositi* (1649). About 1665 Newton also noted the connection between the area under the hyperbola and logarithms and included this relation in his *Method of Fluxions*. He expanded $1/(1 + x)$ by the binomial theorem and integrated term by term to obtain

$$\log_e (1 + x) = x - \frac{x^2}{2} + \frac{x^3}{3} - \cdots.$$

Nicholas Mercator, using Gregory's results, gave the same series independently (though he did not state it explicitly) in his *Logarithmotechnia* of 1668. Other men soon found series which, as we would put it, converged more rapidly. The work on the quadrature of the hyperbola and its relation to the logarithm function was done by many men, and much of it was communicated in letters, so that it is hard to trace the order of discovery and to assign credit.

Up to about 1650 no one believed that the length of a curve could equal exactly the length of a line. In fact, in the second book of *La Géométrie*, Descartes says the relation between curved lines and straight lines is not nor ever can be known. But Roberval found the length of an arch of the cycloid. The architect Christopher Wren (1632–1723) rectified the cycloid by showing

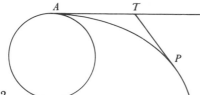

Figure 17.12

(Fig. 17.12) that arc $PA = 2PT$.[6] William Neile (1637–70) also obtained (1659) the length of an arch and, using a suggestion of Wallis, rectified the semicubical parabola ($y^3 = ax^2$).[7] Fermat, too, calculated some lengths of curves. These men usually used an inscribed polygon to approximate the curve, found the sum of the segments, then let the number of segments become infinite as each got smaller. James Gregory (1638–75), a professor at St. Andrews and Edinburgh (whose work was known slightly to his contemporaries but not known generally until a memorial volume, edited by H. W. Turnbull, appeared in 1939), gave in his *Geometriae Pars Universalis* (Universal Part of Geometry, 1668) a method of rectifying curves.

Further results on rectification were obtained by Christian Huygens (1629–95). In particular, he gave the length of arc of the cissoid. He also contributed to the work on areas and volumes and was the first to give results on the areas of surfaces beyond that of the sphere. Thus he obtained the areas of portions of the surfaces of the paraboloid and hyperboloid. Huygens obtained all these results by purely geometric methods, though he did use arithmetic, as Archimedes did occasionally, to obtain quantitative answers.

The rectification of the ellipse defied the mathematicians. In fact, James Gregory asserted that the rectification of the ellipse and the hyperbola could not be achieved in terms of known functions. For a while mathematicians were discouraged from further work on this problem and no new results were obtained until the next century.

We have been discussing the chief contributions of the predecessors of Newton and Leibniz to the four major problems that motivated the work on the calculus. The four problems were regarded as distinct; yet relationships among them had been noted and even utilized. For example, Fermat had used the very same method for finding tangents as for finding the maximum value of a function. Also, the problem of the rate of change of a function with respect to the independent variable and the tangent problem were readily seen to be the same. In fact, Fermat's and Barrow's method of finding tangents is merely the geometrical counterpart of finding the rate of change. But the major feature of the calculus, next to the very concepts of the

6. The method was published by Wallis in *Tractatus Duo* (1659 = *Opera*, 1, 550–69). Wren gave only the result.
7. Neile's work was published by Wallis in the reference in footnote 6.

derivative and of the integral as a limit of a sum, is the fact that the integral can be found by reversing the differentiation process or, as we say, by finding the antiderivative. Much evidence of this relationship had been encountered, but its significance was not appreciated. Torricelli saw in special cases that the rate problem was essentially the inverse of the area problem. It was, in fact, involved in Galileo's use of the fact that the area under a velocity-time graph gives distance. Since the rate of change of distance must be velocity, the rate of change of area, regarded as a "sum," must be the derivative of the area function. But Torricelli did not see the general point. Fermat, too, knew the relationship between area and derivative in special cases but did not appreciate its generality or importance. James Gregory, in his *Geometriae* of 1668, proved that the tangent and area problems are inverse problems but his book went unnoticed. In *Geometrical Lectures*, Barrow had the relationship between finding the tangent to a curve and the area problem, but it was in geometrical form, and he himself did not recognize its significance.

Actually an immense amount of knowledge of the calculus had accumulated before Newton and Leibniz made their impact. A survey of even the one book by Barrow shows a method of finding tangents, theorems on the differentiation of the product and quotient of two functions, the differentiation of powers of x, the rectification of curves, change of variable in a definite integral, and even the differentiation of implicit functions. Though in Barrow's case the geometric formulation made the discernment of the general ideas difficult, in Wallis's *Arithmetica Infinitorum* comparable results were in algebraic form.

One wonders then what remained to be achieved in the way of major new results. The answer is greater generality of method and the recognition of the generality of what had already been established in particular problems. The work on the calculus during the first two thirds of the century lost itself in details. Also, in their efforts to attain rigor through geometry, many men failed to utilize or explore the implications of the new algebra and coordinate geometry, and exhausted themselves in abortive subtle reasonings. What ultimately fostered the necessary insight and the attainment of generality was the arithmetical work of Fermat, Gregory of St. Vincent, and Wallis, the men whom Hobbes criticized for substituting symbols for geometry. James Gregory stated in the preface to *Geometriae* that the true division of mathematics was not into geometry and arithmetic but into the universal and the particular. The universal was supplied by the two all-embracing minds, Newton and Leibniz.

3. *The Work of Newton*

Great advances in mathematics and science are almost always built on the work of many men who contribute bit by bit over hundreds of years; eventually one man sharp enough to distinguish the valuable ideas of his predecessors from the welter of suggestions and pronouncements, imaginative enough to fit the bits into a new account, and audacious enough to build a master plan takes the culminating and definitive step. In the case of the calculus, this was Isaac Newton.

Newton (1642–1727) was born in the hamlet of Woolsthorpe, England, where his mother managed the farm left by her husband, who died two months before Isaac was born. He was educated at local schools of low educational standards and as a youth showed no special flair, except for an interest in mechanical devices. Having passed entrance examinations with a deficiency in Euclidean geometry, he entered Trinity College of Cambridge University in 1661 and studied quietly and unobstrusively. At one time he almost changed his course from natural philosophy (science) to law. Apparently receiving very little stimulation from his teachers, except possibly Barrow, he experimented by himself and studied Descartes's *Géométrie*, as well as the works of Copernicus, Kepler, Galileo, Wallis, and Barrow.

Just after Newton finished his undergraduate work the university was closed down because the plague was widespread in the London area. He left Cambridge and spent the years 1665 and 1666 in the quiet of the family home at Woolsthorpe. There he initiated his great work in mechanics, mathematics, and optics. At this time he realized that the inverse square law of gravitation, a concept advanced by others, including Kepler, as far back as 1612, was the key to an embracing science of mechanics; he obtained a general method for treating the problems of the calculus; and through experiments with light he made the epochal discovery that white light, such as sunlight, is really composed of all colors from violet to red. "All this," Newton said later in life, "was in the two plague years of 1665 and 1666, for in those days I was in the prime of my age for invention, and minded mathematics and philosophy [science] more than at any other time since."

Newton said nothing about these discoveries. He returned to Cambridge in 1667 to secure a master's degree and was elected a fellow of Trinity College. In 1669 Isaac Barrow resigned his professorship and Newton was appointed in Barrow's place as Lucasian professor of mathematics. Apparently he was not a successful teacher, for few students attended his lectures; nor was the originality of the material he presented noticed by his colleagues. Only Barrow and, somewhat later, the astronomer Edmond Halley (1656–1742) recognized his greatness and encouraged him.

At first Newton did not publish his discoveries. He is said to have had an abnormal fear of criticism; De Morgan says that "a morbid fear of opposition from others ruled his whole life." When in 1672 he did publish his work on light, accompanied by his philosophy of science, he was severely criticized by most of his contemporaries, including Robert Hooke and Huygens, who had different ideas on the nature of light. Newton was so taken aback that he decided not to publish in the future. However, in 1675 he did publish another paper on light, which contained his idea that light was a stream of particles—the corpuscular theory of light. Again he was met by a storm of criticism and even claims by others that they had already discovered these ideas. This time Newton resolved that his results would be published after his death. Nonetheless he did publish subsequent papers and several famous books, the *Principia*, the *Opticks* (English edition 1704, Latin edition 1706), and the *Arithmetica Universalis* (1707).

From 1665 on he applied the law of gravitation to planetary motion; in this area the works of Hooke and Huygens influenced him considerably. In 1684 his friend Halley urged him to publish his results, but aside from his reluctance to publish Newton lacked a proof that the gravitational attraction exerted by a solid sphere acts as though the sphere's mass were concentrated at the center. He says, in a letter to Halley of June 20, 1686, that until 1685 he suspected that it was false. In that year he showed that a sphere whose density varies only with distance to the center does in fact attract an external particle as though the sphere's mass were concentrated at its center, and agreed to write up his work.

Halley then assisted Newton editorially and paid for the publication. In 1687 the first edition of the *Philosophiae Naturalis Principia Mathematica* (*The Mathematical Principles of Natural Philosophy*) appeared. There were two subsequent editions, in 1713 and 1726, the second edition containing improvements. Though the book brought Newton great fame, it was very difficult to understand. He told a friend that he had purposely made it difficult "to avoid being bated by little smatterers in mathematics." He no doubt hoped in this way to avoid the criticism that his earlier papers on light had received.

Newton was also a major chemist. Though there are no great discoveries associated with his work in this area, one must take into account that chemistry was then in its infancy. He had the correct idea of trying to explain chemical phenomena in terms of ultimate particles, and he had a profound knowledge of experimental chemistry. In this subject he wrote one major paper, "De natura acidorum" (written in 1692 and published in 1710). In the Philosophical Transactions of the Royal Society of 1701, he published a paper on heat that contains his famous law on cooling. Though he read the works of alchemists, he did not accept their cloudy and mystical views. The chemical and physical properties of bodies could, he believed, be accounted for in terms of the size, shape, and motion of the ultimate particles; he rejected the alchemists' occult forces, such as sympathy, antipathy, congruity, and attraction.

In addition to his work on celestial mechanics, light, and chemistry, Newton worked in hydrostatics and hydrodynamics. Beyond his superb experimental work on light, he experimented on the damping of pendulum motion by various media, the fall of spheres in air and water, and the flow of water from jets. Like most men of the time Newton constructed his own equipment. He built two reflecting telescopes, even making the alloy for the frames, molding the frames, making the mountings, and polishing the lenses.

After serving as a professor for thirty-five years Newton became depressed and suffered a nervous breakdown. He decided to give up research and in 1695 accepted an appointment as warden of the British Mint in London. During his twenty-seven years at the mint, except for work on an occasional problem, he did no research. He became president of the Royal Society in 1703, an office he held until his death; he was knighted in 1705.

It is evident that Newton was far more engrossed in science than in mathematics and was an active participant in the problems of his time. He considered the chief value of his scientific work to be its support of revealed religion and was, in fact, a learned theologian, though he never took orders. He thought scientific research hard and dreary but stuck to it because it gave evidence of God's handiwork. Like his predecessor Barrow, Newton turned to religious studies later in life. In *The Chronology of Ancient Kingdoms Amended,* he tried to date accurately events described in the Bible and other religious documents by relating them to astronomical events. His major religious work was the *Observations Upon the Prophecies of Daniel and the Apocalypse of St. John.* Biblical exegesis was a phase of the rational approach to religion that was popular in the Age of Reason; Leibniz, too, took a hand in it.

So far as the calculus is concerned, Newton generalized the ideas already advanced by many men, established full-fledged methods, and showed the interrelationships of several of the major problems described above. Though he learned much as a student of Barrow, in algebra and the calculus he was more influenced by the works of Wallis. He said that he was led to his discoveries in analysis by the *Arithmetica Infinitorum;* certainly in his own work on the calculus he made progress by thinking analytically. However, even Newton thought the geometry was necessary for a rigorous proof.

In 1669 Newton circulated among his friends a monograph entitled *De Analysi per Aequationes Numero Terminorum Infinitas* (On Analysis by Means of Equations with an Infinite Number of Terms); it was not published until 1711. He supposes that he has a curve and that the area z (Fig. 17.13) under this curve is given by

$$(5) \qquad\qquad z = ax^m,$$

where m is integral or fractional. He calls an infinitesimal increase in x, the moment of x, and denotes it by o, a notation used by James Gregory and the equivalent of Fermat's E. The area bounded by the curve, the x-axis, the y-axis, and the ordinate at $x + o$ he denotes by $z + oy$, oy being the moment of area. Then

$$(6) \qquad\qquad z + oy = a(x + o)^m.$$

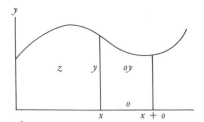

Figure 17.13

He applies the binomial theorem to the right side, obtaining an infinite series when m is fractional, subtracts (5) from (6), divides through by o, neglects those terms that still contain o, and obtains

$$y = max^{m-1}.$$

Thus, in our language, the rate of change of area at any x is the y-value of the curve at that value of x. Conversely, if the curve is $y = max^{m-1}$, the area under it is $z = ax^m$.

In this process Newton not only gave a general method for finding the instantaneous rate of change of one variable with respect to another (z with respect to x in the above example), but showed that area can be obtained by reversing the process of finding a rate of change. Since areas had also been expressed and obtained by the summation of infinitesimal areas, Newton also showed that such sums can be obtained by reversing the process of finding a rate of change. This fact, that summations (more properly, limits of sums) can be obtained by reversing differentiation, is what we now call the fundamental theorem of the calculus. Though it was known in special cases and dimly foreseen by Newton's predecessors, he saw it as general. He applied the method to obtain the area under many curves and to solve other problems that can be formulated as summations.

After showing that the derivative of the area is the y-value and asserting that the converse is true, Newton gave the rule that, if the y-value be a sum of terms, then the area is the sum of the areas that result from each of the terms. In modern terms, the indefinite integral of a sum of functions is the sum of the integrals of the separate functions.

His next contribution in the monograph carried further his use of infinite series. To integrate $y = a^2/(b + x)$, he divided a^2 by $b + x$ and obtained

$$y = \frac{a^2}{b} - \frac{a^2 x}{b^2} + \frac{a^2 x^2}{b^3} - \frac{a^2 x^3}{b^4} + \cdots.$$

Having obtained this infinite series, he finds the integral by integrating term by term so that the area is

$$\frac{a^2 x}{b} - \frac{a^2 x^2}{2b^2} + \frac{a^2 x^3}{3b^3} - \frac{a^2 x^4}{4b^4} + \cdots.$$

He says of this infinite series that a few of the initial terms are exact enough for any use, provided that b be equal to x repeated some few times.

Likewise, to integrate $y = 1/(1 + x^2)$ he uses the binomial expansion to write

$$y = 1 - x^2 + x^4 - x^6 + x^8 - \cdots$$

and integrates term by term. He notes that if, instead, y is taken to be $1/(x^2 + 1)$, then by binomial expansion one would obtain

$$y = x^{-2} - x^{-4} + x^{-6} - x^{-8} + \cdots$$

and now one can integrate term by term. He then remarks that when x is small enough the first expansion is to be used; but when x is large, the second is to be used. Thus he was somewhat aware that what we call convergence is important, but had no precise notion about it.

Newton realized that he had extended term by term integration to infinite series but says in the *De Analysi*:

> And whatever the common Analysis performs by Means of Equations of a finite Number of Terms (provided that can be done) this can always perform the same by Means of infinite Equations so that I have not made any question of giving this the name of Analysis likewise. For the reasonings in this are no less certain than in the other; nor the equations less exact; albeit we Mortals whose reasoning powers are confined within narrow limits, can neither express, nor so conceive all the Terms of these Equations, as to know exactly from thence the quantities we want.

Thus far in his approach to the calculus Newton used what may be described as the method of infinitesimals. Moments are infinitely small quantities, indivisibles or infinitesimals. Of course the logic of what Newton did is not clear. He says in this work that his method is "shortly explained rather than accurately demonstrated."

Newton gave a second, more extensive, and more definitive exposition of his ideas in the book *Methodus Fluxionum et Serierum Infinitarum*, written in 1671 but not published until 1736. In this work he says he regards his variables as generated by the continuous motion of points, lines, and planes, rather than as static aggregates of infinitesimal elements, as in the earlier paper. A variable quantity he now called a fluent and its rate of change, the fluxion. His notation is $\dot{x}$ and $\dot{y}$ for fluxions of the fluents x and y. The fluxion of $\dot{x}$ is $\ddot{x}$, etc. The fluent of which x is the fluxion is $\dot{x}$, and the fluent of the latter is $\ddot{x}$.

In this second work Newton states somewhat more clearly the fundamental problem of the calculus: Given a relation between two fluents, find the relation between their fluxions, and conversely. The two variables whose relation is given can represent any quantities. However, Newton thinks of them as changing with time because it is a useful way of thinking, though, he points out, not necessary. Hence if o is an "infinitely small interval of time," then $\dot{x}o$ and $\dot{y}o$ are the indefinitely small increments in x and y or the moments of x and y. To find the relation between $\dot{y}$ and $\dot{x}$, suppose, for example, the fluent is $y = x^n$. Newton first forms

$$y + \dot{y}o = (x + \dot{x}o)^n,$$

and then proceeds as in the earlier paper. He expands the right side by using the binomial theorem, subtracts $y = x^n$, divides through by o, neglects all terms still containing o, and obtains

$$\dot{y} = nx^{n-1}\dot{x}.$$

In modern notation this result can be written

$$\frac{dy}{dt} = nx^{n-1}\frac{dx}{dt},$$

and since $dy/dx = (dy/dt)/(dx/dt)$, Newton, in finding the ratio of dy/dt to dx/dt or $\dot{y}$ to $\dot{x}$, has found dy/dx.

The method of fluxions is not essentially different from the one used in the *De Analysi*, nor is the rigor any better; Newton drops terms such as $\dot{x}\dot{x}o$ and $\dot{x}\dot{x}o\dot{x}o$ (he writes $\dot{x}^3oo$) on the ground that they are infinitely small compared to the one retained. However, his point of view in the *Method of Fluxions* is somewhat different. The moments $\dot{x}o$ and $\dot{y}o$ change with time o, whereas in the first paper the moments are ultimate fixed bits of x and z. This newer view follows the more dynamic thinking of Galileo; the older used the static indivisible of Cavalieri. The change served, as Newton put it, only to remove the harshness from the doctrine of indivisibles; however, the moments $\dot{x}o$ and $\dot{y}o$ are still some sort of infinitely small quantities. Moreover, $\dot{x}$ and $\dot{y}$, which are the fluxions or derivatives with respect to time of x and y, are never really defined; this central problem is evaded.

Given a relation between $\dot{x}$ and $\dot{y}$, finding the relation between x and y is more difficult than merely integrating a function of x. Newton treats several types: (1) when $\dot{x}$, $\dot{y}$, and x or y are present; (2) when $\dot{x}$, $\dot{y}$, x, and y are present; (3) when $\dot{x}$, $\dot{y}$, $\dot{z}$, and the fluents are present. The first type is the easiest and, in modern notation, calls for solving $dy/dx = f(x)$. Of the second type, Newton treats $\dot{y}/\dot{x} = 1 - 3x + y + x^2 + xy$ and solves it by a successive approximation process. He starts with $\dot{y}/\dot{x} = 1 - 3x + x^2$ as a first approximation, obtains y as a function of x, introduces this value of y on the right side of the original equation, and continues the process. Newton describes what he does but does not justify it. Of the third type, he treats $2\dot{x} - \dot{z} + \dot{y}x = 0$. He assumes a relation between x and y, say $x = y^2$, so that $\dot{x} = 2\dot{y}y$. Then the equation becomes $4\dot{y}y - \dot{z} - \dot{y}y^2 = 0$, from which he gets $2y^2 + (y^3/3) = z$. Thus, if the third type is regarded as a partial differential equation, Newton obtains only a particular integral.

Newton realized that in this paper he had presented a general method. In a letter to John Collins, dated December 10, 1672, wherein he gives the facts of his method and one example, he says,

> This is one particular, or rather corollary, of a general method, which extends itself, without any troublesome calculations, not only to the drawing of tangents to any curved lines, whether geometrical or mechanical... but also to resolving other abstruser kinds of problems about the crookedness, areas, lengths, centres of gravity of curves, etc.; nor is it... limited to equations which are free from surd quantities. This method I have interwoven with that other of working in equations, by reducing them to infinite series.

Newton emphasized the use of infinite series because thereby he could treat functions such as $(1 + x)^{3/2}$, whereas his predecessors had been limited on the whole to rational algebraic functions.

In his *Tractatus de Quadratura Curvarum* (Quadrature of Curves), a third paper on the calculus, written in 1676 but published in 1704, Newton says he has abandoned the infinitesimal or infinitely small quantity. He now criticizes the dropping of terms involving o for, he says,

> in mathematics the minutest errors are not to be neglected.... I consider mathematical quantities in this place not as consisting of very small parts, but as described by a continual motion. Lines are described, and thereby generated, not by the apposition of parts, but by the continued motion of points; superficies by the motion of lines; solids by the motions of superficies; angles by the rotation of the sides; portions of time by continued flux....
>
> Fluxions are, as near as we please, as the increments of fluents generated in times, equal and as small as possible, and to speak accurately, they are in the prime ratio of nascent increments; yet they can be expressed by any lines whatever, which are proportional to them.

Newton's new concept, the method of prime and ultimate ratio, amounts to this. He considers the function $y = x^n$. To find the fluxion of y or x^n, let x "by flowing" become $x + o$. Then x^n becomes

$$(x + o)^n = x^n + nox^{n-1} + \frac{n^2 - n}{2} o^2 x^{n-2} + \cdots.$$

The increases of x and y, namely, o and $nox^{n-1} + \dfrac{n^2 - n}{2} o^2 x^{n-2} + \cdots$ are to each other as (dividing both by o)

$$1 \text{ to } nx^{n-1} + \frac{n^2 - n}{2} ox^{n-2} + \cdots.$$

"Let now the increments vanish and their last proportion will be"

$$1 \text{ to } nx^{n-1}.$$

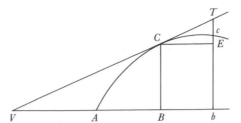

Figure 17.14

Then the fluxion of x is to the fluxion of x^n as 1 to nx^{n-1} or, as we would say today, the rate of change of y with respect to x is nx^{n-1}. This is the prime ratio of the nascent increments. Of course the logic of this version is no better than that of the preceding two; nevertheless Newton says this method is in harmony with the geometry of the ancients and that it is not necessary to introduce infinitely small quantities.

Newton also gave a geometrical interpretation. Given the data in Figure 17.14, suppose bc moves to BC so that c coincides with C. Then the curvilinear triangle CEc is "in the last form" similar to triangle CET, and its "evanescent" sides will be proportional to CE, ET, and CT. Hence the fluxions of the quantities AB, BC, and AC are, in the last ratio of their evanescent increments, proportional to the sides of the triangle CET or triangle VBC.

In the *Method of Fluxions* Newton made a number of applications of fluxions to differentiating implicit functions and to finding tangents of curves, maxima and minima of functions, curvature of curves, and points of inflection of curves. He also obtained areas and lengths of curves. In connection with curvature, he gave the correct formula for the radius of curvature, namely,

$$r = \frac{(1 + \dot{y}^2)^{3/2}}{\ddot{y}}$$

where $\dot{x}$ is taken as 1. He also gave this same quantity in polar coordinates. Finally, he included a brief table of integrals.

Newton did not publish his basic papers in the calculus until long after he had written them. The earliest printed account of his theory of fluxions appeared in Wallis's *Algebra* (2nd ed. in Latin, 1693), of which Newton wrote pages 390 to 396. Had he published at once he might have avoided the controversy with Leibniz on the priority of discovery.

Newton's first publication involving his calculus is the great *Mathematical Principles of Natural Philosophy*.[8] So far as the basic notion of the

8. The third edition was translated into English by Andrew Motte in 1729. This edition, revised and edited by Florian Cajori, was published by the University of California Press in 1946.

calculus, the fluxion, or, as we say, the derivative, is concerned, Newton makes several statements. He rejects infinitesimals or ultimate indivisible quantities in favor of "evanescent divisible quantities," quantities which can be diminished without end. In the first and third editions of the *Principia* Newton says, "Ultimate ratios in which quantities vanish are not, strictly speaking, ratios of ultimate quantities, but limits to which the ratios of these quantities, decreasing without limit, approach, and which, though they can come nearer than any given difference whatever, they can neither pass over nor attain before the quantities have diminished indefinitely."[9] This is the clearest statement he ever gave as to the meaning of his ultimate ratio. Apropos of the preceding quotation, he also says, "By the ultimate velocity is meant that with which the body is moved, neither before it arrives at its last place, when the motion ceases, nor after; but at the very instant when it arrives. . . . And, in like manner, by the ultimate ratio of evanescent quantities is to be understood the ratio of quantities, not before they vanish, nor after, but that with which they vanish."

In the *Principia* Newton used geometrical methods of proof. However, in what are called the Portsmouth Papers, containing unpublished work, he used analytical methods to find some of the theorems. These papers show that he also obtained analytically results beyond those he was able to translate into geometry. One reason he resorted to geometry is believed to be that the proofs would be more understandable to his contemporaries. Another is that he admired Huygens's geometrical work immensely and hoped to equal it. In these geometrical proofs Newton uses the basic limit processes of the calculus. Thus the area under a curve is considered essentially as the limit of the sum of the approximating rectangles, just as in the calculus today. However, instead of calculating such areas, he uses this concept to compare areas under different curves.

He proves that, when AR and BR (Fig. 17.15) are the perpendiculars to the tangents at A and B of the arc ACB, the ultimate ratio, when B approaches and coincides with A, of any two of the quantities chord AB, arc ACB, and AD, is 1. Hence he says in Corollary 3 to Lemma 2 of Book I, "And therefore in all our reasoning about ultimate ratios, we may freely use any one of these lines for any other." He then proves that when B approaches and coincides with A, the ratio of any two triangles (areas) RAB, $RACB$, and RAD will be 1. "And hence in all reasonings about ultimate ratios, we may use any one of these triangles for any other." Also, (Fig. 17.16) let BD and CE be perpendicular to AE (which is not necessarily tangent to arc ABC at A). When B and C approach and coincide with A, the ultimate ratio of the areas ACE and ABD will equal the ultimate ratio of AE^2 to AD^2.

The *Principia* contains a wealth of results, some of which we shall note.

9. Third edition, p. 39.

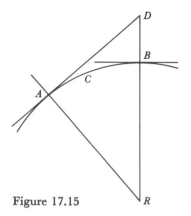

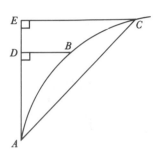

Figure 17.15   Figure 17.16

Though the book is devoted to celestial mechanics, it has enormous importance for the history of mathematics, not only because Newton's own work on the calculus was motivated in large part by his overriding interest in the problems treated therein, but because the *Principia* presented new topics and approaches to problems that were explored during the next hundred years in the course of which an enormous amount of analysis was created.

The *Principia* is divided into three books.[10] In a prefatory section Newton defines concepts of mechanics such as inertia, momentum, and force, then states the three famous axioms or laws of motion. In his words, they are:

Law I. Every body continues in its state of rest, or of uniform motion in a right line, unless it is compelled to change that state by forces impressed upon it.

Law II. The change [in the quantity] of motion is proportional to the motive power impressed; and is made in the direction of the right line in which that force is impressed.

By quantity of motion Newton means, as he has explained earlier, the mass times the velocity. Hence the change in motion, if the mass is constant, is the change in velocity, that is, the acceleration. This second law is now often written as $F = ma$, when the force F is in poundals, the mass m is in pounds, and the acceleration a is in feet per second per second. Newton's second law is really a vector statement; that is, if the force has components in, say, three mutually perpendicular directions, then each component causes an acceleration in its own direction. Newton did use the vector character of force in particular problems, but the full significance of the vector nature of

10. All references are to the edition mentioned in note 8.

the law was first fully recognized by Euler. This law incorporates the key change from the mechanics of Aristotle, which affirmed that force causes velocity. Aristotle had also affirmed that a force is needed to maintain velocity. Law I denies this.

Law III. To every action there is always opposed an equal reaction. . . .

We shall not digress into the history of mechanics except to note that the first two laws are more explicit and somewhat generalized statements of the principles of motion previously discovered and advanced by Galileo and Descartes. The distinction between mass, that is, the resistance a body offers to a change in its motion, and weight, the force gravity exerts on the mass of any object, is also due to these men; and the vector character of force generalizes Galileo's principle that the vertical and horizontal motions of a projectile, for example, can be treated independently.

Book I of the *Principia* begins with some theorems of the calculus, including the ones involving ultimate ratios cited above. It then discusses motion under central forces, that is, forces that always attract the moving object to one (fixed) point (the sun in practice), and proves in Proposition 1 that equal areas are swept out in equal time (which encompasses Kepler's law of areas). Newton considers next the motion of a body along a conic section and proves (Props. 11, 12, and 13) that the force must vary with the inverse square of the distance from some fixed point. He also proves the converse, which contains Kepler's first law. After some treatment of centripetal force, he deduces Kepler's third law (Prop. 15). There follow two sections devoted to properties of the conic sections. The principal problem is the construction of conics that satisfy five given conditions; in practice these are usually observational data. Then, given the time an object has been in motion along a conic section, he determines its velocity and position. He takes up the motion of the apse lines, that is, the lines joining the center of attraction (at one focus) to the maximum or minimum distance of a body moving along a conic that is itself rotating at some rate about the focus. Section 10 is devoted to the motion of bodies along surfaces with special reference to pendulum motion. Here Newton gives due acknowledgment to Huygens. In connection with the accelerating effect of gravity on motions, he investigates geometrical properties of cycloids, epicycloids, and hypocycloids and gives the length of the epicycloid (Prop. 49).

In Section 11 Newton deduces from the laws of motion and the law of gravitation the motion of two bodies, each attracting the other in accordance with the gravitational force. Their motion is reduced to the motion of one around the fixed second body. The moving body traverses an ellipse.

He then considers the attraction exerted by spheres and spheroids of uniform and varying density on a particle. He gives (Sec. 12, Prop. 70) a geometrical proof that a thin homogeneous spherical shell exerts no force on a particle in its interior. Since this result holds for a thin shell, it holds for a sum of such shells, that is, for a shell of finite thickness. (He proves later [Prop. 91, Cor. 3] that the same result holds for a homogeneous ellipsoidal shell, that is, a shell contained between two similar ellipsoidal surfaces, similarly placed.) Proposition 71 shows that the attraction of a thin homogeneous spherical shell on an *external* particle is equivalent to the attraction that would be exerted if the mass of the shell were concentrated at the center, so that the shell attracts the external particle toward the center and with a force varying inversely as the square of the distance from the center. Proposition 73 shows that a solid homogeneous sphere attracts a particle inside with a force proportional to the particle's distance from the center. As for the attraction that a solid homogeneous sphere exerts on an external point, Proposition 74 shows that it is the same as if the mass of the sphere were concentrated at its center. Then if two spheres attract each other, the first attracts every particle of the second as if the mass of the first were concentrated at its center. Thus the first sphere becomes a particle attracted by the distributed mass of the second; hence the second sphere can also be treated as a particle with its mass concentrated at its center. Thus both spheres can be treated as particles with their masses concentrated at their respective centers. All these results, original with Newton, are extended to spheres whose densities are spherically symmetric and to other laws of attraction in addition to the inverse square law.

Newton next takes up the motion of three bodies, each attracting the other two, and obtains some approximate results. The problem of the motion of three bodies has been a major one since Newton's time and has not as yet been solved exactly.

The second book of the *Principia* is devoted to the motion of bodies in resisting media such as air and liquids. It is the beginning of the subject of hydrodynamics. Newton assumes in some problems that the resistance of the medium is proportional to the velocity and in others to the square of the velocity of the moving body. He considers what shape a body must have to encounter least resistance (see Chap. 24, sec. 1). He also considers the motion of pendulums and projectiles in air and in fluids. A section is devoted to the theory of waves in air (e.g., sound waves) and he obtains a formula for the velocity of sound in air. He also treats the motion of waves in water. Newton continues with a description of experiments he made to determine the resistance fluids offer to bodies moving in them. One major conclusion is that the planets move in a vacuum. In this book Newton broke entirely new ground; however, the definitive work on fluid motion was yet to be done

Book III, entitled *On the System of the World*, contains the application of the general theory developed in Book I to the solar system. It shows how the sun's mass can be calculated in terms of the earth's mass, and that the mass of any planet having a satellite can be found in the same way. He calculates the average density of the earth and finds it to be between 5 and 6 times that of water (today's figure is about 5.5).

He shows that the earth is not a true sphere but an oblate spheroid and calculates the flattening; his result is that the ellipticity of the oblate spheroid is 1/230 (the figure today is 1/297). From the observed oblateness of any planet, the length of its day is then calculated. Using the amount of flattening and the notion of centripetal force, Newton computes the variation of the earth's gravitational attraction over the surface and thus the variation in the weight of an object. He proves that the attractive force of a spheroid is not the same as if the spheroid's mass were concentrated at its center.

He then accounts for the precession of the equinoxes. The explanation is based on the fact that the earth is not spherical but bulges out along the equator. Consequently the gravitational attraction of the moon on the earth does not effectively act on the center of the earth but forces a periodic change in the direction of the earth's axis of rotation. The period of this change was calculated by Newton and found to be 26,000 years, the value obtained by Hipparchus by inference from observations available to him.

Newton explained the main features of the tides (Book I, Prop. 66, Book III, Props. 36, 37). The moon is the main cause; the sun, the second. Using the sun's mass he calculated the height of the solar tides. From the observed heights of the spring and neap tides (sun and moon in full conjunction or full opposition) he determined the lunar tide and made an estimate of the mass of the moon. Newton also managed to give some approximate treatment of the effect of the sun on the moon's motion around the earth. He determined the motion of the moon in latitude and longitude; the motion of the apse line (the line from the center of the earth to the maximum distance of the moon); the motion of the nodes (the points in which the moon's path cuts the plane of the earth's orbit; these points regress, that is, move slowly in a direction opposite to the motion of the moon itself); the evection (a periodic change in the eccentricity of the moon's orbit); the annual equation (the effect on the moon's motion of the daily change in distance between the earth and the sun); and the periodic change in the inclination of the plane of the moon's orbit to the plane of the earth's orbit. There were seven known irregularities in the motion of the moon and Newton discovered two more, the inequalities of the apogee (apse line) and of the nodes. His approximation gave only half of the motion of the apse line. Clairaut in 1752 improved the calculation and obtained the full 3° of rotation of the apse line; however, much later John Couch Adams found the correct calculation in Newton's papers. Finally Newton showed that the comets must be moving under the gravitational attraction of the sun because their paths, determined on the basis of observations, are conic sections. Newton devoted a great deal of time to the problem of the moon's motion because, as we noted in the preceding chapter, the knowledge was needed to improve the method of determining longitude. He worked so hard on this problem that he complained it made his head ache.

4. The Work of Leibniz

Though his contributions were quite different, the man who ranks with Newton in building the calculus is Gottfried Wilhelm Leibniz (1646–1716). He studied law and, after defending a thesis on logic, received a Bachelor of Philosophy degree. In 1666, he wrote the thesis *De Arte Combinatoria* (On the Art of Combinations),[11] a work on a universal method of reasoning; this completed his work for a doctorate in philosophy at the University of Altdorf and qualified him for a professorship. During the years 1670 and 1671 Leibniz wrote his first papers on mechanics, and, by 1671, had produced his calculating machine. He secured a job as an ambassador for the Elector of Mainz and in March of 1672 went to Paris on a political mission. This visit brought him into contact with mathematicians and scientists, notably Huygens, and stirred up his interest in mathematics. Though he had done a little reading in the subject and had written the paper of 1666, he says he knew almost no mathematics up to 1672. In 1673 he went to London and met other scientists and mathematicians, including Henry Oldenburg, at that time secretary of the Royal Society of London. While making his living as a diplomat, he delved further into mathematics and read Descartes and Pascal. In 1676 Leibniz was appointed librarian and councillor to the Elector of Hanover. Twenty-four years later the Elector of Brandenburg invited Leibniz to work for him in Berlin. While involved in all sorts of political maneuvers, including the succession of George Ludwig of Hanover to the English throne, Leibniz worked in many fields and his side activities covered an enormous range. He died neglected in 1716.

In addition to being a diplomat, Leibniz was a philosopher, lawyer, historian, philologist, and pioneer geologist. He did important work in logic, mechanics, optics, mathematics, hydrostatics, pneumatics, nautical science, and calculating machines. Though his profession was jurisprudence, his work in mathematics and philosophy is among the best the world has produced. He kept contact by letter with people as far away as China and Ceylon. He tried endlessly to reconcile the Catholic and Protestant faiths. It was he who proposed, in 1669, that a German Academy of Science be founded; finally the Berlin Academy was organized in 1700. His original recommendation had been for a society to make inventions in mechanics and discoveries in chemistry and physiology that would be useful to mankind; Leibniz wanted knowledge to be applied. He called the universities "monkish" and charged that they possessed learning but no judgment and were absorbed in trifles.

11. Published 1690 = *Die philosophische Schriften*, 4, 27–102.

Instead he urged the pursuit of real knowledge—mathematics, physics, geography, chemistry, anatomy, botany, zoology, and history. To Leibniz the skills of the artisan and the practical man were more valuable than the learned subtleties of the professional scholars. He favored the German language over Latin because Latin was allied to the older, useless thought. Men mask their ignorance, he said, by using the Latin language to impress people. German, on the other hand, was understood by the common people and could be developed to help clarity of thought and acuteness of reasoning.

Leibniz published papers on the calculus from 1684 on, and we shall say more about them later. However, many of his results, as well as the development of his ideas, are contained in hundreds of pages of notes made from 1673 on but never published by him. These notes, as one might expect, jump from one topic to another and contain changing notation as Leibniz's thinking developed. Some are simply ideas that occurred to him while reading books or articles by Gregory of St. Vincent, Fermat, Pascal, Descartes, and Barrow or trying to cast their thoughts into his own way of approaching the calculus. In 1714 Leibniz wrote *Historia et Origo Calculi Differentialis*, in which he gives an account of the development of his own thinking. However, this was written many years after he had done his work and, in view of the weaknesses of human memory and the greater insight he had acquired by that time, his history may not be accurate. Since his purpose was to defend himself against an accusation of plagiarism, he might have distorted unconsciously his account of the origins of his ideas.

Despite the confused state of Leibniz's notes we shall examine a few, because they reveal how one of the greatest intellects struggled to understand and create. By 1673 he was aware of the important direct and inverse problem of finding tangents to curves; he was also quite sure that the inverse method was equivalent to finding areas and volumes by summations. The somewhat systematic development of his ideas begins with notes of 1675. However, it seems helpful, in order to understand his thinking, to note that in his *De Arte Combinatoria* he had considered sequences of numbers, first differences, second differences, and higher-order differences. Thus for the sequence of squares

$$0, 1, 4, 9, 16, 25, 36,$$

the first differences are

$$1, 3, 5, 7, 9, 11$$

and the second differences are

$$2, 2, 2, 2, 2, 2.$$

Leibniz noted the vanishing of the second differences for the sequence of natural numbers, the third differences for the sequence of squares, and so on.

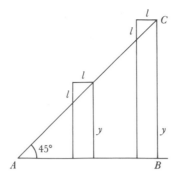

Figure 17.17

He also observed, of course, that if the original sequence starts from 0, the sum of the first differences is the last term of the sequence.

To relate these facts to the calculus he had to think of the sequence of numbers as the y-values of a function and the difference of any two as the difference of two nearby y-values. Initially he thought of x as representing the order of the term in the sequence and y as representing the value of that term.

The quantity dx, which he often writes as a, is then 1 because it is the difference of the orders of two successive terms, and dy is the actual difference in the values of two successive terms. Then using omn. as an abbreviation for the Latin *omnia*, to mean sum, and using l for dy, Leibniz concludes that omn. $l = y$, because omn. l is the sum of the first differences of a sequence whose terms begin with 0 and so gives the last term. However, omn. yl presents a new problem. Leibniz obtains the result that omn. yl is $y^2/2$ by thinking in terms of the function $y = x$. Thus, as Figure 17.17 shows, the area of triangle ABC is the sum of the yl (for "small" l) and it is also $y^2/2$. Leibniz says, "Straight lines which increase from nothing each multiplied by its corresponding element of increase form a triangle." These few facts already appear, among more complicated ones, in papers of 1673.

In the next stage he struggled with several difficulties. He had to make the transition from a discrete series of values to the case where dy and dx are increments of an arbitrary function y of x. Since he was still tied to sequences, wherein x is the order of the term, his a or dx was 1; so he inserted and omitted a freely. When he made the transition to the dy and dx of any function, this a was no longer 1. However, while still struggling with the notion of summation he ignored this fact.

Thus in a manuscript of October 29, 1675, Leibniz starts with

(7)
$$\text{omn.}\, yl = \overline{\text{omn.}\overline{\text{omn.}\, l\frac{l}{a}}},$$

which holds because y itself is omn. l. Here he divides l by a to preserve dimensions. Leibniz says that (7) holds, whatever l may be. But, as we saw in connection with Figure 17.17,

(8) $$\text{omn. } yl = \frac{y^2}{2}.$$

Hence from (7) and (8)

(9) $$\frac{y^2}{2} = \text{omn.}\overline{\text{omn. } l}\frac{l}{a}.$$

In our notation, he has shown that

$$\frac{y^2}{2} = \int \left\{ \int dy \right\} \frac{dy}{dx} = \int y \frac{dy}{dx}.$$

Leibniz says that this result is admirable.

Another theorem of the same kind, which Leibniz derived from a geometrical argument, is

(10) $$\text{omn. } xl = x \text{ omn. } l - \text{omn.omn. } l,$$

where l is the difference in values of two successive terms of a sequence and x is the number of the term. For us this equation is

$$\int x \, dy = xy - \int y \, dx.$$

Now Leibniz lets l itself in (10) be x, and obtains

$$\text{omn. } x^2 = x \text{ omn. } x - \text{omn.omn. } x.$$

But omn. x, he says, is $x^2/2$ (he has shown that omn. yl is $y^2/2$). Hence

$$\text{omn. } x^2 = x \frac{x^2}{2} - \text{omn. } \frac{x^2}{2}.$$

By transposing the last term he gets

$$\text{omn. } x^2 = \frac{x^3}{3}.$$

In this manuscript of October 29, 1675, Leibniz decided to write $\int$ for omn., so that

$$\int l = \text{omn. } l \quad \text{and} \quad \int x = \frac{x^2}{2}.$$

The symbol $\int$ is an elongated S for "sum."

Leibniz realized rather early, probably from studying the work of Barrow, that differentiation and integration as a summation must be inverse processes; so area, when differentiated, must give a length. Thus, in the same manuscript of October 29, Leibniz says, "Given l and its relation to x, to find $\int l$." Then, he says, "Suppose that $\int l = ya$. Let $l = ya/d$. [Here he puts d in the denominator. It would mean more to us if he wrote $l = d(ya)$.] Then just as $\int$ will increase, so d will diminish the dimensions. But $\int$ means a sum, and d, a difference. From the given y we can always find y/d or l, that is, the difference of the y's. Hence one equation may be transformed into the other; just as from the equation

$$\overline{\int c \int \overline{l^2}} = \frac{c \int \overline{l^3}}{3a^3},$$

we can obtain the equation

$$c \int \overline{l^2} = \frac{c \int \overline{l^3}}{3a^3 d}."$$

In this early paper Leibniz seems to be exploring the *operations* of $\int$ and d and sees that they are inverses. He finally realizes that $\int$ does not raise dimension nor d lower it, because $\int$ is really a summation of rectangles, and so a sum of areas. Thus he recognizes that, to get back to dy from y, he must form the difference of y's or take the differential of y. Then he says, "But $\int$ means a sum and d a difference." This may have been a later insertion. Hence a couple of weeks afterwards, in order to get from y to dy, he changes from dividing by d to taking the differential of y, and writes dy.

Up to this point Leibniz had been thinking of the y-values as values of terms of a sequence and of x usually as the order of these terms, but now, in this paper, says, "All these theorems are true for series in which the differences of the terms bear to the terms themselves a ratio that is less than any assignable quantity." That is, dy/y may be less than any assignable quantity.

In a manuscript dated November 11, 1675, entitled "Examples of the inverse method of tangents," Leibniz uses $\int$ for the sum and x/d for difference. He then says x/d is dx, the difference of two consecutive x-values, but apparently here dx is a constant and equal to unity.

From barely intelligible arguments such as the above, Leibniz asserted the fact that *integration as a summation process is the inverse of differentiation*. This idea is in the work of Barrow and Newton, who obtained areas by antidifferentiation, but it is first expressed as a relation between summation and differentiation by Leibniz. Despite this outright assertion, he was by no means clear as to how to obtain an area from what one might loosely write as $\sum y \, dx$—that is, how to obtain an area under a curve from a set of rectangles. Of course this difficulty beset all the seventeenth-century workers. Not possessing a clear concept of a limit, or even clear notions about area, Leibniz thought of the latter sometimes as a sum of rectangles so small and so

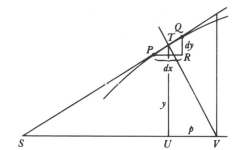

Figure 17.18

numerous that the difference between this sum and the true area under the curve could be neglected, and at other times as a sum of the ordinates or y-values. This latter concept of area was common, especially among the indivisibilists, who thought that the ultimate unit of area and the y-value were the same.

With respect to differentiation, even after recognizing that dy and dx can be arbitrarily small quantities, Leibniz had yet to overcome the fundamental difficulty that the ratio dy/dx is not quite the derivative in our sense. He based his argument on the characteristic triangle, which Pascal and Barrow had also used. This triangle (Fig. 17.18) consists of dy, dx, and the chord PQ, which Leibniz also thought of as *the curve between* P *and* Q *and part of the tangent at* T. Though he speaks of this triangle as indefinitely small, he maintains nevertheless that it is similar to a definite triangle, namely, the triangle STU formed by the subtangent SU, the ordinate at T, and the length of tangent ST. Hence dy and dx are ultimate elements, and their ratio has a definite meaning. In fact, he uses the argument that, from the similar triangles PRQ and SUT, $dy/dx = TU/SU$.

In the manuscript of November 11, 1675, Leibniz shows how he can solve a definite problem. He seeks the curve whose subnormal is inversely proportional to the ordinate. In Figure 17.18, the normal is TV and the subnormal p is UV. From the similarity of triangles PRQ and TUV, he has

$$\frac{dy}{dx} = \frac{p}{y}$$

or

$$p \, dx = y \, dy.$$

But the curve has the given property

$$p = \frac{b}{y},$$

where b is the proportionality constant. Hence

$$dx = \frac{y^2}{b} \, dy.$$

Then

$$\int dx = \int \frac{y^2}{b} \, dy$$

or

$$x = \frac{y^3}{3b}.$$

Leibniz also solved other inverse tangent problems.

In a paper of June 26, 1676, he realizes that the best method of finding tangents is to find dy/dx, where dy and dx are differences and dy/dx is the quotient. He ignores $dx \cdot dx$ and higher powers of dx.

By November of 1676, he is able to give the general rules $dx^n = nx^{n-1} \, dx$ for integral and fractional n and $\int x^n = x^{n+1}/n + 1$, and says, "The reasoning is general, and it does not depend upon what the progressions of the x's may be." Here x still means the order of the terms of a sequence. In this manuscript he also says that to differentiate $\sqrt{a + bz + cz^2}$, let $a + bz + cz^2 = x$, differentiate $\sqrt{x}$, and multiply by dx/dz. This is the chain rule.

By July 11, 1677, Leibniz could give the correct rules for the differential of sum, difference, product, and quotient of two functions and for powers and roots, but no proofs. In the manuscript of November 11, 1675, he had struggled with $d(uv)$ and $d(u/v)$, and thought that $d(uv) = du \, dv$.

In 1680, dx has become the difference of abscissas and dy the differences in the ordinates. He says, "... now these dx and dy are taken to be infinitely small, or the two points on the curve are understood to be a distance apart that is less than any given length...." He calls dy the "momentaneous increment" in y as the ordinate moves along the x-axis. But PQ in Figure 17.18 is still considered part of a straight line. It is "an element of the curve or a side of the infinite-angled polygon that stands for the curve...." He continues to use the usual differential form. Thus, if $y = a^2/x$, then

$$dy = -\frac{a^2}{x^2} \, dx.$$

He also says that differences are the opposite to sums. Then, to get the area under a curve (Fig. 17.19), he takes the sum of the rectangles and says one can neglect the remaining "triangles, since they are infinitely small com-

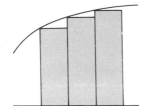

Figure 17.19

pared to the rectangles . . . thus I represent in my calculus the area of the figure by $\int y\, dx.$ . . . ." He also gives, for the element of arc,

$$ds = \sqrt{dx^2 + dy^2};$$

and, for the volume of a solid of revolution obtained by revolving a curve around the x-axis,

$$V = \pi \int y^2\, dx.$$

Despite prior statements that dx and dy are small differences, he still talks about sequences. He says, "Differences and sums are the inverses of one another, that is to say, the sum of the differences of a series [sequence] is a term of the series, and the difference of the sums of a series is a term of the series, and I enumerate the former thus, $\int dx = x$, and the latter thus, $d\int x = dx$." In fact, in a manuscript written after 1684, Leibniz says his method of infinitesimals has become widely known as the calculus of differences.

Leibniz's first publication on the calculus is in the *Acta Eruditorum* of 1684.[12] In this paper the meaning of dy and dx is still not clear. He says in one place, let dx be any arbitrary quantity, and dy is defined by (see Fig. 17.18)

$$dy : dx = y : \text{subtangent}.$$

This definition of dy presumes some expression for the subtangent; hence the definition is not complete. Moreover, Leibniz's definition of tangent as a line joining two infinitely near points is not satisfactory.

He also gives in this paper the rules he had obtained in 1677 for the differential of the sum, product, and quotient of two functions and the rule for finding $d(x^n)$. In this last case he sketches the proof for positive integral n but says the rule is true for all n; for the other rules he gives no proofs. He makes applications to finding tangents, maxima and minima, and points of inflection. This paper, six pages long, is so unclear that the Bernoulli brothers called it "an enigma rather than an explication."[13]

12. *Acta Erud.*, 3, 1684, 467–73 = *Math. Schriften*, 5, 220–26.
13. Leibniz: *Math. Schriften*, 3, Part 1, 5.

In a paper of 1686[14] Leibniz gives

$$y = \sqrt{2x - x^2} + \int \frac{dx}{\sqrt{2x - x^2}}$$

as the equation of the cycloid. His point here is to show that by his methods and notation some curves can be expressed as equations not obtainable in other ways. He reaffirms this in his *Historia* where he says that his dx, ddx (second difference), and the sums that are the inverses of these differences can be applied to all functions of x, not excepting the mechanical curves of Vieta and Descartes, which Descartes had said have no equations. Leibniz also says that he can include curves that Newton could not handle even with his method of series.

In the 1686 paper as well as in subsequent papers,[15] Leibniz gave the differentials of the logarithmic and exponential functions and recognized exponential functions as a class. He also treated curvature, the osculating circle, and the theory of envelopes (see Chap. 23). In a letter to John Bernoulli of 1697, he differentiated under the integral sign with respect to a parameter. He also had the idea that many indefinite integrals could be evaluated by reducing them to known forms and speaks of preparing tables for such reductions—in other words, a table of integrals. He tried to define the higher-order differentials such as ddy (d^2y) and $dddy$ (d^3y), but the definitions were not satisfactory. Though he did not succeed, he also tried to find a meaning for $d^\alpha y$ where α is any real number.

With respect to notation, Leibniz worked painstakingly to achieve the best. His dx, dy, and dy/dx are, of course, still standard. He introduced the notation log x, d^n for the nth differential, and even d^{-1} and d^{-n} for $\int$ and the nth iteration of summation, respectively.

In general Leibniz's work, though richly suggestive and profound, was so incomplete and fragmentary that it was barely intelligible. Fortunately, the Bernoulli brothers, James and John, who were immensely impressed and stirred by Leibniz's ideas, elaborated his sketchy papers and contributed an immense number of new developments we shall discuss later. Leibniz agreed that the calculus was as much theirs as his.

5. *A Comparison of the Work of Newton and Leibniz*

Both Newton and Leibniz must be credited with seeing the calculus as a new and general method, applicable to many types of functions. After their work, the calculus was no longer an appendage and extension of Greek geometry, but an independent science capable of handling a vastly expanded range of problems.

14. *Acta Erud.*, 5, 1686, 292–300 = *Math. Schriften*, 5, 226–33.
15. *Acta Erud.*, 1692, 168–71 = *Math. Schriften*, 5, 266–69; *Acta Erud.*, 1694 = *Math. Schriften*, 5, 301–6.

Both also arithmetized the calculus; that is, they built on algebraic concepts. The algebraic notation and techniques used by Newton and Leibniz not only gave them a more effective tool than geometry, but also permitted many different geometric and physical problems to be treated by the same technique. A major change from the beginning to the end of the seventeenth century was the algebraicization of the calculus. This is comparable to what Vieta had done in the theory of equations and Descartes and Fermat in geometry.

The third vital contribution that Newton and Leibniz share is the reduction to antidifferentiation of area, volume, and other problems that were previously treated as summations. Thus the four main problems—rates, tangents, maxima and minima, and summation—were all reduced to differentiation and antidifferentiation.

The chief distinction between the work of the two men is that Newton used the infinitely small increments in x and y as a means of determining the fluxion or derivative. It was essentially the limit of the ratio of the increments as they became smaller and smaller. On the other hand, Leibniz dealt directly with the infinitely small increments in x and y, that is, with differentials, and determined the relationship between them. This difference reflects Newton's physical orientation, in which a concept such as velocity is central, and Leibniz's philosophical concern with ultimate particles of matter, which he called monads. As a consequence, Newton solved area and volume problems by thinking entirely in terms of rate of change. For him differentiation was basic; this process and its inverse solved all calculus problems, and in fact the use of summation to obtain an area, volume, or center of gravity rarely appears in his work. Leibniz, on the other hand, thought first in terms of summation, though of course these sums were evaluated by antidifferentiation.

A third distinction between the work of the two men lies in Newton's free use of series to represent functions; Leibniz preferred the closed form. In a letter to Leibniz of 1676, Newton stressed the use of series even to solve simple differential equations. Though Leibniz did use infinite series, he replied that the real goal should be to obtain results in finite terms, using the trigonometric and logarithmic functions where algebraic functions would not serve. He recalled to Newton James Gregory's assertion that the rectification of the ellipse and hyperbola could not be reduced to the circular and logarithmic functions and challenged Newton to determine by the use of series whether Gregory was correct. Newton replied that by the use of series he could decide whether some integrations could be achieved in finite terms, but gave no criteria. Again, in a letter of 1712 to John Bernoulli, Leibniz objected to the expansion of functions into series and stated that the calculus should be concerned with reducing its results to quadratures (integrations) and, where necessary, quadratures involving transcendental functions.

There are differences in their manner of working. Newton was empirical, concrete, and circumspect, whereas Leibniz was speculative, given to generalizations, and bold. Leibniz was more concerned with operational formulas to produce a calculus in the broad sense; for example, rules for the differential of a product or quotient of functions, his rule for $d^n(uv)$ (u and v being functions of x), and a table of integrals. It was Leibniz who set the canons of the calculus, the system of rules and formulas. Newton did not bother to formulate rules, even when he could easily have generalized his concrete results. He knew that if $z = uv$, then $\dot{z} = u\dot{v} + v\dot{u}$, but did not point out this general result. Though Newton initiated many methods, he did not stress them. His magnificent applications of the calculus not only demonstrated its value but, far more than Leibniz's work, stimulated and determined almost the entire direction of eighteenth-century analysis. Newton and Leibniz differed also in their concern for notation. Newton attached no importance to this matter, while Leibniz spent days choosing a suggestive notation.

6. *The Controversy over Priority*

Nothing of Newton's work on the calculus was published before 1687, though he had communicated results to friends during the years 1665 to 1687. In particular, he had sent his tract *De Analysi* in 1669 to Barrow, who had sent it to John Collins. Leibniz visited Paris in 1672 and London in 1673 and communicated with some of the people who knew Newton's work. However, he did not publish on the calculus until 1684. Hence the question of whether Leibniz had known the details of what Newton did was raised, and Leibniz was accused of plagiarism. However, investigations made long after the deaths of the two men show that Leibniz was an independent inventor of major ideas of the calculus, though Newton did much of his work before Leibniz did. Both owe much to Barrow, though Barrow used geometrical methods almost exclusively. The significance of the controversy lies not in the question of who was the victor but rather in the fact that the mathematicians took sides. The Continental mathematicians, the Bernoulli brothers in particular, sided with Leibniz, while the English mathematicians defended Newton. The two groups became unfriendly and even bitter toward each other; John Bernoulli went so far as to ridicule and inveigh against the English.

As a result, the English and Continental mathematicians ceased exchanging ideas. Because Newton's major work and first publication on the calculus, the *Principia*, used geometrical methods, the English continued to use mainly geometry for about a hundred years after his death. The Continentals took up Leibniz's analytical methods and extended and improved them. These proved to be far more effective; so not only did the English mathematicians fall behind, but mathematics was deprived of contributions that some of the ablest minds might have made.

7. Some Immediate Additions to the Calculus

The calculus is of course the beginning of that most weighty part of mathematics generally referred to as analysis. We shall be following the important developments of this field in succeeding chapters; we might note here, however, some additions that were made immediately after the basic work of Newton and Leibniz.

In his *Arithmetica Universalis* (1707) Newton established a theorem on the upper bound to the real roots of polynomial equations. The theorem says: A number a is an upper bound of the real roots of $f(x) = 0$ if, when a is substituted for x, it gives to $f(x)$ and to all its derivatives the same sign.

In his *De Analysi* and *Method of Fluxions*, he gave a general method of approximating the roots of $f(x) = 0$, which was published in Wallis's *Algebra* of 1685. In his tract *Analysis Aequationum Universalis* (1690), Joseph Raphson (1648–1715) improved on this method; though he applied it only to polynomials, it is much more broadly useful. It is this modification that is now known as Newton's method or the Newton-Raphson method. It consists in first choosing an approximation a. Then calculate $a - f(a)/f'(a)$. Call this b, and calculate $b - f(b)/f'(b)$. Call this last result c, and so forth. The numbers a, b, c, . . . are successive approximations to the root. (The notation is modern.) Actually the method does not necessarily give better and better approximations to the root. J. Raymond Mourraille showed in 1768 that a must be chosen so that the curve of $y = f(x)$ is convex toward the axis of x in the interval between a and the root. Much later Fourier discovered this fact independently.

In his *Démonstration d'une méthode pour résoudre les égalitéz de tous les dégrez* (16191), Michel Rolle (1652–1719) gave the famous theorem now named after him, namely, that if a function is 0 at two values of x, say, a and b, then the derivative is 0 at some value of x between a and b. Rolle stated the theorem but did not prove it.

After Newton and Leibniz the two most important founders of the calculus were the Bernoulli brothers, James and John. James (= Jakob = Jacques) Bernoulli (1655–1705) was self-taught in mathematics and so matured slowly in that subject. At the urging of his father he studied for the ministry, but eventually turned to mathematics, and in 1686 became a professor at the University of Basle. His chief interests thereafter were mathematics and astronomy. When, in the late 1670s, he began to work on mathematical problems, Newton's and Leibniz's work was still unknown to him. He too learned from Descartes's *La Géométrie*, Wallis's *Arithmetica Infinitorum* and Barrow's *Geometrical Lectures*. Though he took much from

Barrow, he put it into analytical form. He gradually became familiar with Leibniz's work, but because so little of the latter appeared in print, much of what James did overlapped Leibniz's results. Actually he, like the other mathematicians of the time, did not fully understand Leibniz's work.

James's activity is closely linked with that of his younger brother John (= Johann = Jean, 1667–1748). John was sent into business by his father but turned to medicine, while learning mathematics from his brother. He became a professor of mathematics at Groningen in Holland and then succeeded his brother at Basle.

Both James and John corresponded constantly with Leibniz, Huygens, other mathematicians, and each other. All these men worked on many common problems suggested in letters or posed as challenges. Since results, too, were in those days often communicated in letters with or without subsequent publication, the matter of priority is complicated. Sometimes credit was claimed for a result that was announced even though no proof was given at that time. The question is further complicated by the peculiar relationships that developed. John was extremely anxious to secure fame and began to compete with his brother; soon each was challenging the other on problems. John did not hesitate to use unscrupulous means to appear to be the discoverer of results he got from others, including his brother. James was very sensitive and reacted in kind. Each published papers that owed much to the other without acknowledging the origins of their ideas. John actually became a vitriolic critic of his brother, and Leibniz tried to mediate between the two. Though James had said earlier, while praising Barrow, that Leibniz's work should not be depreciated, he became more and more distrustful of Leibniz. Moreover, he resented Leibniz's superior insights and thought Leibniz was arrogant in pointing out that he had done things James thought were original with himself. He became convinced that Leibniz sought only to belittle his work and was favoring John in the disputes between the brothers. When Nicholas Fatio de Duillier (1664–1753) gave Newton credit for creating the calculus and became embroiled in controversy with Leibniz, James wrote letters to Fatio opposing Leibniz.

As to the Bernoullis' work in the calculus, they, too, tackled problems such as finding the curvature of curves, evolutes (envelopes of the normals to a curve), inflection points, the rectification of curves, and other basic calculus topics. The results of Newton and Leibniz were extended to spirals of various sorts, the catenary, and the tractrix, which was defined as the curve (Fig. 17.20) for which the ratio PT to OT is a constant. James also wrote five major papers on series (Chap. 20, sec. 4), which extended Newton's use of series to integrate complicated algebraic functions and transcendental functions. In 1691 both James and John gave the formula for the radius of curvature of a curve. James called it his "golden theorem" and wrote it as

$$z = dx\, ds : ddy = dy\, ds : ddx$$

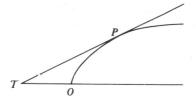

Figure 17.20

where z is the radius of curvature. If we divide numerator and denominator of each ratio by ds^2 we get

$$z = \frac{dx/ds}{d^2y/ds^2} = \frac{dy/ds}{d^2x/ds^2},$$

which are more familiar forms. James also gave the result in polar coordinates.

John produced a now-famous theorem for obtaining the limit approached by a fraction whose numerator and denominator approach 0. This theorem was incorporated by Guillaume F. A. l'Hospital (1661–1704), a pupil of John, in an influential book on the calculus, the *Analyse des infiniment petits* (1696), and is now known as L'Hospital's rule.

8. The Soundness of the Calculus

From the very introduction of the new methods of finding rates, tangents, maxima and minima, and so forth, the proofs were attacked as unsound. Cavalieri's use of indivisible ultimate elements and his arguments shocked those who still respected logical rigor. To their criticism Cavalieri responded that the contemporary geometers had been freer with logic than he—for example, Kepler, in his *Stereometria Doliorum*. These geometers, he continued, had been content in their calculation of areas to imitate Archimedes' method of summing lines, but had failed to give the complete proofs that the great Greek had used to make his work rigorous. They were satisfied with their calculations, provided only that the results were useful. Cavalieri felt justified in adopting the same point of view. He said that his procedures could lead to new inventions and that his method did not at all oblige one to consider a geometrical structure as composed of an infinite number of sections; it had no other object than to establish correct ratios between areas or volumes. But these ratios preserved their sense and value whatever opinion one might have about the composition of a continuum. In any case, said Cavalieri, "rigor is the concern of philosophy and not of geometry."

Fermat, Pascal, and Barrow recognized the looseness of their work on summation but believed that one could make precise proofs in the manner of Archimedes. Pascal, in *Letters of Dettonville* (1659), affirmed that the infinitesimal geometry and classical Greek geometry were in agreement. He concluded, "What is demonstrated by the true rules of indivisibles could be demonstrated also with the rigor and the manner of the ancients." Further, he said the method of indivisibles must be accepted by any mathematicians who pretend to rank among geometers. It differs only in language from the method of the ancients. Nevertheless, Pascal, too, had ambivalent feelings about rigor. At times he argued that the heart intervenes to assure us of the correctness of mathematical steps. The proper "finesse," rather than geometrical logic, is what is needed to do the correct work, just as the religious appreciation of grace is above reason. The paradoxes of geometry as used in the calculus are like the apparent absurdities of Christianity; and the indivisible in geometry has the same relation to the finite as man's justice has to God's.

The defenses Cavalieri and Pascal offered applied to the summation of infinitely small quantities. As to the derivative, early workers such as Fermat and Roberval thought they had a simple algebraic process that had a very clear geometric interpretation and so could be justified by geometrical arguments. Actually Fermat was careful not to assert general theorems when he advanced any idea he could not justify by the method of exhaustion. Barrow argued only geometrically and, despite his attacks on the algebraists for their lack of rigor, was less scrupulous about the soundness of his geometrical arguments.

Neither Newton nor Leibniz clearly understood nor rigorously defined his fundamental concepts. We have already observed that both vacillated in their definitions of the derivative and differentials. Newton did not really believe that he had departed from Greek geometry. Though he used algebra and coordinate geometry, which were not to his taste, he thought his underlying methods were but natural extensions of pure geometry. Leibniz, however, was a man of vision who thought in broad terms, like Descartes. He saw the long-term implications of the new ideas and did not hesitate to declare that a new science was coming to light. Hence he was not too concerned about the lack of rigor in the calculus.

In response to criticism of his ideas, Leibniz made various, unsatisfactory replies. In a letter to Wallis of March 30, 1690[16] he said:

> It is useful to consider quantities infinitely small such that when their ratio is sought, they may not be considered zero but which are rejected as often as they occur with quantities incomparably greater. Thus if we have $x + dx$, dx is rejected. But it is different if we seek the difference between $x + dx$ and x. Similarly we cannot have $x\, dx$ and $dx\, dx$ standing together. Hence if we are to differentiate xy we write $(x + dx)(y + dy) - xy = x\, dy + y\, dx + dx\, dy$. But here $dx\, dy$ is to be rejected as incomparably less than $x\, dy + y\, dx$. Thus in any particular case, the error is less than any finite quantity.

16. Leibniz: *Math. Schriften*, 4, 63.

As to the ultimate meanings of dy, dx and dy/dx, Leibniz remained vague. He spoke of dx as the difference in x values between two infinitely near points and of the tangent as the line joining such points. He dropped differentials of higher order with no justification, though he did distinguish among the various orders. The infinitely small dx and dy were sometimes described as vanishing or incipient quantities, as opposed to quantities already formed. These indefinitely small quantities were not zero, but were smaller than any finite quantity. Alternatively he appealed to geometry to say that a higher differential is to a lower one as a point is to a line [17] or that dx is to x as a point to the earth or as the radius of the earth to that of the heavens. The ratio of two infinitesimals he thought of as a quotient of inassignables or of indefinitely small quantities, but one which could nevertheless be expressed in terms of definite quantities such as the ratio of ordinate to subtangent.

A flurry of attacks and rebuttals was initiated in books of 1694 and 1695 by the Dutch physician and geometer Bernard Nieuwentijdt (1654–1718). Although he admitted that in general the new methods led to correct results, he criticized the obscurity and pointed out that sometimes the methods led to absurdities. He complained that he could not understand how the infinitely small quantities differed from zero and asked how a sum of infinitesimals could be finite. He also challenged the meaning and existence of differentials of higher order and the rejection of infinitely small quantities in portions of the arguments.

Leibniz, in a draft of a reply to Nieuwentijdt, probably written in 1695, and in an article in the *Acta Eruditorum* of 1695,[18] gives various answers. He speaks of "overprecise" critics and says that excessive scrupulousness should not cause us to reject the fruits of invention. He then says his method differs from Archimedes' only in the expressions used, but that his own are better adapted to the art of discovery. The words "infinite" and "infinitesimal" signify merely quantities that one can take as large or as small as one wishes in order to show that the error incurred is less than any number that can be assigned—in other words, that there is no error. One can use these ultimate things—that is, infinite and infinitely small quantities—as a tool, much as algebraists used imaginary roots with great profit.

Leibniz's argument thus far was that his calculus used only ordinary mathematical concepts. But since he could not satisfy his critics, he enunciated a philosophical principle known as the law of continuity, which was practically the same as one already stated by Kepler. In 1687, in a letter to Pierre Bayle,[19] Leibniz expressed this principle as follows: "In any supposed transition, ending in any terminus, it is permissible to institute a general reasoning, in which the final terminus may also be included." To support

17. *Math. Schriften*, 5, 322 ff.
18. *Acta Erud.*, 1695, 310–16 = *Math. Schriften*, 5, 320–28.
19. *Math. Schriften*, 5, 385.

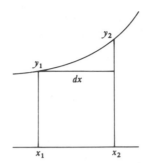

Figure 17.21

this principle he gives, in an unpublished manuscript of about 1695, the example of including under one argument ellipses and parabolas, though the parabola is a limiting case of the ellipse when one focus moves off to infinity. He then applies the principle to the calculation of $dy:dx$ for the parabola $y = x^2/a$. After obtaining

$$dy:dx = (2x + dx):a,$$

he says, "Now, since by our postulate it is permissible to include under the one general reasoning the case also in which [Fig. 17.21] the ordinate x_2y_2 is moved up nearer and nearer to the fixed ordinate x_1y_1 until it ultimately coincides with it, it is evident that in this case dx becomes equal to 0 and should be neglected...." Leibniz does not say what meaning should be given to the dx that appears at the left side of the equation.

Of course, he says, things that are absolutely equal have a difference that is absolutely nothing; therefore a parabola is not an ellipse.

> Yet a state of transition may be imagined, or one of evanescence, in which indeed there has not yet arisen exact equality or rest.... but in which it is passing into such a state that the difference is less than any assignable quantity; also that in this state there will still remain some difference, some velocity, some angle, but in each case one that is infinitely small....
>
> For the present, whether such a state of instantaneous transition from inequality to equality ... can be sustained in a rigorous or metaphysical sense, or whether infinite extensions successively greater and greater, or infinitely small ones successively less and less, are legitimate considerations, is a matter that I own to be possibly open to question....
>
> It will be sufficient if, when we speak of infinitely great (or, more strictly, unlimited) or of infinitely small quantities (i.e., the very least of those within our knowledge), it is understood that we mean quantities that are indefinitely great or indefinitely small, i.e., as great as you please, or as small as you please, so that the error that any one may assign may be less than a certain assigned quantity.

On these suppositions, all the rules of our algorithm, as set out in the *Acta Eruditorum* for October 1684, can be proved without much trouble.

Leibniz then goes over these rules. He introduces the quantities $(d)y$ and $(d)x$ and carries out the usual processes of differentiation with them. These he calls assignable or definite nonvanishing quantities. After obtaining the final result, he says, we can replace $(d)y$ and $(d)x$ by the evanescent or unassignable quantities dy and dx, by making "the supposition that the ratio of the evanescent quantities dy and dx is equal to the ratio of $(d)y$ and $(d)x$, because this supposition can always be reduced to an undoubtable truth."

Leibniz's principle of continuity is certainly not a mathematical axiom today, but he emphasized it and it became important later. He gave many arguments that are in accordance with this principle. For example, in a letter to Wallis,[20] Leibniz defended his use of the characteristic triangle as a form without magnitude, the form remaining after the magnitudes had been reduced to zero, and challengingly asked, "Who does not admit a form without magnitude?" Likewise, in a letter to Guido Grandi,[21] he said the infinitely small is not a simple and absolute zero but a relative zero, that is, an evanescent quantity which yet retains the character of that which is disappearing. However, Leibniz also said, at other times, that he did not believe in magnitudes truly infinite or truly infinitesimal.

Leibniz, less concerned with the ultimate justification of his procedures than Newton, felt that it lay in their effectiveness. He stressed the procedural or algorithmic value of what he had created. Somehow he had confidence that if he formulated clearly the rules of operation and these were properly applied, reasonable and correct results would be obtained, however doubtful might be the meanings of the symbols involved.

It is apparent that neither Newton nor Leibniz succeeded in making clear, let alone precise, the basic concepts of the calculus: the derivative and the integral. Not being able to grasp these properly, they relied upon the coherence of the results and the fecundity of the methods to push ahead without rigor.

Several examples may illustrate the lack of clarity even among the great immediate successors of Newton and Leibniz. John Bernoulli wrote the first text on the calculus in 1691 and 1692. The portion on the integral calculus was published in 1742;[22] the part on the differential calculus, *Die Differentialrechnung*, was not published until 1924. However, the Marquis de l'Hospital did publish a slightly altered French version (already referred to) under his own name in 1696. Bernoulli begins the *Differentialrechnung* with three postulates. The first reads: "A quantity which is diminished or increased by an

20. *Math. Schriften*, 4, 54.
21. *Math. Schriften*, 4, 218.
22. *Opera Omnia*, 3, 385–558.

infinitely small quantity is neither increased nor decreased." His second postulate is: "Each curved line consists of infinitely many straight lines, these themselves being infinitely small." In his reasoning he followed Leibniz and used infinitesimals. Thus to obtain dy from $y = x^2$, he uses e for dx and gets $(x + e)^2 - x^2$, or $2xe + e^2$, and then just drops e^2. Like Leibniz, he used vague analogies to explain what differentials were. Thus, he says, the infinitely large quantities are like astronomical distances and the infinitely small are like animalcules revealed by the microscope. In 1698 he argued that infinitesimals must exist.[23] One has only to consider the infinite series $1, 1/2, 1/4, \ldots$. If one takes 10 terms, then $1/10$ exists; if one takes 100 terms, then $1/100$ exists. Corresponding to the infinite number of terms there is the infinitesimal.

A few men, Wallis and John Bernoulli among them, tried to define the infinitesimal as the reciprocal of ∞, the latter being a definite number to them. Still others acted as though what was incomprehensible needed no further explanation. For most of the seventeenth-century men, rigor was not a matter of concern. What they often said could be rigorized by the method of Archimedes could actually not have been rigorized by an Archimedes; this is particularly true of the work on differentiation, which had no parallel in Greek mathematics.

Actually the new calculus was introducing concepts and methods that inaugurated a radical departure from earlier work. With the work of Newton and Leibniz, the calculus became a totally new discipline that required foundations of its own. Though they were not fully aware of it, the mathematicians had turned their backs on the past.

Germs of the correct new concepts can be found even in the seventeenth-century literature. Wallis, in the *Arithmetica Infinitorum*, advanced the arithmetical concept of the limit of a function as a number approached by the function so that the difference between this number and the function could be made less than any assignable quantity and would vanish ultimately when the process was continued to infinity. His wording is loose but contains the right idea.

James Gregory in his *Vera Circuli et Hyperbolae Quadratura* (1667) explicitly pointed out that the methods used to obtain areas, volumes, and lengths of curves involved a new process, the limit process. Moreover, he added, this operation was distinct from the five algebraic operations of addition, subtraction, multiplication, division, and extraction of roots. He put the method of exhaustion into algebraic form and recognized that the successive approximations obtained by using rectilinear figures circumscribed about a given area or volume and those obtained by using inscribed rectilinear figures both converged to the same "last term." He also noted that

23. Leibniz: *Math. Schriften*, **3**, Part 2, 563 ff.

this limit process yields irrationals not obtainable as roots of rationals. But these insights of Wallis and Gregory were ignored in their century.

The foundations of the calculus remained unclear. Adding to the confusion was the fact that the proponents of Newton's work continued to speak of prime and ultimate ratios, while the followers of Leibniz used the infinitely small non-zero quantities. Many of the English mathematicians, perhaps because they were in the main still tied to the rigor of Greek geometry, distrusted all the work on the calculus. Thus the century ended with the calculus in a muddled state.

Bibliography

Armitage, A.: *Edmond Halley*, Thomas Nelson and Sons, 1966.

Auger, L.: *Un Savant méconnu: Gilles Persone de Roberval (1602–1675)*, A. Blanchard, 1962.

Ball, W. W. R.: *A Short Account of the History of Mathematics*, Dover (reprint), 1960, pp. 309–70.

Baron, Margaret E.: *The Origins of the Infinitesimal Calculus*, Pergamon Press, 1969.

Bell, Arthur E.: *Newtonian Science*, Edward Arnold, 1961.

Boyer, Carl B.: *The Concepts of the Calculus*, Dover (reprint), 1949.

————: *A History of Mathematics*, John Wiley and Sons, 1968, Chaps. 18–19.

Brewster, David: *Memoirs of the Life, Writings and Discoveries of Sir Isaac Newton*, 2 vols., 1855, Johnson Reprint Corp., 1965.

Cajori, Florian: *A History of the Conceptions of Limits and Fluxions in Great Britain from Newton to Woodhouse*, Open Court, 1919.

————: *A History of Mathematics*, Macmillan, 1919, 2nd ed., pp. 181–220.

Cantor, Moritz: *Vorlesungen über Geschichte der Mathematik*, 2nd ed., B. G. Teubner, 1900 and 1898, Vol. 2, pp. 821–922; Vol. 3, pp. 150–316.

Child, J. M.: *The Geometrical Lectures of Isaac Barrow*, Open Court, 1916.

————: *The Early Mathematical Manuscripts of Leibniz*, Open Court, 1920.

Cohen, I. B.: *Isaac Newton's Papers and Letters on Natural Philosophy*, Harvard University Press, 1958.

Coolidge, Julian L.: *The Mathematics of Great Amateurs*, Dover (reprint), 1963, Chaps. 7, 11, and 12.

De Morgan, Augustus: *Essays on the Life and Work of Newton*, Open Court, 1914.

Fermat, Pierre de: *Œuvres*, Gauthier-Villars, 1891–1912, Vol. 1, pp. 133–79, Vol. 3, pp. 121–56.

Gibson, G. A.: "James Gregory's Mathematical Work," *Proc. Edinburgh Math. Soc.*, 41, 1922/23, 2–25.

Huygens, C.: *Œuvres complètes*, 22 vols., Société Hollandaise des Sciences, Nyhoff, 1888–1950.

Leibniz, G. W.: *Œuvres*, Firmin-Didot, 1859–75.

————: *Mathematische Schriften*, ed. C. I. Gerhardt, 7 vols., Ascher-Schmidt, 1849–63. Reprinted by Georg Olms, 1962.

More, Louis T.: *Isaac Newton*, Dover (reprint), 1962.

Montucla, J. F.: *Histoire des mathématiques*, Albert Blanchard (reprint), 1960, Vol. 2, pp. 102–77, 348–403; Vol. 3, pp. 102–38.

Newton, Sir Isaac: *The Mathematical Works*, ed. D. T. Whiteside, 2 vols., Johnson Reprint Corp., 1964–67. Vol. 1 contains translations of the three basic papers on the calculus.

———: *Mathematical Papers*, ed. D. T. Whiteside, 4 vols., Cambridge University Press, 1967–71.

———: *Mathematical Principles of Natural Philosophy*, ed. Florian Cajori, 3rd ed., University of California Press, 1946.

———: *Opticks*, Dover (reprint), 1952.

Pascal, B.: *Œuvres*, Hachette, 1914–21.

Scott, Joseph F.: *The Mathematical Work of John Wallis*, Oxford University Press, 1938.

———: *A History of Mathematics*, Taylor and Francis, 1958, Chaps. 10–11.

Smith, D. E.: *A Source Book in Mathematics*, Dover (reprint), 1959, pp. 605–26.

Struik, D. J.: *A Source Book in Mathematics, 1200–1800*, Harvard University Press, 1969, pp. 188–316, 324–28.

Thayer, H. S.: *Newton's Philosophy of Nature*, Hafner, 1953.

Turnbull, H. W.: *The Mathematical Discoveries of Newton*, Blackie and Son, 1945.

———: *James Gregory Tercentenary Memorial Volume*, Royal Society of Edinburgh, 1939.

Turnbull, H. W. and J. F. Scott: *The Correspondence of Isaac Newton*, 4 vols., Cambridge University Press, 1959–1967.

Walker, Evelyn: *A Study of the* Traité des indivisibles *of Gilles Persone de Roberval*, Columbia University Press, 1932.

Wallis, John: *Opera Mathematica*, 3 vols., 1693–99, Georg Olms (reprint), 1968.

Whiteside, Derek T.: "Patterns of Mathematical Thought in the Seventeenth Century," *Archive for History of Exact Sciences*, 1, 1961, pp. 179–388.

Wolf, Abraham: *A History of Science, Technology and Philosophy in the 16th and 17th Centuries*, 2nd ed., George Allen and Unwin, 1950, Chaps. 7–14.

Abbreviations

Journals whose titles have been written out in full in the text are not listed here.

Abh. der Bayer. Akad. der Wiss. Abhandlungen der Königlich Bayerischen Akademie der Wissenschaften (München)

Abh. der Ges. der Wiss. zu Gött. Abhandlungen der Königlichen Gesellschaft der Wissenschaften zu Göttingen

Abh. König. Akad. der Wiss., Berlin Abhandlungen der Königlich Preussischen Akademie der Wissenschaften zu Berlin

Abh. Königlich Böhm. Ges. der Wiss. Abhandlungen der Königlichen Böhmischen Gesellschaft der Wissenschaften

Abh. Math. Seminar der Hamburger Univ. Abhandlungen aus dem Mathematischen Seminar Hamburgischen Universität

Acta Acad. Sci. Petrop. Acta Academiae Scientiarum Petropolitanae

Acta Erud. Acta Eruditorum

Acta Math. Acta Mathematica

Acta Soc. Fennicae Acta Societatis Scientiarum Fennicae

Amer. Jour. of Math. American Journal of Mathematics

Amer. Math. Monthly American Mathematical Monthly

Amer. Math. Soc. Bull. American Mathematical Society, Bulletin

Amer. Math. Soc. Trans. American Mathematical Society, Transactions

Ann. de l'Ecole Norm. Sup. Annales Scientifiques de l'Ecole Normale Supérieure

Ann. de Math. Annales de Mathématiques Pures et Appliquées

Ann. Fac. Sci. de Toulouse Annales de la Faculté des Sciences de Toulouse

Ann. Soc. Sci. Bruxelles Annales de la Société Scientifique de Bruxelles

Annali di Mat. Annali di Matematica Pura ed Applicata

Annals of Math. Annals of Mathematics

Astronom. Nach. Astronomische Nachrichten

Atti Accad. Torino Atti della Reale Accademia delle Scienze di Torino

Atti della Accad. dei Lincei, Rendiconti Atti della Reale Accademia dei Lincei, Rendiconti

Brit. Assn. for Adv. of Sci. British Association for the Advancement of Science

Bull. des Sci. Math. Bulletin des Sciences Mathématiques

Bull. Soc. Math. de France Bulletin de la Société Mathématique de France

Cambridge and Dublin Math. Jour. Cambridge and Dublin Mathematical Journal

Comm. Acad. Sci. Petrop. Commentarii Academiae Scientiarum Petropolitanae

Comm. Soc. Gott. Commentationes Societatis Regiae Scientiarum Gottingensis Recentiores

Comp. Rend. Comptes Rendus

Corresp. sur l'Ecole Poly. Correspondance sur l'Ecole Polytechnique

Encyk. der Math. Wiss. Encyklopädie der Mathematischen Wissenschaften

Gior. di Mat. Giornale di Matematiche

Hist. de l'Acad. de Berlin Histoire de l'Académie Royale des Sciences et des Belles-Lettres de Berlin

Hist. de l'Acad. des Sci., Paris Histoire de l'Académie Royale des Sciences avec les Mémoires de Mathématique et de Physique

Jahres, der Deut. Math.-Verein. Jahresbericht der Deutschen Mathematiker-Vereinigung

Jour. de l'Ecole Poly. Journal de l'Ecole Polytechnique

Jour. de Math. Journal de Mathématiques Pures et Appliquées

Jour. des Sçavans Journal des Sçavans

Jour. für Math. Journal für die Reine und Angewandte Mathematik

Jour. Lon. Math. Soc. Journal of the London Mathematical Society

Königlich Sächsischen Ges. der Wiss. zu Leipzig Berichte über die Verhandlungen der Königlich Sächsischen Gesellschaft der Wissenschaften zu Leipzig

Math. Ann. Mathematische Annalen

Mém. de l'Acad. de Berlin See *Hist. de l'Acad. de Berlin*

Mém. de l'Acad. des Sci., Paris See *Hist. de l'Acad. des Sci., Paris;* after 1795, Mémoires de l'Academie des Sciences de l'Institut de France

Mém. de l'Acad. Sci. de St. Peters. Mémoires de l'Académie Impériale des Sciences de Saint-Petersbourg

Mém. des sav. étrangers Mémoires de Mathématique et de Physique Présentés à l'Académie Royal des Sciences, par Divers Sçavans, et Lus dans ses Assemblées

Mém. divers Savans See *Mém. des sav. étrangers*

Misc. Berolin. Miscellanea Berolinensia; also as *Hist. de l'Acad. de Berlin (q.v.)*

Misc. Taur. Miscellanea Philosophica-Mathematica Societatis Privatae Taurinensis (published by Accademia della Scienze di Torino)

Monatsber. Berliner Akad. Monatsberichte der Königlich Preussischen Akademie der Wissenschaften zu Berlin

N.Y. Math. Soc. Bull. New York Mathematical Society, Bulletin

Nachrichten König. Ges. der Wiss. zu Gött. Nachrichten von der Königlichen Gesellschaft der Wissenschaften zu Göttingen

Nou. Mém. de l'Acad. Roy. des Sci., Bruxelles Nouveaux Mémoires de l'Académie Royale des Sciences, des Lettres, et des Beaux-Arts de Belgique

Nouv. Bull. de la Soc. Philo. Nouveau Bulletin de la Société Philomatique de Paris

Nouv. Mém. de l'Acad. de Berlin Nouveaux Mémoires de l'Académie Royale des Sciences et des Belles-Lettres de Berlin

Nova Acta Acad. Sci. Petrop. Nova Acta Academiae Scientiarum Petropolitanae

Nova Acta Erud. Nova Acta Eruditorum

Novi Comm. Acad. Sci. Petrop. Novi Commentarii Academiae Scientiarum Petropolitanae

Phil. Mag. The Philosophical Magazine

Philo. Trans. Philosophical Transactions of the Royal Society of London

Proc. Camb. Phil. Soc. Cambridge Philosophical Society, Proceedings

Proc. Edinburgh Math. Soc. Edinburgh Mathematical Society, Proceedings

Proc. London Math. Soc. Proceedings of the London Mathematical Society

Proc. Roy. Soc. Proceedings of the Royal Society of London

Proc. Royal Irish Academy Proceedings of the Royal Irish Academy

Quart. Jour. of Math. Quarterly Journal of Mathematics

Scripta Math. Scripta Mathematica

Sitzungsber. Akad. Wiss zu Berlin Sitzungsberichte der Königlich Preussischen Akademie der Wissenschaften zu Berlin

Sitzungsber. der Akad. der Wiss., Wien Sitzungsberichte der Kaiserlichen Akademie der Wissenschaften zu Wien. Mathematisch-Naturwissenschaftlichen Klasse

Trans. Camb. Phil. Soc. Cambridge Philosophical Society, Transactions

Trans. Royal Irish Academy Transactions of the Royal Irish Academy

Zeit. für Math. und Phys. Zeitschrift für Mathematik und Physik

Zeit. für Physik Zeitschrift für Physik

Name Index

iii

Subject Index